Springer Wien New York

www.robarch2012.org
chair@robarch2012.org

Editors:
Sigrid Brell-Çokcan, Johannes Braumann
Association for Robots in Architecture
www.robotsinarchitecture.org

Funded by KUKA Robotics and
the Association for Robots in Architecture

SpringerWienNewYork is part of
Springer Science+Business Media
springer.at

Editing:
James Roderick O'Donovan
Design Concept and Cover:
Toledo i Dertschei
Layout:
Marko Tomicic
Printing:
Holzhausen Druck GmbH, 1140 Wien, Austria

Printed on acid-free and chlorine-free bleached paper
SPIN 86175717

With 445 coloured figures.

Library of Congress Control Number: 2012953005
ISBN 978-3-7091-1464-3 SpringerWienNewYork

Sigrid Brell-Çokcan
Johannes Braumann (Eds.)

Rob | Arch 2012

Robotic Fabrication in Architecture, Art, and Design

Springer Wien New York

Introduction

Keynotes

Workshops

Projects

Research

Industry

Sigrid Brell-Çokcan, Johannes Braumann

Introduction

Rob|Arch: Robotic Fabrication in Architecture, Art, and Design

Towards Robots in Architecture

Architects have been fascinated by robots for many decades, from "Chantier de Construction Électrique", Villemard's utopian vision of an architect building a house with robotic labor in 1910, to the design of buildings that are robots themselves, such as Archigram's Walking City. In the 1980s and 1990s it briefly seemed as if robots had finally arrived in architecture, when the Japanese construction industry started using highly customized robots for high-rise construction. However, amid the turmoil of Japan's financial problems in the 1990s these experiments were discontinued. Many later robotic projects were performed in a purely virtual environment, as architects were unable to transform their theories into a physical output.

Today, architects, artists and designers are again approaching the topic of robotic fabrication but with a different strategy: Instead of utopian proposals like Archigram's or highly specialized robots like the ones that were used in Japan, the current focus of architectural robotics is industrial robots. These robotic arms have six degrees of freedom and are widely used in industry, especially for automotive production lines. What makes robotic arms so interesting for the creative industry is their multi-functionality and their low price: instead of having to develop specialized machines, a multifunctional robot arm can be equipped with a wide range of end-effectors, similar to a human hand using various tools. Furthermore, due to their prevalence in industry, these robots are not prototypical machines, but certified, reliable, and increasingly affordable, today costing 70% less than the average price in the 1990s.

General research into industrial robots has been going on since the 1950s as an interdisciplinary effort involving mostly mechanical and electrical engineers, as well as computer scientists and mathematicians to deal with various aspects, from kinematic calculations to the design of efficient motors. This has led to a wide range of industrial robots, from desktop-sized small robots with a carrying weight of a few kilograms to massive machines capable of lifting a car chassis.

Therefore, architectural research into robotics is not so much directed at reinventing machines for architectural fabrication, but rather at re-using industrial robots as a well-established basis and adapting them for architectural purposes by developing custom software interfaces and end-effectors.

Pioneering Work

While the use of industrial robots in the construction industry was explored by researchers as early as the 1980s, pioneering work was done at ETH Zurich by Fabio Gramazio and Matthias Kohler, whose projects such as the Gantenbein Vineyard Façade showed that robotic arms are not only capable of replicating human labor, but can perform fabrication strategies that are outside the scope of human labor.

That was in 2006. In the past six years, more than 20 architecture faculties around the globe have acquired industrial robots and are actively researching new and innovative uses for these multifunctional machines, among them the University of Stuttgart, whose research pavilions have been published worldwide by architectural and mainstream media.

At the end of 2010, the Association for Robots in Architecture was founded, with the goal of making industrial robots accessible to the creative industry. We pursue that goal with a dual strategy, on the one hand by developing custom tools for accessible robot control, which later resulted in e.g. KUKA|prc, and on the other hand by acting as an open platform for artists, designers, researchers, technicians, and corporations involved in creative robotic fabrication.

The idea of organizing the first international conference dedicated to robots in architecture, art, and design emerged in mid-2011 and has since then met with an extremely positive feedback from both universities and industry partners.

Rob|Arch

Robotic fabrication in architecture, art, and design is a relatively young discipline, whose focus is on applied research, performed on the one hand by young designers, artists and researchers from the "digital generation" and on the other by innovative firms and startups, researching applications that go beyond typical industry solutions. This is reflected in the structure of this book, which does not consist solely of full-length scientific papers but has four distinct sections: workshop papers, research papers, project papers, and industry papers.

Workshop Papers

One of the centerpieces of the Rob|Arch conference is the robot workshops, organized by ETH Zurich, University of Stuttgart, TU Delft, TU Vienna, TU Graz, Harvard GSD, SciArc, and HAL/Robots in Architecture. For the first time, these internationally recognized institutions are opening their robotic labs and allowing participants to take part in their exciting research.

These workshops are not recapitulations of existing work, but contain new ideas that were developed for this conference and are published in this book. Stuttgart's workshop contribution builds upon the joining technology that was initially developed for the research pavilion, and the influence of biomimetic design strategies, while the ETH's workshop paper shows how their robotic bricklaying algorithm has evolved into an accessible design tool. New interfaces are also a significant topic for most of the other workshop paper: Thibault Schwartz presents a versatile tool for the visual programming of ABB robots, while TU Vienna links the fabrication process with real-time data captured from a camera and SciArc explores cooperating robotic arms. Harvard GSD's workshop focuses on ceramics as a material, while the remaining two workshops deal with subtractive fabrication methods: TU Graz explores robotic milling of foam glass, while TU Delft uses wire cutting for the rapid generation of three-dimensional free-forms.

Research Papers

Research papers are full scientific papers that were reviewed by the scientific committee and show a wide range of robotic applications that go far beyond today's industrial applications.

Despite the large spectrum of applications, we can still identify a common ground linking these research projects. Aggregations can be observed in multiple contributions, from uniform black balls that are robotically glued together to form organic structures, via macro-scale granulates, to metallic molecules that are shaped by robot-mounted electromagnets. New robotic end-effectors are also explored in various papers, such as for the robotic bending of metal facades, the shaping of clay, or simply for holding a tile long enough at an arbitrary position in three-dimensional space until the bonding material sets.

The third common topic is augmented reality and non-physical fabrication, where on the one hand gestural interfaces and head-mounted displays assist in the design and fabrication of physical objects, while on the other hand robots are programmed to paint with light or even to use light for the visualization of non-visible radiation - controlled by a robotic arm.

Project Papers

The project paper section contains innovative robotic projects from a wide range of robot users, from mechanical engineers to artists. The deformation of metal is explored in very different ways, from the bending of metal rods - either for creating three-dimensional objects for the Venice Biennale, or as reinforcement for non-standard concrete structures - to the three-dimensional deforming of sheet metal with a spherical tool.

Another area is the robotic application of materials, such as the weaving of spider-silk-like nylon strands, the shaping of plaster, and the extrusion of recycled plastic for furniture design. Furthermore, custom end-effectors, e.g. chainsaws and jigsaws, are explored, along with potential uses of industrial robots for large-scale architectural projects.

Industry Papers

In addition to the workshop, research, and project papers, Rob|Arch's innovative industry partners were invited to submit papers that showcase their most recent developments in the context of robotic fabrication in architecture, art, and design. Robot manufacturers present their newest series of industrial robots, alongside software systems that allow the direct loading of CNC code or the simultaneous control of multiple robots. Various interesting and innovative robotic fabrication methods are discussed, from the robotic winding of composite materials, to automated steel beam construction, high-end subtractive fabrication, and new programming strategies. The industry papers in particular show that robots are already being used for various tasks in the building industry and will soon become valuable state-of-the-art tools for the creative industry.

Outlook

The contributions presented in this book show that robotic fabrication in architecture, art, and design has evolved from being a small, specialized, and exclusive field of research, to a large community where robots are no longer used simply for milling or welding – as they were in the past decades – but as multifunctional machines that can perform an extremely wide range of tasks, from replacing human labor to performing tasks that would be impossible for the human hand.

New interfaces, developed by architects and designers themselves, enable the creative industry to control robots out of common Computer Aided Design (CAD) software, instead of having to rely on engineering-focused, specialized robotic software. This customization, not only of the end-effectors, but also of the software interfaces, allows architects and designers to move beyond industry-standard robotic applications towards highly optimized and customized machines. Architects, artists and designers have advanced from being mere "users" of robots, and have successfully emerged as recognized developers and trendsetters in robotic fabrication.

We are extremely grateful to our supporters, conference partners, and workshop hosts, as well as all the authors, for making this significant event happen. Special thanks go to Rob|Arch's main sponsor KUKA, represented by Alois Buchstab, for their steady support, not only of this conference, but of innovative and creative robotic projects in general. We would also like to acknowledge the fast growing community of "Robots in Architecture" who share their expertise, knowledge and passion of robots to meet the Association´s goal of making robots accessible to a wide range of new users.

Robotic fabrication in architecture, art, and design has gained great significance within the space of just a few years. As a central node of the creative robotic community, Rob|Arch will continue to carry this momentum. We strongly believe that, this time, robots are here to stay.

Sigrid Brell-Çokcan
Johannes Braumann
Rob|Arch 2012 Chairs
Association for Robots in Architecture

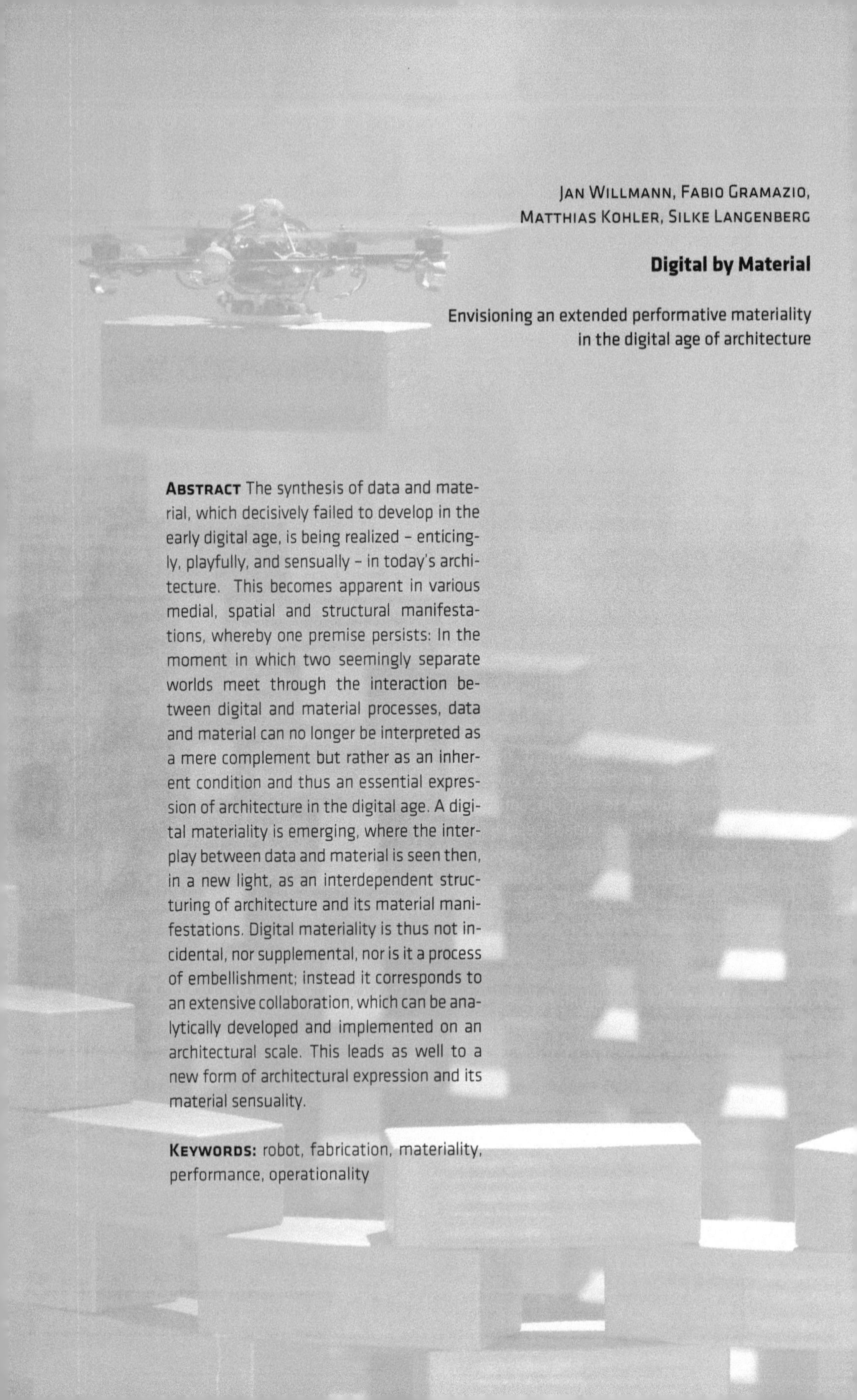

Jan Willmann, Fabio Gramazio,
Matthias Kohler, Silke Langenberg

Digital by Material

Envisioning an extended performative materiality in the digital age of architecture

Abstract The synthesis of data and material, which decisively failed to develop in the early digital age, is being realized – enticingly, playfully, and sensually – in today's architecture. This becomes apparent in various medial, spatial and structural manifestations, whereby one premise persists: In the moment in which two seemingly separate worlds meet through the interaction between digital and material processes, data and material can no longer be interpreted as a mere complement but rather as an inherent condition and thus an essential expression of architecture in the digital age. A digital materiality is emerging, where the interplay between data and material is seen then, in a new light, as an interdependent structuring of architecture and its material manifestations. Digital materiality is thus not incidental, nor supplemental, nor is it a process of embellishment; instead it corresponds to an extensive collaboration, which can be analytically developed and implemented on an architectural scale. This leads as well to a new form of architectural expression and its material sensuality.

Keywords: robot, fabrication, materiality, performance, operationality

The Digitalisation of Materiality

Today, at the threshold between the mechanical and digital age, it appears that a large part of contemporary architecture is determined by algorithmically established design procedures in which the constructive and building implementation is of insufficient significance and appears secondary; it is resolved only upon completion of the architectural design. With *Digital Materiality* [1] something entirely different is introduced: instead of realizing a design, an image, or a drawing, a comprehensive design and building process is conceived. Here, the central issue is not the design of a form, rather it is the design of a production process that is informed essentially and in equal measure by the constructive organization and the implementation. Thereby conceptual commonalities between the construction of a building component and the programming of a computer become apparent; just like a computer program that conducts different operations in logical order, constructive principles can be determined that define the production of architectural components as interrelated production steps [2].

The architectural creative will is nevertheless maintained in the setting of essential parameters and dependencies as well as by the actual design of a comprehensive building system. The creative will unfolds even more fully through the constructive collaboration of highly diverse parameters of the design and its materialization. Thus *Digital Materiality* is characterized by material precision and clarity; it is uncoupled, however, from formal guidelines and relocated to another (constructive) level. It is a design and construction process controlled in all its details by the architect, a fundamental balancing or weighing of real possibilities, so to speak, during the process of making (Fig. 1). Conversely, we are not talking about building systems that can be configured endlessly in the virtual space of a computer [3]. Rather the constructive logic of programming and the material realization are linked to each other. The structural production process that emerges in *Digital Materiality* is no longer that of the construction site or the workshop but rather a design process according to specific guidelines of the architect. The digital construction process reveals itself to

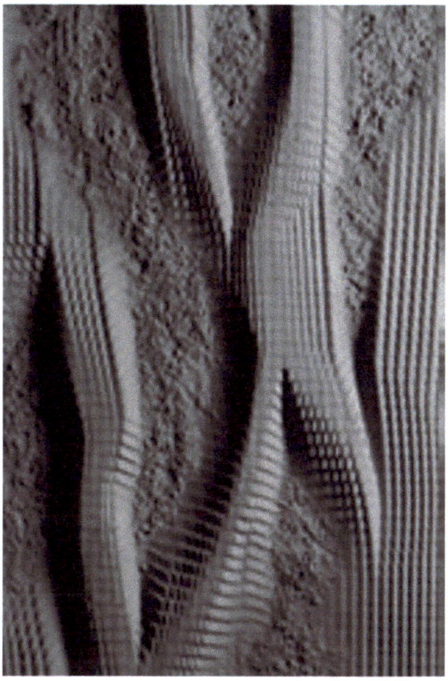

Figure 1 Robotically fabricated concrete element, casted at full scale in concrete (Procedural Landscapes, Gramazio & Kohler, ETH Zurich, 2011, in cooperation with Prof. Girot, ILA, ETH Zurich und Yael Girot, Atelier Girot).

Figure 2 Geometric precision of an algorithmically designed and robotically fabricated sand landscape (Procedural Landscapes, Gramazio & Kohler, ETH Zurich, 2011, in cooperation with Prof. Girot, ILA, ETH Zurich und Yael Girot, Atelier Girot)

be a constructive structuring, disengaged from the formal, the result of a "demystified" understanding of digital technologies and a freer, more autonomous use of the computer [4].

Consequently, *Digital Materiality* allows one to combine the abilities and deficiencies of human beings and machines to deliberate advantage. In the digital age this means that, while the machine with its numerical logic can rule over an infinitely large quantity of numbers, only human beings with their cognitive abilities and intuitive approaches can recognize meaning in them (Fig. 2). The result is an added architectural value through the "interactive connection" of the human and the machine, who are not equal but rather "equivalent" partners [5]. The added value points the way to completely new possibilities for a future constructional reality – not just quantitatively but also qualitatively. Thus *Digital Materiality* is far more than a mere rhetorical figure in the digital discourse; it represents an architectural vision [6].

New Modular Capacities in Space

The central problem is to what extent the difference between data and material, or digital and analogous realities, can be maintained, since *digital materiality* would seem to dispense with the frequently discussed dichotomies of programming and construction, of human beings and machines. Thus it recalls what Gottfried Semper pointed out long ago – the constructional requirements of architecture can be deduced primarily from different cultural and material models. According to Semper, the architectural result is formed by its own history, that is, by the process of its origin, the process of its making. It appears relieved of its original characteristic style of form and appearance, nearly emancipated. As Semper puts it, despite all these influences and transformations, in the end the different characteristics should remain recognizable and "owe their origin to the combined engineering arts in a primitive architectonic installation."[7]

Representing many other projects by Gramazio & Kohler, the project *The Fragile Structure* (Fig. 3) demonstrates these principles [8]. It makes apparent that *digital materiality* develops its greatest potential whenever the number of single components that stand in relation to one another is particularly high, although these linkages are not random but rather build on each other and are conditional on rules; they are material structurings assembled with an essentially open set of rules [9]. This creates specific constructive and aesthetic interdependencies and potentials, through which complex design processes can be developed not only in immediate dependency on the material used in each case but also in relation to their material sensuality. As part of the project, a robot freely stacks more than one thousand geometrically discrete wood elements without additional fasteners, so that the issue of inherent stability takes on a decisive role in the design [10]. In addition, *The Fragile Structure* is assembled by a mobile robotic unit. This makes it possible to efficiently and precisely build a structure whose dimensions exceed by many times those of the conventional work area of an industrial robot. It is also possible, during the process of installation, to adapt the size of the architectural structure to its surrounding environment [11].

The project *The Fragile Structure* was built in a parking garage because the spatial situation closely resembles that of a construction site; a slanted floor, interior supports, and a restricted ceiling height provided essential characteristics for adapting the structure to specific surroundings. Through the mobile robotic unit equipped with additional sensor technology, it was possible on one hand to recognize the environment and its geometric deviations compared with the idealized computer-planned situation. On the other hand, these data specific to the actual position could be immediately entered into the building process [12].

In *The Fragile Structure*, curved planes seamlessly merge into each other and span the distance between ceiling and floor; an interaction is set up between the rhythmic repetition of the additively assembled wood elements and their delicate dissolution into the spatially-adapted, self-supporting entirety. Complex visual phenomena are thus engendered, whereby the transparent appearance of this structure is due to a specific constructive dissolution, which, in turn, can be attributed to the fragility of the entirety and the geometry of the individual wood elements. Because of the porous assembly of the wood elements, a complex visual effect appears on the surface depending on the viewer's perspective

Figure 3 In-situ robotic fabrication of a complex modular building structure (*The Fragile Structure*, Gramazio & Kohler, ETH Zurich, 2012)

and on light conditions (Fig 4). Of course this is evident only at a distance; if one steps closer the illusion dissolves, leaving only a multiplicity of geometrically discrete elements. This means that one is dealing not only with the appearance of a material compound structure, but also with a visual and, in the original meaning of the word, a "virtual" event, that is, with a kind of "virtual materiality" (Fig. 5). Consequently, with regard to *The Fragile Structure*, it can be stated that such an architectural structure and its effect do not just inform each other, rather a further architectural potential of this structural and visual interdependency can be explored and differentiated through the use of the robot.

The Fragile Structure was initially developed by means of quite different variations. Early on, numerous robotically assembled prototypes were built, which allowed distinctly different building systems to be developed and validated in very rapid sequences – in an "evolutionary physical way". Moreover, the specific objective of an "untethered" structure that approaches a state of equilibrium made it difficult to accomplish this development work by hand. Through its constructional clarity and differentiated articulation, *The Fragile Structure* shows how concrete materialistic-empirical research is granted an important place in the digital age. Its value accrues because this kind of experimentation and repetition, programming and constructing, not only forms the small units that comprise various complex compounds, linkages and aggregations; rather within such a connection of data and material, these can now also be implemented in a controlled way within a specific spatial environment.

The Extended Operationality of Architectural Utopia

As becomes apparent in *Flight Assembled Architecture*, this implementation can go far beyond the scale of individual building components or structures. The project at the Fonds régional d'art contemporain (FRAC) Centre in Orléans represents the first architectural installation in the world to be built

Figure 4 Additive robotic assembly of discrete wood elements without additional fasteners (*The Fragile Structure*, Gramazio & Kohler, ETH Zurich, 2012)

Figure 5 Plastic deformation of the final structure in contrast to the firm materiality of the surrounding parking garage (*The Fragile Structure*, Gramazio & Kohler, ETH Zurich, 2012)

by flying robots [13]. In this case, several "quadrocopters" developed by Raffaello D'Andrea put together more than 1,500 elements to create a six meter tall complex vertical structure (Fig. 6). The flight behavior of the quadrocopters is based on the algorithmic translation of digital design data, according to which they land on a platform where they pick up individual elements and assemble them according to an assigned construction sequence. Thus a geometrically differentiated entity is created where the individual building layers are mutually offset and unite to present a six hundred meter tall building that paves the way for entirely new scales of digitally fabricated architectures.

Flight Assembled Architecture is not only an architectural installation, but rather a vertical urban utopia – a *Vertical Village* [14]. With 180 floors and a usable space of 1.3 million square meters, the *Vertical Village* provides living space for more than 30,000 inhabitants (Fig. 7). With its porous structure it creates the largest possible diversity of urban living [15]. Consisting of vertical core structures and staggered-module chains, the *Vertical Village* employs a grid-like organisation that allows a great degree of freedom to vary the arrangements of the modules. However, the varying arrangement does not run horizontally as in gridded city, rather it is turned vertically and is closed to form a circular unit. This results in a geometric compound that is the basis for the particularly constructive, self-stabilizing properties of the entire structure. Moreover, in the transition from an ideal urban plan to a spatially differentiated and highly condensed urbanity, it also strives for nothing less than a revision of the organisational

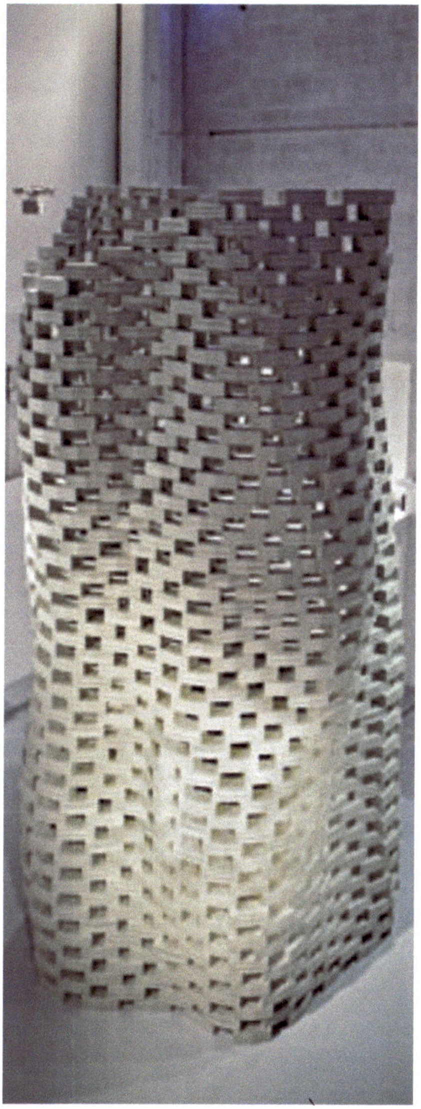

Figure 6 *Flight Assembled Architecture*. Installation assembled from 1,500 building modules at the FRAC Centre Orléans (*Flight Assembled Architecture*, Gramazio & Kohler and Raffaello D'Andrea in collaboration with ETH Zurich, 2011, image by François Lauginie)

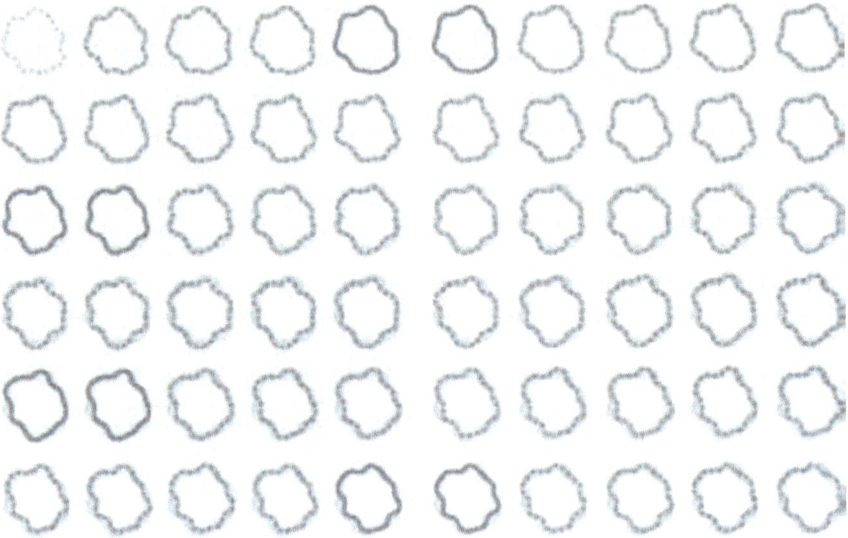

Figure 7 Sixty different building layers, creating the self-stabilizing, porous arrangement of the *Vertical Village* (Flight Assembled Architecture, Gramazio & Kohler and Raffaello D'Andrea in collaboration with ETH Zurich, 2011)

Figure 8 A quadrocopter placing a polystyrene module (*Flight Assembled Architecture*, Gramazio & Kohler and Raffaello D'Andrea in collaboration with ETH Zurich, 2011, image by François Lauginie)

scheme of a city in the 21st century. The monotonous and often unbearable density of earlier times becomes an engine for a newly discovered urbanity.

Flight Assembled Architecture also represents a technological "intensification." Here the use of flying robots not only provides for the architectural implementation of design data, their accumulation and processing, it also results in implementation of an actual, built installation. Far beyond that, the "quadrocopters" correspond to a model of thinking; they are a kind of a "conceptional door opener" that can free one from the constraint of the present and facilitate instead a radical architectural utopia that excludes neither the possibility of material experiments nor a possible built reality of the future. Similar to the case of the industrial robot, which underwent a breakthrough in industrial automation several decades before its first application in architecture, the flying "variation", too, represents an already established device that has been the focus of many research endeavors and is available on the commercial market. Similar to the industrial robot, the quadrocopter has a "generic nature"; it can be variously adapted and applied for dealing with architectural scalings, their basic requirements and degrees of flexibility [16]. It is important to stress, however, that the quadrocopter has the capacity to leave the conventional work area of an industrial robot; the airspace not only corresponds to an architectural environment, it also becomes an all-determining design paradigm [17].

From its essential tendency to combine different technologies, perspectives, and potentials *Flight Assembled Architecture* generates for the viewer a near "state of hovering" between a real architectural installation and a utopia (Fig. 8). Thus the installation – although on a scale a hundred times smaller than the projected *Vertical Village* – calls into question the supposedly distinct border between utopia and realities. Beyond that, the focus on the creation of a parallel reality, which is recorded with matching precision in *Flight Assembled Architecture*, thus becomes a systematic expansion of both the imaginable and the real [18]. At the same time, the boundaries between installation and building process, between architecture and robotics, increasingly dissolve here and themselves become an instrument that fathoms anew the "borders between the real and that which is conceivable" and present *Digital Materiality* in a new architectural "perspectivity." This indicates how we may understand and investigate robot-based design processes in the digital age of architecture. It registers that empirical and at the same time "speculative" character, without which the relation to architectural research would remain disengaged and distanced from both reality and future.

Thus *Flight Assembled Architecture* is not restricted to a pure "projection or imagining of the future," rather it is propelled by a concrete "logic of making." In the history of architecture there has always been an impetus to tame the new and thus to transform a multiplicity of possibilities and risks into concrete realities. Consequently, utopia in architecture – beginning with ancient descriptions of ideal states and cities, biblical approximations, the first architectural theory by Vitruvius, the idealized medieval representations of cities, the ideal city, architecture of the revolution up to the early socialists, modernists, and postmodern subversity – has its firm place in the history of the built and planned environment [19]. With this premise, the installation *Flight Assembled Architecture* shows that – as in all projects by Gramazio & Kohler – it is not a distinct architectural drawing or a pure picture-like vision that stands in the foreground; above all, it is a matter of pointing out, comprehending, and implementing an architectural process with all its spatial, functional, and aesthetic consequences. Thereby *Flight Assembled Architecture* opens up a radical material practice and comprehensive interdisciplinarity; at precisely this moment utopia becomes research into the future.

The "Interactive Connection" of Man and Machine

In navigating through the world of digital design and fabrication, the question always arises as to why these tasks cannot be accomplished without a robot. Indeed, if viewed from a global perspective one must concede that it is easier for human beings than for a robot to produce a simple brick wall. What might be called the "operationality of the robot" [20] pertains in situations of a certain complexity – when one does not want to continue to carry old paradigms as empty shells through the field of architecture. Every material entity that can be represented and grasped only with the help of a robot becomes at the same time the reason why the robot facilitates what human beings cannot do. A robot unfolds its potential precisely where an increasing number of complex relations and individual requirements justify its use, without leaving human beings out entirely [21].

As the following project, *Spatial Aggregations*, demonstrates, the question arises of how to fundamentally deal with architectural complexity (Figs. 9, 10), that is, the problem of the relation of spatial differentiality and functional performance. Initially this corresponds less to the efficiency, precision, and flexibility facilitated by the robot, it relates rather to the connection with architecturally complex tasks and artefacts [22]. Only there, where these tasks and artefacts are developed and materialized does the possibility emerge for representing, understanding, and developing

Figure 9 Robot-made complex spatial structure, assembled from a large number of generic rods (*Spatial Aggregations*, Gramazio & Kohler, ETH Zurich, 2012).

Figure 10 Scaled robotic fabrication (*Spatial Aggregations*, Gramazio & Kohler, ETH Zurich, 2012)

Figure 11 The connection of man and machine – mechanic precision and human intervention during the robotic assembly process (*Spatial Aggregations*, Gramazio & Kohler, ETH Zurich, 2012)

them further. For the architectural implementation of the project *Spatial Aggregations*, simple rod elements were selected. These are connected point-like with each other, and a large quantity is assembled in a geometrically differentiated manner. This results in statically redundant, spatially differentiated load-bearing structures, which – in contrast to traditional framework constructions – are individually adaptable and capable of assuming diverse configurations. Through robot-based fabrication, it becomes possible to produce these rod structures without recalibration and auxiliary structures; that is, the robot grabs a generic rod element, shortens and marks it before positioning it in space, according to the programming data from the already built structure in space. In this sense, the fundamental installation sequence exerts a decisive influence on the architectural design and building process; essential here is both the spatial positioning in the construction sequence as well as the connection of the individual elements.

This process requires new decision-making processes and extended degrees of flexibility; most of all it requires an intensive collaboration of human beings and machines, because the assembly of the rod elements is in no way a fully autonomous procedure. On the contrary, human beings become part of the mechanical process by inserting individual rod elements and installing them according to the previously applied markings. In this instance the robot merely positions them in space. The tolerances introduced through the human intervention are in the meantime compensated for by the other elements put in place by the robot. This example demonstrates the potential for future adaptive and recursive processes in digital design and construction procedures. The individual elements of *Spatial Aggregations* fit adaptively as described to form a coherent, differentiated and nevertheless harmonious whole so that even more unique and highly resolved spatial structures can be built. The goal therefore is not so much the pure, automatized materialisation of a concrete, previously defined state, but rather a procedural investigation into the cooperation between the human and the machine during a complex constructive assembly process.

While experimental research in architecture in the 1970s was still entirely concerned with "natural"[23], self-organizing or purely industrial-modular building systems, a very interesting shift emerges here: *Spatial Aggregations* is probably an important trial not least of all because the interaction between human intervention and digital fabrication procedures can be directly connected, allowing for the mechanical logic of the robot to work jointly with the cognitive intelligence of human beings (Fig. 11). In this respect, the benefit gained from the robot lies in its structure-generating differentiation of a large number of generic elements.

In *Spatial Aggregations*, the structures assembled by the robot can be not only quantified as a whole, each individual element has become qualifiable because human interaction, control and correction can be effectively architecturally integrated. It is, particularly, these cognitive design decisions that do justice to a comprehensive idea of construction in connection with the robot. What emerges are highly resolved, spatially complex aggregations, the results of which can be predicted or simulated only conditionally; they become accessible only through real experiments and actual materialization processes.

In this context, it must be stated that human beings will in no way be relativized by the robot; instead they will establish themselves as leading figures in a constructive reality between programming and fabrication. The implied turn towards an associative logic and human interaction could prove to be so far-reaching that the central issue of our debate would consequently be the opposite; that is, through the use of the robot, the human experiences a far-reaching re-conceptualization in the "force field" [24] of the architectural information age.

The Return of the Machine

It is perhaps this conceptual connection of human beings and machines that imbues *Digital Materiality* with its expression and makes possible the introduction of the robot into the architectural discipline. As a "multiple tool," the robot allows one to execute diverse applications in a rapid and precise way, but above all, to work directly at the immediate interface between digital and material spheres and thus to exert decisive influence on the programming and consequently on the design. Since the beginning of the 1990s, the robot has indeed become a primary tool of industrial and standardized forms of production, which throughout the entire 20th century have been influential in our understanding of contemporary society and its stimuli for the design disciplines. However, the development towards an increasingly reflexive, individual and global "stratification" [25] of cultural forms paradoxically represents an additional, almost complementary "turning point" [26]. This explains why, in the future, the robot will be granted more rather than less significance: Because the robot masters not only the language of unity but also that of diversity (Fig. 12).

Unquestionably, the hardly noted dispute about the division of labor belongs in this context, that is, the relation of the transformation of labor and the renewal of architectural formation in the transition from the mechanical to the industrial age [27]. Particularly here, the use of the robot makes clear that instead of perceiving the robot in the context of industrialization or unleashed capitalist production of goods, it rather must be seen as an expression of a more fundamental process of the digital age: as an expression of a new digital "workability" that is, as a fabric of diverse relations that has not only dispensed with the difference between authorship and the actual producer, the production of an original and its copy, but it has also materialized in an architectural reality, which has already begun.

As if this potential had always been present in the "DNA" of the robot, now comes a breakthrough that makes the robot the suitable tool, not only of a standardized but also of an individual and global world of production. Its "generic" properties handle the most diverse tasks with consistent efficiency, precision, and flexibility while always remaining open for additional adaptations, extensions and tasks.

The same is true for architecture: the robot attains significance for the architectural discipline because it allows for the implementation of individual work processes instead of uniform, repetitive building sequences, and it realizes them on an architectural scale. Thereby the industrial robot connects the (old) world of industrial logic with the (new) world of the information age, making it possible – between efficiency and precision – to grant the general primacy of individualization, even in technology (Fig. 13).

The concepts of the industrial division of labor, widespread to this day, are based on individual work sequences, on spatial and temporal surveys of human being-machine systems, which build largely on empirical knowledge. Frederick Winslow Taylor was the first to aim for clear rules and instructions for dealing with complex issues of the division of labor, exchangeability of single parts and efficient mass production. Although production processes became considerably more precise and less expensive – the same is true for standardization, distribution, and the repair of industrial artifacts that were produced in this way – reference was always made to the implementation of predefined objects and sequential procedures [28]. However, in the post-mechanical age, the fixed adherence to sequential production of architectural artifacts and their limited variation have been dispensed with, so that "productivity" and "specialization" are no longer necessarily

Figure 12 The world's first architectural robotic laboratory for non-standard assembly processes at ETH Zurich (Gramazio & Kohler, ETH Zurich, 2006)

contradictory, rather they realistically depict the interconnectedness of information and technology. The sequential categorization of architectural production thus falls away, and with it the classic division of labor of the discipline. It is precisely here that the robot in architecture makes an important contribution, so that the sometimes dialectic and equally marginal influence of digital technologies on architecture now corresponds to a "reflexive" form, through which it gains considerable significance. In this process diverse influences and disciplines enrich each other and enter into a mutual connection of diverse information and environments – less because of increased sales or efficiency potentials and more from an awareness of a culturally strengthened architectural production capacity [29].

Towards an Extended Performative Materiality in Architecture

For our present discussion, the great achievement is that the questions of efficiency, precision and flexibility can be simultaneously reinterpreted as a question of how to deal fundamentally with building. What is sometimes inconceivable in our current modes of thinking is that the robot-facilitated approach to a comprehensive technological fabrication capability corresponds in no way to a devaluation of human complexity; on the contrary, human capabilities can be considerably expanded through the "operationality" of the robot. Thus monitoring and control of complex material processes are not only improved, they can be implemented in a differentiated

Figure 13 Robotic fabrication laboratory for the design of robotically fabricated high rise buildings, SEC Future Cities Laboratory (FCL), Singapore (Gramazio & Kohler, SEC Singapore, 2012, image by Bas Princen)

way and thereby targeted for architectural purposes. Within these conceptual goals, the robot no longer is tied to the making of things, it also connects with the thinking of things. According to Mario Carpo, the consequence is that the division between the acts of designing and producing, which has existed since the Renaissance, dissolves. Thus, the "operationality" of the robot is not related exclusively to the material act of producing, of material operation, and of pure implementation, but rather equally to the way architecture is intellectually conceived, programmed, and designed. Carpo's thesis gains analytic acuity when it is modified to say, conversely: programming can be interpreted as an "anthropological" form of designing, constructing, and materialization, so that it is ultimately questionable whether – in the synthesis of programming and robot-based fabrication – the intrinsic "self-referentiality" of human beings and machine becomes generally visible. Thus it could be provisionally stated: the robot, outside of reasonable or sensible assessment, is a fascinating instrument in architecture, particularly because it facilitates – far removed from any determinism – the discovery of new constructive and spatial worlds, which in turn provide new insights for further discoveries. It remains to be seen how the robot will develop in the future. However, one thing can already be stated: In no way does operationality of data and material aim merely at digital aesthetics; it is far more than a short-lived chapter of the digital age. Rather it is a "perspectivity" that facilitates – from concrete, technology-based examination of computer programming to fabrication with the aid of robots – an open, complex and tangible asset of architecture. The thrust of digital advancement in design and manufacture can be thus investigated and included in the content of the discipline; it becomes possible to spatially and materially relate these developments and thereby to make them culturally significant.

References

[1] The merger of the terms "digital" and "materiality" can be traced to the essay "Digital Materiality in Architecture" in Fabio Gramazio, Matthias Kohler, *Digital Materiality in Architecture*, Lars Müller Publishers, Baden, 2008. Combining seemingly ambivalent concepts – the digital and the material – architecture is enabled to generate new constructive and sensual realities where data and material, programming and construction are interwoven by the techniques of digital fabrication. This allows not only to control the architectural manufacturing process through design data and therefore to "inform" material, but also to express a new sensuality in the digital age of architecture, being characterized by an unusually large number of precisely arranged elements, a sophisticated level of detail, and the simultaneous presence of different scales of formation.

[2] Central to this is an additive principle, which allows the assembly of complex architectural structures from single elements, and to control and manipulate them so that new kinds of spatial and functional configurations can arise.

[3] Cache, B, 2004, 'Towards and Associative Architecture', in Leach, N, Turnbull, D and Williams C, (eds), *Digital Tectonics*, Wiley & Sons, pp. 103–109.

[4] Gramazio, F and Kohler, M 2008, *Digital Materiality in Architecture*, Lars Müller Publishers, Baden, pp. 8.

[5] If the aforementioned similarities between the machine and the robot were further differentiated, however, this, while interesting, would be going too far in this context. Because it is not that the well-known properties of mechanical processes and procedures (e.g. seriality) to some extent are the "mirror image" of the use of the robot as well as justifying its importance in architecture.

[6] For more information, see Professorship for Architecture and Digital Fabrication, ETH Zurich, http://www.dfab.arch.ethz.ch (01.06.2012).

[7] Semper, G 1878, *Der Stil in den technischen und tektonischen Künsten oder praktische Ästhetik*, Band 1, Bruckmann, Munich.

[8] The project *The Fragile Structure* was developed 2012 at the ETH Zurich and realised with the support of Schilliger Holz AG (Project leader: Luka Piskorec; team: Volker Helm, Selen Ercan, Thomas Cadalbert; students: Petrus Aejmelaeus Lindström, Leyla Ilman, David Jenny, Michi Keller, Beat Lüdi) For more information, see Professorship for Architecture and Digital Fabrication, ETH Zurich, http://www.dfab.arch.ethz.ch/web/e/lehre/225.html (01.07.2012).

[9] Kolarevic, B 2003, *Architecture in the Digital Age: Design and Manufacturing*, Spon Press, New York.

[10] Interestingly, this debate is directly related to the modularity of such systems. *The Fragile Structure* demonstrates that such a modular approach on the one hand incorporates the logic of traditional serial systems, on the other hand, however, it articulates a wholly new states of affairs where such a generic and multiple construction systems I dimension of architecture. This "reversal" is, however, less radical as it is generally propagated within the contemporary architectural discourse on mass customization. For, although the customization and increasingly technological nature of construction make possible the integration of new freedoms and complexities, it also in parallel to ever new conventions, standardizations and simplifications, even when these are at first out of direct visibility.

[11] This in turn means that the traditional use of industrial ground robots in constant environments is questioned by this project, in that a mobile robot unit is made capable of manufacturing digitally informed and geometrically distinct building components, to a certain extent "in-situ". For more information, see "In situ robotic fabrication" (Echord/EU funded re search project), Professorship for Architecture and Digital Fabrication, http://www.dfab.arch.ethz.ch/index.php?lang=e&this_page=forschung&this_page_old=&this_type=&this_year=&this_id=198 (01.07.2012).

[12] For this purpose, the mobile robot unit mainly addresses cognitive characteristics: It recognizes its own position, the surrounding and building materials and processes the information gained. Moreover, manually produced components can be combined-with components manufactured by the robot. As a result, the robotic system responds to construction tolerances and is able to adapt to changing conditions autonomously. Research and development of this mobile robot unit is advanced by the ECHORD project within the Seventh Framework Programme of the European Union in order to create new use cases and develop the necessary technologies. For more information, see In-situ robotic fabrication (Project leader: Volker Helm; team: Dr. Ralph Bärtschi, Tobias Bonwetsch, Selen Ercan, Ryan Luke Johns, Dominik Weber), Professorship for Architecture and Digital Fabrication, http://www.dfab.arch.ethz.ch/index.php?lang=e&this_page=forschung&this_page_old=&this_type=&this_year=&this_id=198 (01.07.2012)

[13] This project is based on a collaboration of the Professorship for Architecture and Digital Fabrication (Prof. Gramazio, Prof. Kohler) and the Institute for Dynamic Systems and Control (Prof. Raffaello D'Andrea), both of ETH Zurich.

[14] Gramazio, F, Kohler, M and D'Andrea, R 2012, *Flight Assembled Architecture*, Editions Hyx, Orleans.

[15] The question of the diversity and accessibility of urban spaces and their contents becomes a central theme of the *Vertical Village*; in as much as the embedding of four gigantic continuous public double-rings with a combined length of 1 km each, are found not, as usually is the case, on the lowest floor, but are spread across the entire height of the building volume, creating heterogeneous city structures. The public space thus spreads across the entire height. Together with the inner courtyard – with a diameter of over 300 meters, certainly comparable with a valley in a landscape – this creates the possibility of an urban generosity and permeability, which treats public life with all it offers, less as uniform, horizontal and insular and more as an essential feature. At the same time, a completely unique form of intimacy takes place. For through the sheer size and structure of the Vertical Village, the inhabitants and their comings and goings are only roughly visible from outside, whilst remaining recognisable, and thus create an intimate presence within the *Vertical Village*.

[16] For more information, see Institute for Dynamic Systems and Control, http://www.idsc.ethz.ch (01.07.2012).

[17] Kohler, M 2012, 'Aerial Architecture', in: *LOG 25*, New York.

[18] Tönnesmann, A 2007, *Die Kunst der Renaissance*, C.H. Beck, Munich, pp. 22-23.

[19] Alison, J and Brayer, MA 2007, *Future city: experiment and utopia in architecture*, Thames & Hudson, London.

[20] Kohler, M, Gramazio F and Willmann, J 2011, "Die Operationalität von Daten und Material im Digitalen Zeitalter", in: *Die Zukunft des Bauens*, Detail - Institut für internationale Architektur-Dokumentation, Munich, pp. 8–17.

[21] Conversely, it would make less sense to use robots producing standardized building components, even when this would be technically possible. For an architectural use of robots for such things, the essential complexity would be missing; quite apart from the fact that the human represents with craft skills developed over millennia, and the highly advanced technology of mass production, a far more efficient framework within which simple and similar components can be manufactured.

[22] The project *Spatial Aggregations* was developed 2012 at the ETH Zurich and realised with the support of REHAU (Project leader: Luka Piskorec; team: Thomas Cadalbert; students: Petrus Aejmelaeus-Lindström, David Jenny, Gabriela Schär, Ripple Chauhan, Evangelos Pantazis, Stylianos Psaltis, Rahil Shah, Stella Azariadi, Ivana Damjanovic, Hjalmar Schmid, Lukas Mersch, Katharina Schwiete, Enzo Valerio, Andreas Kissel, Kulshresth Patel, Christian Grewe-Rellmann, Sonja Cheng, Joe Liao, Yushi Sasada, Tarika Sajnani, Janki Vyas, Bo Li, Yuji Mukaiyama, James Yeo, Eveline Job, Joséphine Simonian). For more information, see http://www.dfab.arch.ethz.ch/web/e/lehre/228.html (01.07.2012).

[23] Otto, F 1988, *Gestaltwerdung. Zur Formentstehung in Natur, Technik und Baukunst*, Müller, Cologne, pp. 5.

[24] Bourdieu, P 1997, *Zur Soziologie der symbolischen Formen*, Suhrkamp, Frankfurt/M.

[25] Beck, U, Giddens, A and Lash, S 1994, *Reflexive Modernization. Politics, Tradition and Aesthetics in the Modern Social Order*, Cambridge University Press, Cambridge.

[26] What has not remained unknown is that Konrad Wachsmann's *Wendepunkt im Bauen* (1959) deals with a similar theme, as does the work of Pier Luigi Nervi or also Felix Candela. The essential aspect is that Wachsmann had recognised early on the conceptual effects of industrialised production processes for architecture, and anticipated these for the digital age. Within this "Marxist" perspective, it is Wachsmann's distinction between technology and the art of building, which is essential to the debate conducted here, insofar as through the robot the "natural sense for material and joints" that Wachsmann postulated appears to be newly articulated and experiences a new "turning point", in the age of the viable individual and digital production of architecture.

[27] Aureli, PV 2011, *The Possibility of an Absolute Architecture*, MIT Press, Cambridge, MA., pp. 251.

[28] Taylor, FW 1913, *The Principles of Scientific Management*, Harper & Brothers, London.

[29] DeLanda, M 2006, *A New Philosophy of Society. Assemblage Theory and Social Complexity*, Continuum, New York.

Achim Menges

Morphospaces of Robotic Fabrication

From theoretical morphology to design computation and digital fabrication in architecture

Abstract The research presented in this paper investigates the possible transfer of the concept of morphospaces from theoretical morphology in biology to the realm of robotic fabrication and design computation in architecture. This investigation is concerned with the search for suitable methods of differentiating between the geometrically possible and robotically fabricable in integrative computational design processes, a critical component for further developing a morphogenetic approach to design. In the first, second and third part of the paper, the relevant aspects of morphogenetic design in architecture, theoretical morphology in biology and the related distinction between empirical and theoretical morphospaces are introduced. In the fourth and fifth part, the transfer of the concept of theoretical morphospaces from biology to design computation and robotic fabrication is introduced and explained along with the research on constructing machinic morphospaces for robotic production for robotically fabricated plate structures with finger joint connections.

Keywords: robotic fabrication, machinic morphospaces, computational design, digital morphogenesis, theoretical morphology

Introduction

Over the last two decades the logics and economics of serial production in the industrial prefabrication of building elements has eroded rapidly in the face of an increasing availability of computer-aided manufacturing and digital fabrication processes in the building sector. Whereas in industrial prefabrication the product was the one and singular outcome of a specific automated machine, the first wave of digital fabrication in architecture based on the introduction of computer-numeric control of long-established analogue machines (CNC mills, CNC saws, CNC joinery machines, etc.) led to a first significant increase of variability in production processes.

The second wave of digital fabrication currently underway entails a transition from job-specific computer controlled machinery to more generic production robots. This generic character of the basic robotic hardware – that only becomes specific when equipped with a particular effector and tool - enables the design of new fabrication processes prior or in parallel to a specific project, and thus potentially challenges the conventional hierarchy and sequences still predominant in design and fabrication in today's architectural practice. The research presented in this paper forms part of a larger research undertaking that investigates possible convergences of computational form generation and computer-aided materialisation in architecture through integrative design computation: an approach that has been termed morphogenetic design (Hensel, Menges et al. 2004 / Hensel, Menges et al 2006).

1 Morphogenetic Design

Contemporary architectural design is still characterised by a clear separation and hierarchical conception of the creation of form, space and structure and its subsequent preparation for materialisation. In contrast the approach presented here seeks to employ computational processes for a higher level of integration of form generation and materialisation (Menges 2011). Analogous to the processes of becoming that derive the complex organisation, versatile structure and articulated shape of natural systems, here the genesis of form is conceptualized as the interaction between system intrinsic materialisation capacities and constraints as well as system external influences and pressures. Exploring the space of the physically producible, this design process enables novel modes of architectural inquiry, functional integration, performative capacity and material resourcefulness.

A more detailed investigation of morphogenetic design computation has been discussed in various other contexts, including [i] the transfer of morphogenetic and evolutionary concepts by process biomimetics (Menges 2012), [ii] its relation to parametric design (Menges 2005), [iii] the relation between computational form and material gestalt (Menges 2008), [iv] the integration of material behavior (Fleischmann et al 2012), [v] the integrative characteristics of the developed design processes (Menges 2006), [vi] the underlying conception of performativity (Hensel and Menges 2008), [vii] the related multidisciplinary design approach (Fleischmann and Menges 2011) and [viii] various specific research project examples (Menges 2010 / Menges et al 2011). This paper will present the research on transferring another methodology from the realm of biology to morphogenetic computational design with a particular focus on robotic manufacturing. Here the main focus is on developing a robust method for describing the morphological variance of building elements together with the machine constraints of their robotic fabrication within one system, so that a non-hierarchical, direct feedback between key parameters of the computational form generation and materialisation can be established.

Thus, the paper will first introduce the notion of theoretical morphology in biology as a systematic approach of describing not-yet-existing form, and relate this to the definition of formal, multidimensional spaces based on morphological parameters, so called morphospaces, which can be adapted to serve as mathematical constructs to explore the space of the geometrically possible and physically producible.

2 Theoretical Morphology in Biology

Theoretical morphology comprises an important part of the discipline of morphology in biology. In contrast to empirical morphological studies it is concerned with the range of forms that biological entities could theoretically take (McGhee 1999). In biology it aims tor investigate why certain natural forms exist but others have not developed, with the ultimate goal of a better understanding of evolution. For this research the concept of theoretical morphology is of particular interest due to two reasons: First, it provides a methodological framework that allows for the transition from an analytic approach to morphology to a generative one, including the possibility of computing conceivable morphologies. Second, it allows the establishment of a distinction between empirical and theoretical morphospaces, which constitutes a critical part of the research on integrating morphogenetic design and robotic fabrication.

In the late 18th century Johann Wolfgang von Goethe invented the notion of morphology, stating that "morphol-

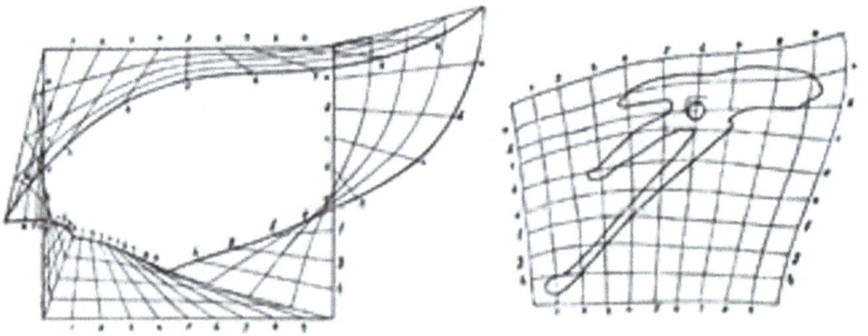

Figure 1a Based on the coordinate systems of the pelvis of Archaeopteryx and the pelvis of Apatornis D'Arcy Thompson derived three intermediate systems as an interpolation (left), which allows a corresponding inscribed pelvis to be generated (right). Source: D'Arcy Thompson (1947). Reproduced with permission of Cambridge University Press.

ogy may be said to include the principles of structured form and the formation and transformation of organic bodies" (Miller 1995). His work provides a profound first contribution to theoretical morphology, in particular his approach to generalizing the combinatorial logic of plant organs and animal bones. These investigations were strongly based on the understanding that morphological diversity develops within the constrictions of specific structures and that the infinity of possible forms is limited in natural variation. For the research presented here a critical moment in the development of early theoretical morphology occurs in D'Arcy Thompson's work. Thompson is generally regarded as a significant contributor to laying the conceptual foundations for contemporary computational design (Weinstock 2004), especially in terms of parametrics and the logics of associative geometry. One particularly relevant side to Thompson's work is the progression of his mathematical concept of biological form and transformation from a merely analytical mode of mapping existing form to one that suggests how his mathematical operations can become generative (Ahlquist and Menges 2011). The research described in his "Theory of Transformation" on educing an undiscovered, intermediate pelvic structure between the Jurassic bird Archaeopteryx and the Cretaceous bird Apatornis marks the point of transition from reconstructing known forms to generating unknown forms with mathematical methods (Thompson 1961) (Fig. 1a). It is interesting to note that Thompson's method has subsequently been adapted and tested in computational approaches (Rasskin-Gutman and Buscalioni 2001) (Fig. 1b). The underlying conceptual framework established by theses early researches is still considered to be of relevance and significant influence for con-

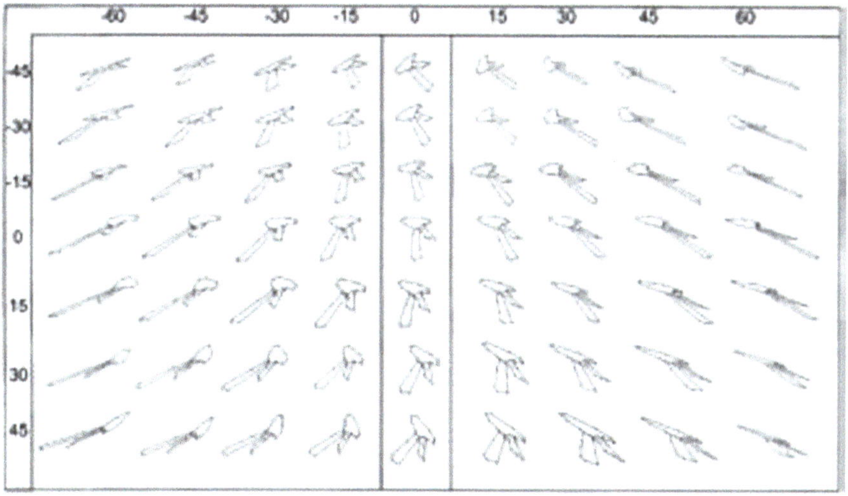

Figure 1b The affine morphospace of the hip outline of the theropod dinosaur Deinioniychus antirrhopus generated by the computer program D'ARCYGRAPH. Source: Rasskin-Gutman, D. and Buscalioni, A. (2001). Reproduced with permission of the Paleontological Society.

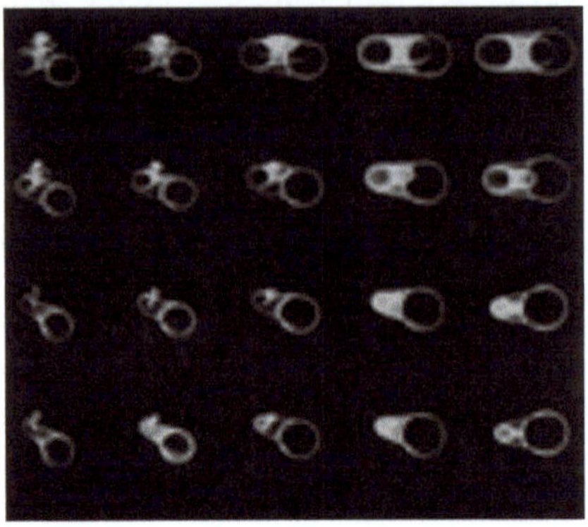

Figure 2a Oscilloscope photographs show the coiling geometries generated by an analog computer. Source: Raup, D. M. and Michelson, A. (1965). Reproduced with permission of Science.

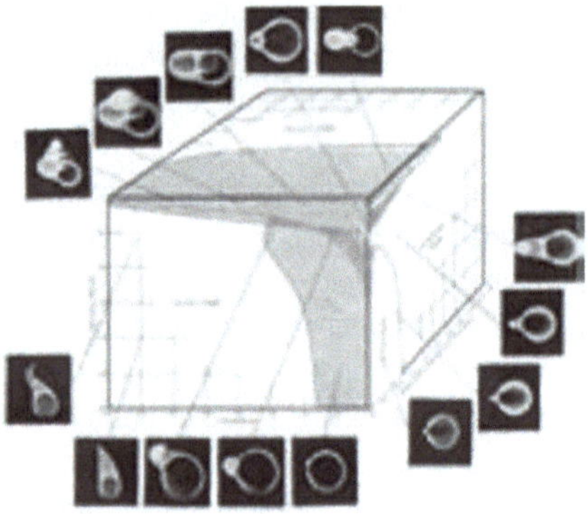

Figure 2b The morphospace of coiled shells is constructed based on three parameters. It shows the difference between the empirical morphospace of existing taxa (grey regions) and the theoretical morphospace (white regions). Source: Raup, D.M. (1966). Reproduced with permission of SEPM Society for Sedimentary Geology.

temporary theoretical morphology (Eble 1999). However Mendelism and Darwinism shifted the attention to other theories in biology. Further progress into modern theoretical morphology only occurred after the modern evolutionary synthesis and the rise of computational methods in the second half of the 20th century. Another critical moment for the research presented here is David Raup's seminal work on pioneering the use of computers for developing the theoretical morphology of coiled shells in the 1960s. Employing both digital computers (IBM 7094 with an Calcomp X-Y plotter) and analog computers (PACE TR-10 with an oscilloscope) he generated an entire spectrum of possible shell forms (Raup 1965) (Fig. 2a).

3 Theoretical Morphospaces and Empirical Morphospaces in Biology

In addition to the mathematical simulation of form, Raup's work also exemplifies the second critical component of theoretical morphology: the construction of morphospaces where the possible and the actual can be mapped and compared (Eble 1999). In order to introduce the relevant distinction between these two kinds of morphospaces the general concept needs to be explained: In evolutionary and developmental biology morphological spaces, or, in short, morphospaces, constitute formal spaces defined by multiple dimensions each corresponding with a variable parameter of morphology. They serve as computational and conceptual tools that allow for describing and relating the vast variance of organismal form in living nature (Mitteroecker and Huttegger 2009). A good example is the above mentioned morphospace of coiled shells developed by Raup (Fig. 2b). The selection of three parameters of the computational generation of shell morphologies defines the three axes of this morphospace: [i] the distance between the cross section and the coiling axes, [ii] the rate of translation of the cross section along the axis per revolution and [iii] the rate of size increase of the shell per revolution (Raup 1966). Another critical aspect exemplified by Raup's morphospace is the distinction between two regions: the parts where existing taxa can be mapped into this space – the empirical morphospace (marked grey in Fig. 2b) – and the entirety of the space – the theoretical morphospace – of which large regions remain empty (the white regions in Fig. 2b).

In morphospace studies the difference between empirical and theoretical descriptions plays a central role. Whereas empirical morphospaces are representations of what has actually come into existence in nature (McGhee 1991), theoretical morphospaces are based on mathematical parameters and represent what could be possible. With their dimensions being "geometric or mathematical abstractions of form" (McGhee 1991), theoretical morphospaces have the "ability to specify nonexistent form" (McGhee 1991). This characteristic suggests a potential overlap between the concept of morphospaces in biology and morphogenetic design computation.

4 Morphospace Concept: Transfer to Computational Design and Robotic Fabrication

In computational design the generation of form is based on algorithmic processes that operate within specific variable ranges of selected parameters. Thus the design of the generative process always precedes the design of a specific result, and the designer's focus extends beyond shape, space and structure towards the underlying generative system (Ahlquist and Menges 2011). The variance range of each parameter of this underlying system can be conceptualized as delineating a multidimensional space very similar to the morphospaces of theoretical morphology in biology. In this way, the capacity of a generative computational design system to derive "nonexistent form" can be directly related to the constraints of machinic processes of actualizing specific shape.

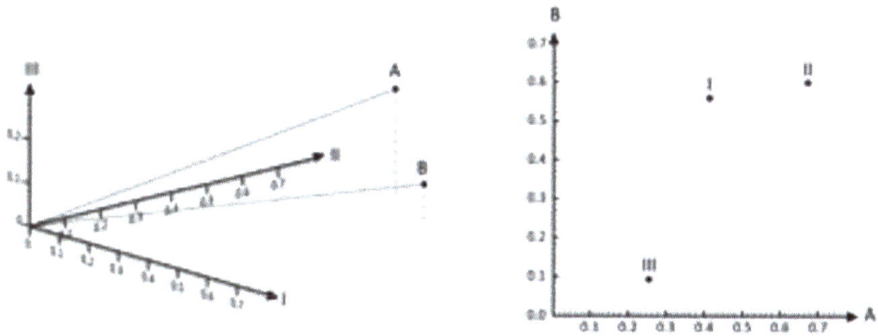

Figure 3 Data for p parameters in n cases can be represented in R-spaces and Q-spaces. Whereas the R-space is spanned, for example, by the three cases I,II and III and the parameter values A and B are two points in this space (left), the three cases are single points in a Q-space defined by the parameter ranges of A and B. Based on: Mitteroecker and Huttegger (2009).

The transfer of concepts from theoretical morphology in biology to morphogenetic design and robotic fabrication in architecture may be of profound relevance, as it allows the representation of the variable parameter range of both form generation and materialisation in one morphospace. This provides a critical methodological step for overcoming the inherent hierarchy of established design processes – including most digital ones – that prioritize the definition of form over its subsequent physical realisation. The concurrent mapping of both [i] the theoretical morphological variance of p shape parameters and [ii] the specific region of manufacturability based on the specific constraints and affordances of a robotic fabrication processes within the resulting space for n possible morphological actualisations lays the foundation for a systematic investigation of a non-hierachical convergence of computational design and digital fabrication.

4.1 Q-space and R-space in Multivariate Statistics

To represent data for p parameters in n cases two different kinds of formal spaces are employed in multivariate statistics (Mardia et al. 1979): [i] the first kind is the n-dimensional R-space (Fig. 3 left), in which variables are represented by p points. In biological theoretical morphology phenomena like morphological integration and modularity are investigated in R-space, as in this space the correspondence between variables can be calculated as the cosine of the angle given by the two vectors that connects the origin with the related points (Mitteroecker and Bookstein 2007).

[ii] The second, more common kind is the so called Q-space (Fig. 3 right), which is the p-dimensional space spanned by the variables. In this space each morphological instance is represented by a single point. Thus morphological similarities and dissimilarities among instances are indicated by decreasing or increasing distances between points. If the Q-space has a Euclidian structure these distances can be computed as the Euclidian distance between Cartesian coordinates of an underlying Euclidian vector space. On the one hand, Euclidian vector space has the advantage that it allows various geometric and algebraic operations, as the vector space underlying Euclidian geometry determines its algebraic structure. Based on a fully defined notion of addition and scalar multiplication, morphological instances can be generated through different transformations and a morphological instance situated between two other morphologies can be generated. On the other hand, the stringent properties and relations characterising Euclidian morphospaces have the disadvantage that they considerably limit the range of variables that can be investigated. Very often in both biology and the architectural research presented here, precisely formalizing parameters results in morphospaces that are "weaker" than Euclidian space. Whereas modern morphometrics regularly employs more abstract spaces based on, for example, Riemannian geometry, "traditional" multivariate morphometrics can also be adapted to handle a wider range of parameters. This is of particular relevance when comparing morphological features expressed through incommensurate units within one morphospace.

4.2 Affine Q-spaces for Incommensurate Parameter Units

Raup's "classical" morphospace of coiled shells can again serve as an interesting example. His parameters cannot be accommodated by a strictly Euclidian structure. While only two of his parameters are at least of the same unit, they are still employed to serve profoundly different purposes in the geometric model and they lack a "natural" scalar relation (Mitteroecker and Huttegger 2009). Nevertheless, they still result in a meaningful morphospace. This example shows how a vector space can be constituted by axes that do not rely on possessing commensurate units. Thus the representation of different parameters such as, for example, volumes, distances and angles, which do not share a common scale, is possible. A space defined by such variables is not a Euclidian space but an affine space. This is the space underlying affine geometry, which is characterised by remaining unchanged when undergoing affine transformations. Geometric properties that are invariantly relative to affine transformations include for example collinearity, incidence relationships and barycentric combinations. The awareness of the variance and invariance of geometric properties is of critical importance for the correct understanding and transfer of such morphospaces to the realm of morphogenetic design and robotic fabrication.

4.3 Machinic Morphospaces and Robotic Production

Manufacturing and fabrication – or more precisely the constraints and affordances of these processes – play an important role in morphogenetic computational design. This aspect of manufacturing-informed design computation requires the development of appropriate design methods capable of navigating the narrow path between under-determining manufacturing specificity, which leads to a lack of rigor and consequently operativeness, and over-constraining machine limits, geometric properties and boundary conditions resulting in premature convergence and lack of exploratory potential. The transfer of the concept of morphospaces from biology to computational design offers the possibility to develop a robust methodological framework for the concurrent mapping of the morphological variability and production constraints of building elements. While other, more recent morphospace – as for example the Kendall Shape space – do exist, this transfer will be explained in the next paragraphs through a project specific, affine Q-space based on the variable range of three parameters derived from robotic fabrication.

5 Constructing a Machinic Morphospace: Example of Robotically Fabricated Polygonal Plywood Plates with Finger Joints

The concept of machinic morphospaces has been investigated in the context of researching robotic prefabrication processes for polygonal plywood plate structures. This research originated from the recognition that robotic fabrication has the potential to extend traditional wood jointing techniques. Finger joints are of particular interest, as they allow for embedding the joint in the actual plates with no need for any additional mechanical or adhesive elements, creating a mono-material form- and force-fitting connection that further enhances the excellent ecological performance of wood structures with regard to a very low level of embodied energy (Alcorn 1996) and positive carbon footprint (Kolb 2008). In traditional industrial processes of fabricating finger joints the set-up of the respective machine limits the possible wood plate connection to one angle and one set of plate thicknesses. In contrast, the robotic fabrication process developed as part of this research project enables the cutting of finger joints for a wide range of connection angles (Fig. 4a) and even differing plate thicknesses. These polygonal plywood plates with robotically fabricated finger joints provided the starting point for the development of a lightweight wood construction system (Fig. 4b). As the resulting wood plate structure consist entirely only of these polygonal plates, the mapping of their possible morphological variance in relation to the specific fabrication set-up presents a critical facet for the ensuing computational design process. To construct a morphospace of plate morphologies a structure similar to the one investigated in theoretical morphology in biology can be developed.

5.1 Definition of Building Element Morphology Parameters

At the onset, the construction of the morphospace requires determining the relevant morphological parameters that comprehensively describe the theoretical morphological variance across a large number of possible plate instances (Fig. 5). The following parameters were identified for the plate

Figure 4a The robotic fabrication of finger joints allows the connection of plywood plates at variable angles without the need for any additional mechanical elements.

morphology: [i] the size of the plate is expressed as the radius of the circle around the polygon centroid through the polygonal vertices. This parameter is called polygon radius and measured in millimetres. The initial parameter range comprises 0 to 1100mm, approximately half the distance between the axis 1 and the axis 7 of the robotic set-up. [ii] The connection angle α between plates is expressed in degrees and can theoretically vary between -180° and +180°, with the range between -180° and 0° representing concave plate connections and the range between 0° and +180° representing convex plate connections. [iii] The shape of the polygon is expressed through

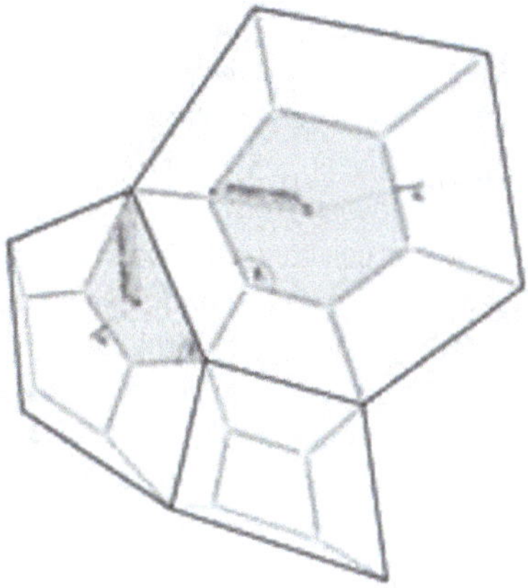

Figure 5 The three relevant parameters of the plate morphology are the polygon radius, the connection angle α, the polygon angle β.

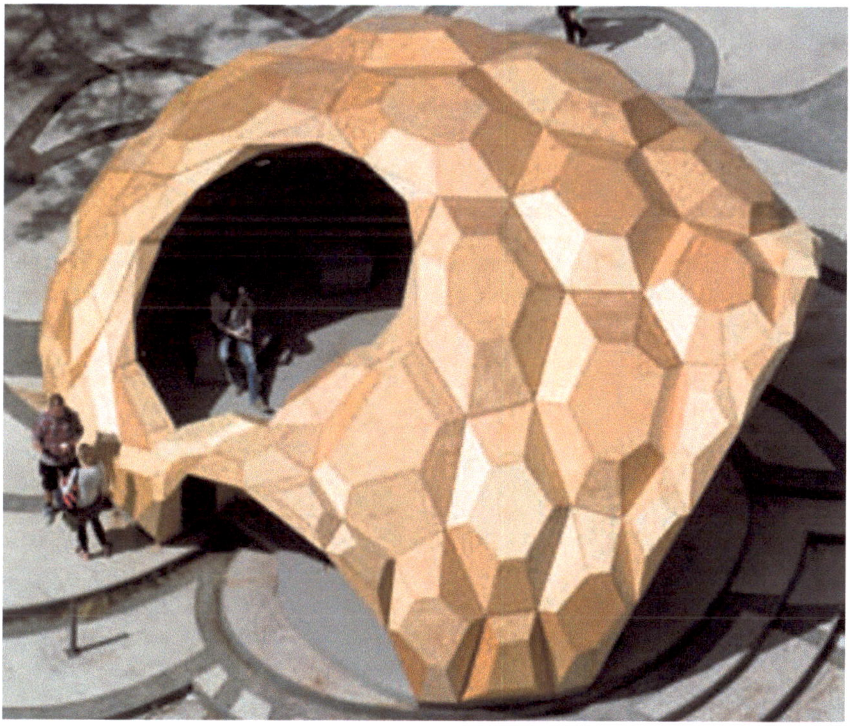

Figure 4b The development of a lightweight system of wood plate structures is based on the morphological variance of the finger joint plates enabled by robotic fabrication.

the angle between the polygon edges. The related parameter, termed polygon angle β, is expressed in degrees and can theoretically vary between -180° and +180°, with the range between -180° and 0° representing concave polygon segments and the range between 0° and +180° representing convex polygon segment. In the case of the research project described here the range of polygons was limited to four- to seven-sided polygons and the parameter of the plate thickness was set to be constant at 6.5 mm in order to demonstrate the system's lightweight potential.

5.2 Definition of Robot Fabrication Set-up

The delineation of the morphospace's region that separates the producible from the geometrically possible is highly set-up specific. In addition to understanding the possible working positions that the TCP (tool centre point) can reach, other constraints of the specific robot kinematic, which comprises the interaction between motors and internal sensors, need to be taken into account, as for example the limited rotation around the A6 axis of various robot types caused by the effector's power and air supply pipes. In the case of this research project the robot set-up for fabricating the finger-jointed plywood plates (Fig. 6) consist of: [i] KUKA KR125/2 robot with a distance of 1210mm between the A3 and A6 axes, [ii] KUKA KPF1-V500V1 turntable as A7 axis, on which the stock piece is mounted, with a distance of 2225mm between the A1 and A7 axes, [iii] HSD ES 350 spindle unit and [iv] Leitz 20/120 Z4 milling tool, which was custom-made by one of the project's industrial partners and used for all fabrication processes. Any change in this set-up will have an effect on the related morphospace.

5.3 Identification of Set-up Specific Fabrication Constraints

A critical step in the morphospace's definition is identifying the robot set-up's specific fabrication constraints and relating them to the morphological parameters of the plates to be fabricated. The connection angle α between plates is defined through the normal vector No of the stock piece to

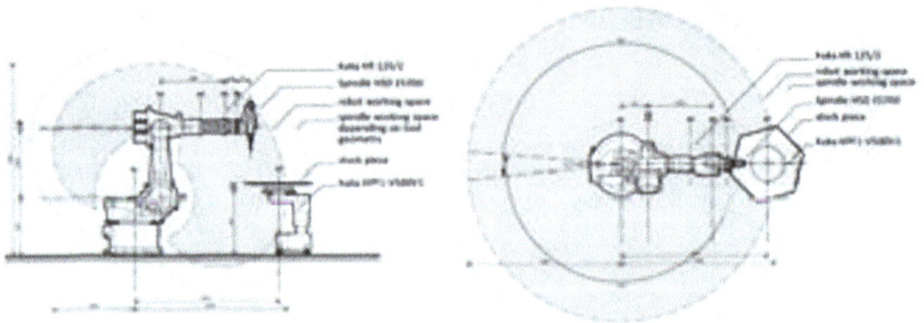

Figure 6 The robot set-up for fabricating the finger-jointed plywood plates consists of a KUKA KR 125/2 robot equipped with a HSD ES 350 spindle unit and a Leitz 20/120 Z4 milling tool as well as a KUKA KPF1-V500V1 turntable.

be fabricated and the normal vector N_1 of the connecting plate. The stock piece to be fabricated is cut to shape by a 3-axis milling cycle and subsequently mounted to the turntable by a quick-release lock. The orientation of the local coordinate system in relation to the robot's A_1 axis is defined by the centroid of the polygon, a second vector point in the same plane and the plate's normal vector N. All plates are mounted on the turntable with the later outside surface pointing upwards. Whereas concave connection angles can be produced with the tool approaching the stock piece from the top, convex connection angles require the effector to be below the plate (Fig. 7 left). The connection angle determines the tool angle of the three fabrication steps: [i] In a first step, the pre-cut plate's outline is trimmed to the exact dimensions and angle α by end milling. [ii] In the second step the mitre angle defined as the by-sector of the angle 180°- α is cut by end milling with the same tool. [iii] In the third step the finger joints are cut by both the tip and the shank of the same tool at the angle orthogonal to α (Fig. 7 right).

The robot set-up constrains the connection angle α to be considerably smaller than then the geometrically possible -180° to +180°. Both the specification of the effector, as for example the dimensions of the spindle, and the length of the tool, limit the range of producible outline-, miter- and joint-angles (Fig. 8). As the specific angle constraints of all three fabrication

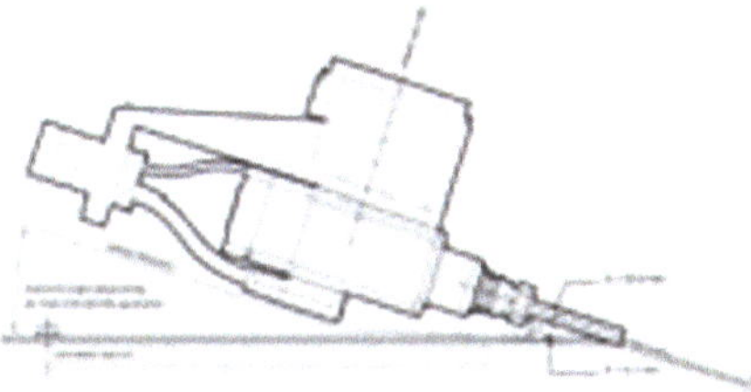

Figure 8 The specific spindle and tool geometry constrains the variable range of the parameter connection angle α.

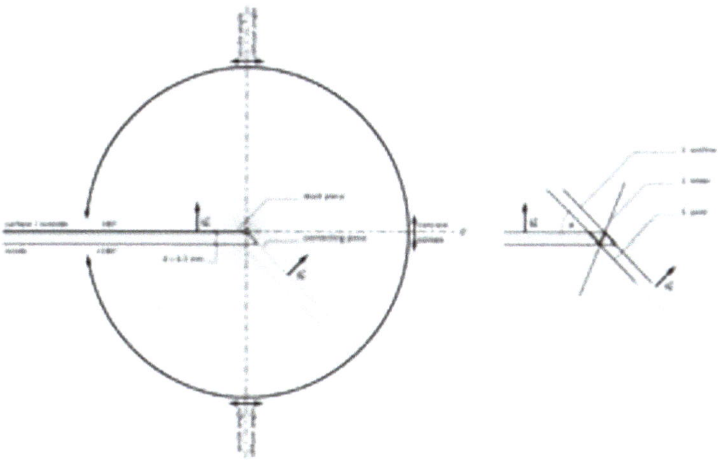

Figure 7 The parameter connection angle α is defined by the normal vectors N_0 and N_1 of adjacent plates (left) and determines the tool angle of the three steps of fabricating the plate outline, mitres and finger joints (right).

processes are a function of α, they can all be expressed through the limits of the variable range of the parameter connection angle α.

The parameter range of the polygon radius reflecting the plate size also directly depends on the specific robot set-up. The distance between robot and turntable – i.e. axis 1 and axis 7 – in conjunction with the length of the robot arm – i.e. the distance between axis 3 and axis 6 and the robot-kinematics – determines the maximum plate size, especially for convex connection angles, as they require the tool to approach the plate from below. The minimum plate size is constrained by the geometry and dimension of the mounting mechanism on the turntable (Fig. 9).

The important limits of the parameter polygon angle β range are only indirectly influenced by the specific robot set-up. This mainly relates to maintaining a minimum distance d that results from the

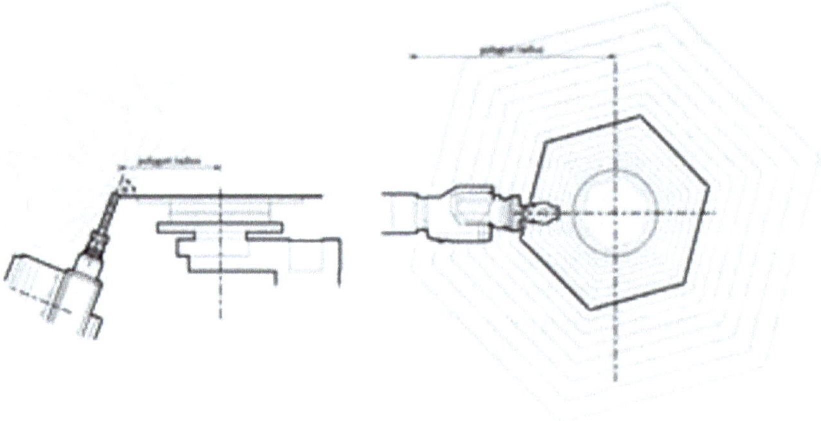

Figure 9 The variable range of the parameter polygon radius reflects the minimum and maximum fabricable plate size.

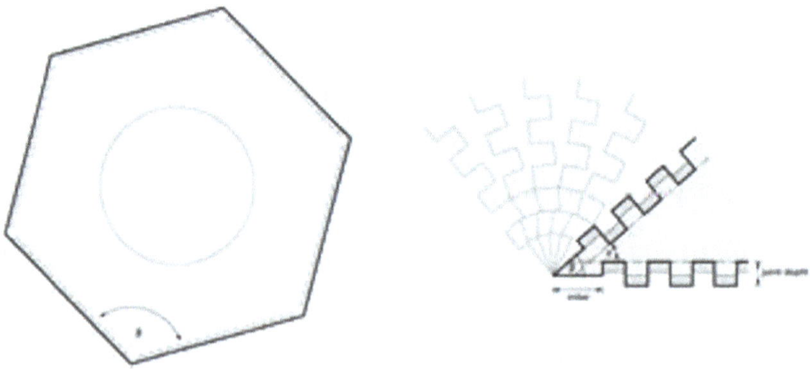

Figure 10 The convex range of the parameter polygon angle β is limited by a minimum distance d between the joints.

joint depth determined by angle α and the convex range of angle β (Fig. 10). Obviously more directly machine-related constraints are the diameter of the mounting mechanism on the turntable, which – together with the maximum reach of the tool – has an effect on the how far β can approach 180° for extreme convex angles. Similarly, the effector dimensions and tool length determines how close β can get to -180° for extreme convex angles.

5. 4 Interdependence of Parameter Min/ Max ranges

The kinematics of the 6-axis robot allow for multiple robot positions to reach the same TCP. Thus the actual numerical limits of the parameters are identified through fabrication simulations (Fig. 11). This process allows not only for establishing the min/max values for each parameter, but also enables the investigation of the interdependence of the parameter ranges. For the subsequent morphospace construction it is of critical importance to understand how one parameter range's limits has an effect on the other, and that this interdependence does usually not constitute a linear relation. For example, the interdependency between the possible plate size expressed through the parameter polygon radius and the producible finger joints expressed through the parameter connection angle α increases non-linearly. Another example is the interdependency between the connection angle α and the polygon angle β: as the finger joint depth increases the closer angle α gets to 0° and +/- 180° (with the shortest finger joints at +/-90°). This has an effect on the maximum value for polygon angle β to prevent the finger joints of adjacent polygon edges from falling below the minimum distance d.

5.5 Construction of Machinic Morphospaces

Following the definition of the morphological parameters of the building element, the identification of set-up specific fabrication constraints and the interdependence of the

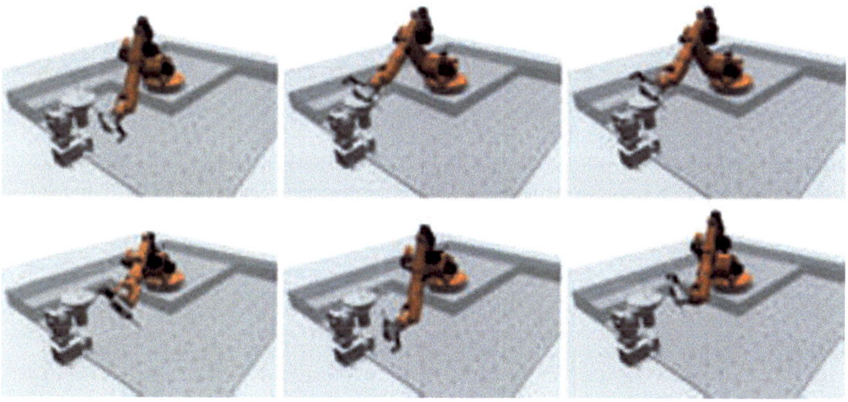

Figure 11 Fabrication simulations determine the minimum (top row) and maximum (bottom row) range of the parameters polygon radius (left), connection angle α (centre) and polygon angle β (right) based on the specific robot-set up.

parameter ranges, a morphospace of robotically fabricated plates with finger joints can be developed. In this case, a Q-type space with three axes representing the [i] polygon radius (0 to 1100 mm), [ii] connection angle α (-180° to +180°) and [iii] polygon angle β (-180° to +180°) was developed. The resulting three-dimensional space describes the theoretical morphospace including all theoretically possible plate morphologies (Fig. 12: white regions). The spatial regions that include all plate morphologies fabricable with the specific 7-axis robot set-up are mapped into this space based on the set-up constraints of the parameter ranges described above (Fig. 12: blue regions). This space, which resides within the theoretical morphospace of a particular building element's entire morphological range but only includes the regions theoretically fabricable with a specific machine, can be termed machinic morphospace (Menges and Schwinn 2012). The concept of machinic morphospace presents significant possibilities for

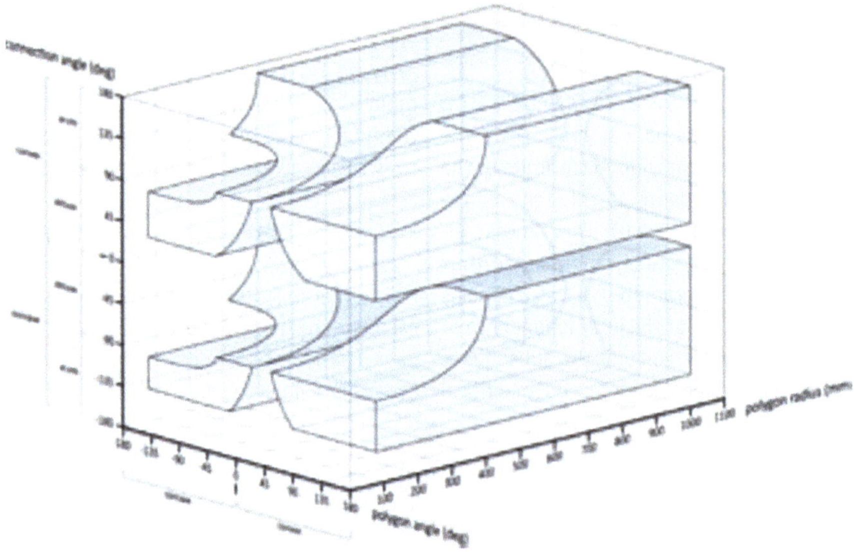

Figure 12 The theoretical morphospace of plate morphologies (white) includes the spatial regions fabricable with the specific 7-axis robot set-up (blue), which is the machinic morphospace based on the min/max ranges of the parameters polygon radius, connection angle α and polygon angle β.

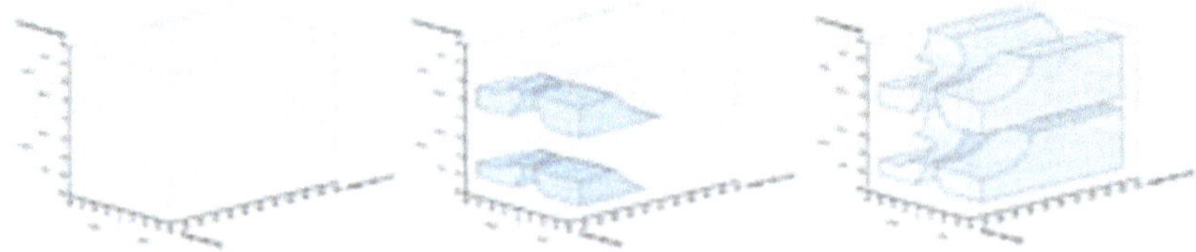

Figure 13 Morphospace of all theoretically possible plate morphologies(left) with morphospace regions fabricable with a 6-axis robot set-up (centre) and a 7-axis robot set-up (left).

morphogenetic design computation: it offers mathematic and non-ambiguous fabrication checks, it allows for concurrent monitoring of producibility during computational form generation and thus provides a critical step in enabling a full design exploration of robotic fabrication without preconceived limits to what may or may not be possible.

5.6 Shifting Machinic Morphospaces

In addition to the design exploration of one robot configuration, the machinic morphospace also allows for rapid adaptations to changes in the specific set-up. For example, within the same morphospace of all theoretically possible plate morphologies (Fig. 13a) both the regions fabricable with a 6-axis robot set-up (Fig. 13b) and the regions producible with the 7-axis robot set-up (Fig. 13c) can be registered and compared. In this case, the related expansion of the machinic morphospace identifies the significant gain of morphological variance enabled by the inclusion of the turntable. In the same way, multiple machinic morphospaces representing changes to variables in the robot set-up (i.e. different tools, different axes positions, different mounting mechanisms, etc) can be mapped into the same theoretical morphospace to register possible shifts and expansions to the space of fabricable building element morphologies.

5.7 Hyperdimensional Morphospaces

The research example project introduced above also implies two other relevant aspects for the transfer of the morphospace concept to morphogenetic design computation in architecture: First, it indicates that even for a mono-material, single element construction system a complex morphospace is required. Taking the interrelation between local building element morphology, regional surface/structure morphology and global building morphology into account, the need for additional parameters becomes obvious. A larger number of possible parameters, which means a larger number of possible dimensions of form, can be employed to construct a hyper-dimensional space of possible morphologies (McGhee 2007), potentially even including multiple scales of morphological articulation or different levels of hierarchy. A second important point is the recognition that multidimensional theoretical morphospaces, even the three-dimensional one introduced

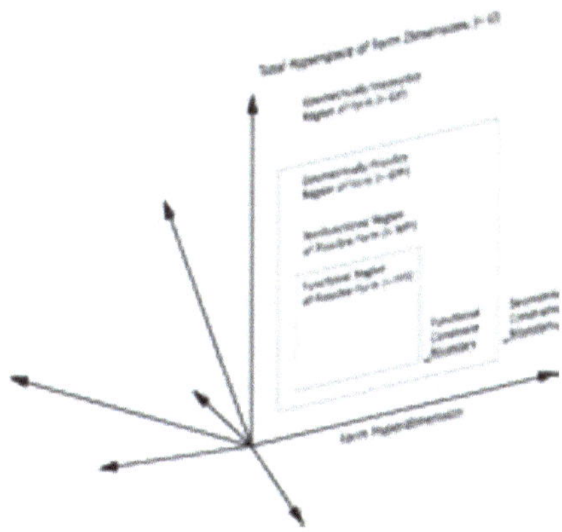

Figure 14 Each dimension of a theoretical hyperdimensional space represents one parameter of form. The total hyperdimensional space is divided into geometrically impossible (GIF) and possible (GPF) regions of form by the geometric constraint boundary. Within the GPF region the functional constraint boundary separates the non-functional (NPF) and functional (FPF) region of form. Based on: McGhee (2007).

above, also include a geometrically impossible region of form (GIF). The geometric constraint boundary separates this region from the geometrically possible region of form (GPF), which in itself is divided into a non-functional region of possible form (NPF) and a functional region of possible form (FPF) by the functional constraint boundary (McGhee 2006) (Fig. 14b). In the research example the FPF region equals the machinic morphospace, which may be subject to yet another set of sub-regions.

5.8 Navigating the Machinic Morphospace

In the theoretical morphospace of robotically fabricated plate morphologies presented, one sub-region of the machinic mophospace is of particular interest: the one that includes element morphologies with a high performance capacity. In other words, one can say that the machinic morphospace still describes a vast number of plate morphologies which includes plate configurations that make more or less sense in regards to architectural, structural or environmental performance. Thus, one challenge of morphogenetic design is not only to maintaining coherency with the FPF region, but also to computationally populate particularly promising areas of it.

In the research project presented above, biomimetics has been identified as a suitable approach for navigating the machinic morphospace. The bottom-up design development of the material system – the plywood plate structure based on robot-fabricated finger joints – and the related machinic morphospace was "filtered" by a top-down biomimetic method. This entails the identification of a technical problem or opportunity (here: the plate system), the search for biological analogies that provide the base for a performance gain of a technical system, the subsequent detection of relevant principles, their abstraction and transfer from biology to design (Knippers and Speck 2012).

The plate skeleton morphology of the sand dollar (Clypeasteroida), a subspecies of the sea urchin (Echinoidea), was recognized as a suitable biological system as it consist of discreet polygonal plates that are connected by finger joint-like calcite protrusions at their edges (Seilacher 1979). The sand dollar's shell has evolved ways to compensate the innate weakness of finger joints with regard to transferring bending moments or tensile forces. It has developed plate morphologies that are stabilized only by shear forces acting along the plates' edges, and thus capitalizing on the inherent structural capacity of finger joints. Eight morphological principles relevant for this structural performance were identified and embedded in algorithmic design rules. By synthesising the principles of performance-oriented design through functional

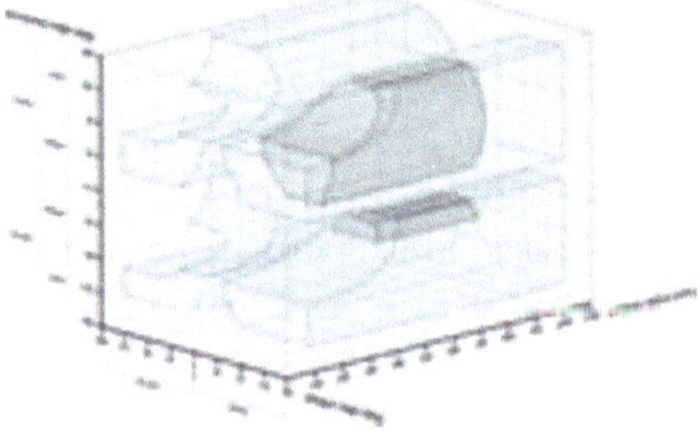

Figure 15 The plate morphologies actually produced during the construction of the prototype lightweight structure all populate a particular region (grey) of the theoretical machinic morphospace of fabricable plate morphologies (light blue).

morphology and the constraints of producibility through machinic morphospaces the computational design process enabled the exploration of complex plate morphologies providing both exceptional structural performance and novel architectural opportunities. This was verified by the construction of a full scale demonstrator, the ICD/ITKE Research Pavilion (Knippers, Menges et al 2012) developed by the Institute for Computational Design and the Institute of Building Structures and Structural Design at the University of Stuttgart. The structure consists of more than 850 geometrically unique, robotically fabricated birch plywood plates with more than 100,000 individual finger joints, which only populate a particular region of the machinic morphospace (Fig. 15). The performance capacity of the plate morphology is demonstrated by the fact that the entire pavilion (Fig. 16) could be built exclusively out of 6.5 mm thin sheets of plywood, resulting in a very materially efficient and lightweight construction system that envelopes a gross spatial volume of 200m³ consuming only 2m³ of wood. For a more comprehensive description of the integrative computational design and robotic fabrication processes of this research project refer to the research paper "Robotically Fabricated Wood Plate Morphologies" (page 48).

6 Conclusion

The transfer of the concept of morphospaces from theoretical morphology in biology to the realm of robotic fabrication and design computation enables the definition of theoretical morphospaces for digitally produced building elements. These multidimensional spaces make it possible to differentiate between regions containing geometrically impossible, geometrically possible and robotically fabricable element morphologies, with the latter region being defined as the machinic morphospace of a specific robot set-up. The rapid adaptability to changes in the robot set-up and a biomimetic method for exploring the machinic morphospace has been explained along a particular project example which shows that machinic morphospaces allow the synthesizing of the processes of form generation and materialisation through robotic fabrication in one integrative, morphogenetic computational design process.

Figure 16 The prototype structure ICD/ITKE Research Pavilion 2011 consists of more than 850 geometrically unique, robotically fabricated plate morphologies with more than 100,000 individual finger joints. The use of 6.5mm plywood only demonstrates the system's structural performance capacity.

The concept of machinic morphospaces also offers promising opportunities for future research. In addition to the development of more complex multidimensional spaces describing a larger number of parameters, further research on agent-based modelling for robotically fabricable construction systems has been identified as a particularly interesting area of investigation. Here, form generating agents can be informed in direct feedback with the concurrent agent-based navigation of multidimensional machinic morphospace, altering the generative agents' behavioural rules based on the corresponding position within the morphospace. Another promising field for employing machinic morphospaces is in the application of real-time, online robot control. In this context machinic morphospaces can provide a robust framework for checking the responsive control of the robot based on data collected at runtime sensing or scanning.

Most importantly, the concept of machinic morphospaces may provide one relevant facet for the design exploration of the dramatically increased production possibilities in architecture offered by robotic fabrication.

Acknowledgements

The author expresses his sincere gratitude to Tobias Schwinn and Oliver David Krieg for their contributions to the content of the paper, and to Oliver David Krieg for his significant help with the illustrations. In addition, he would like to thank the entire project team of the ICD/ITKE Research Pavilion 2011 including all participating students and Steffen Reichert and Tobias Schwinn of the Institute for Computational Design as well as Professor Jan Knippers, Markus Gabler, Riccardo La Magna, and Frederic Waimer of the Institute of Building Structures and Structural Design at the University of Stuttgart.

References

Ahlquist S and Menges A 2011, 'Computational Design Thinking' in Menges, A and Ahlquist, S (eds), Computational Design Thinking, John Wiley and Sons, London, pp. 10-29.

Alcorn A 1996, Embodied Energy Coefficients of Building Materials, Centre for Building Performance Research, Wellington.

Eble GJ 2000, 'Theoretical morphology: state of the art', Paleobiology, 26, pp. 520-528.

Fleischmann M, Knippers J, Lienhard J, Menges A and Schleicher S 2012, 'Material Behaviour: Embedding Physical Properties in Computational Design Processes', Architectural Design, Vol. 82 No. 2, pp. 44-51.

Fleischmann M and Menges A 2011, "ICD/ITKE Research Pavilion: A case study of multi-disciplinary computational design" in Gengnagel C, Kilian A, Palz N and Scheurer F (eds.), *Computational Design Modeling*, Springer, Berlin, pp. 239-248.

Hensel M and Menges A 2008, "Inclusive Performance: Efficiency versus Effectiveness", *Architectural Design*, Vol. 78 No. 2, pp. 54-63.

Hensel M, Menges A and Weinstock M (eds.) 2006, *Techniques and Technologies in Morphogenetic Design*, Wiley Academy, London.

Hensel M, Menges A and Weinstock M (eds.), 2004, *Emergence - Morphogenetic Design Strategies*, Wiley Academy, London.

Knippers J and Speck T 2012, "Design and construction principles in nature and architecture", *Bioinspiration and Biomimetics*, vol. 7, no. 1.

Knippers J, Menges A, La Magna R, Schwinn T, Waimer F 2012, "From Nature to Fabrication: Biomimetic Design Principles for the Production of Complex Spatial Structures", *Proceedings of the Advances in Architectural Geometry 2012 Conference*, Springer, Wien.

Kolb J 2008, *Systems in Timber Engineering: Loadbearing Structures and Component Layers*, Birkhäuser, Basel.

Mardia KV, Kent JT, Bibby JM 1979, *Multivariate Analysis*, Academic Press, London.

McGhee GR Jr. 1991, "Theoretical morphology: the concept and its applications", in Gilinsky, NL and Signor, PW (eds.), *Analytical Paleobiology*, Paleontological Society, Knoxville, pp. 87-102.

McGhee GR Jr. 1999, *Theoretical Morphology: the concept and its applications*, Columbia University Press, New York.

McGhee GR Jr. 2006, "Exploring the spectrum of existent, non-existent and impossible biological form: a research program" in Laublicher, M and Müller, GB (eds.), *Modeling Biology: Structures, Behaviors, Evolution*, MIT Press, Cambridge (MA).

McGhee GR Jr. 2007, *The Geometry of Evolution*, Cambridge University Press, Cambridge.

Menges A 2005, "Pluripotent Components - An Alternative Approach to Parametric Design", *AA Files*, no. 52, pp. 63-74.

Menges A 2006, "Polymorphism", *Architectural Design*, vol. 76, no. 2, pp. 78-87.

Menges A 2008, "Integral Formation and Materialisation: Computational Form and Material Gestalt", in Kolarevic B. and Klinger K. (eds.), *Manufacturing Material Effects: Rethinking Design and Making in Architecture*, Routledge, New York, pp. 195–210.

Menges A 2010, "AA Membrane Canopy in London", *Detail*, vol. 2010, no. 04, pp. 350-353.

Menges A 2011, "Simple Systems – Complex Capacities: Integrative Processes of Computational Morphogenesis in Architecture", *TECHNE Journal of Technology for Architecture and Environment*, no. 02/2011, pp. 68-77.

Menges A 2012, "Biomimetic design processes in architecture: morphogenetic and evolutionary computational design", *Bioinspiration and Biomimetics*, vol. 7, no. 1.

Menges A, Schleicher S, Fleischmann M 2011, "Research Pavilion ICD/ITKE", in Glynn, R and Sheil, B (eds), *Proceedings of the Fabricate: Making Digital Architecture Conference, University College London*, Riverside Architectural Press, Waterloo, pp. 22-27.

Menges A and Schwinn T 2012, "Manufacturing Reciprocities", *Architectural Design*, vol. 82, no. 2, pp. 118-125.

Miller D (ed.) 1995, *Goethe. The collected works.* Volume 12, Scientific studies, Princeton University Press, Princenton.

Mitteroecker P and Bookstein FL 2007, "The Conceptual and Statistical Relationship between Modularity and Morphological Integration", *Systematic Biology*, vol. 56, no. 5, pp. 818-836.

Mitteroecker P and Huttegger S 2009, "The Concept of Morphospaces in Evolutionary and Developmental Biology: Mathematics and Metaphors". *Biological Theory*, vol. 4, no. 1, pp. 54-67.

Raup DM and Michelson A 1965, "Theoretical morphology of the coiled shell", *Science* 147, pp. 1294-1295.

Raup DM 1966, "Geometric analysis of shell coiling: general problems", *Journal of Paleontology*, 40, pp. 1178-1190.

Rasskin-Gutman D and Buscalioni A 2001, "Theoretical Morphology of the Archosaur (Reptilia: Diapsida) Pelvic Girdle", *Paleobiology*, vol. 27, no. 1, pp. 59-78.

Seilacher A 1979, "Constructional Morphology of Sand Dollars", *Paleobiology*, 5(3), 191-221.

Thompson D'Arcy 1961, *On Growth and Form*, Cambridge University Press, Cambridge.

Weinstock M 2004, "Morphogenesis and the Mathematics of Emergence", *Architectural Design*, vol. 74, no. 3, pp. 10-17.

Tobias Schwinn, Oliver David Krieg, Achim Menges

Robotically Fabricated Wood Plate Morphologies

Robotic prefabrication of a biomimetic, geometrically differentiated, lightweight, finger joint timber plate structure

Abstract Due to their relative affordability and ease of use industrial manipulators aka robots have become increasingly common in the field of architectural experimentation and research. Specifically for timber construction, their higher degrees of kinematic freedom and fabricational flexibility, compared to established and process-specific computer numerically controlled (CNC) wood working machines, allow for new design and fabrication strategies or else the reinterpretation and re-appropriation of existing techniques – both of which offer the potential for novel architectural systems. In the case study presented here an investigation into the transfer of morphological principles of a biological role model (Clypeasteroida) is initiated by the robotic implementation of a newly developed finger-joint fabrication process. In the subsequent biomimetic design process the principles are translated into a generative computational design tool incorporating structural constraints as well as those of robotic fabrication leading to a full-scale built prototype.

Keywords: robotic fabrication, biomimetics, parametric design, timber construction, finger joint

Introduction

Wood, one of the predominant building materials in the pre-industrial era, is currently experiencing resurging interest from the construction industry. As a renewable resource and natural CO^2 storage, wood maintains a positive carbon footprint, even if today's heavily industrial wood processing is taken into account (Scheer et al. 2006). To a large extent this interest can therefore be attributed to the sustainability discussion in construction and the search for alternatives to fossil fuel-based resources. But also within the context of computer numerically controlled (CNC) fabrication, wood is proving to be an increasingly sought-after building material (Schindler 2009).

The evolution of CNC technologies has seen an increasing sophistication and refinement in the implementation of application-specific fabrication processes. Within the domain of CNC technologies, industrial manipulators aka robots constitute a relatively novel and somewhat atypical addition to the field. As general purpose fabrication machines, they are not limited to specific fabrication processes and can be applied in a variety of different fabrication contexts (Brell-Çokcan and Braumann 2010). Over the last decade, they also became increasingly prevalent in the field of architectural experimentation and research due to their relative affordability, flexibility, and ease of use (Bechthold 2010).

Traditional pre-industrial wood jointing techniques are characterized by the localized geometric requirements of each specific connection resulting in customized, albeit labor-intensive, mono-material joints that can be highly adaptive (Schindler 2010). In contrast, engineered timber connections today usually rely on fasteners which pose additional challenges not only from the point of view of material behavior but also from an aesthetic and recycling standpoint.

Specifically in timber construction, industrial robots can provide higher degrees of kinematic freedom and fabricational flexibility in comparison to established and process-specific CNC wood working machines, and therefore offer the opportunity for new design and fabrication strategies or else the reinterpretation and re-appropriation of existing techniques - both of which offer the potential for novel architectural systems (Menges 2011).

Research Objectives

The aim of the presented research is to activate the respective advantages and potentials in robotic fabrication and traditional wood jointing techniques as part of a new lightweight material system in architecture. The potential for morphological differentiation enabled by robotic fabrication in combination with the robotic re-interpretation of a highly performative and geometrically complex mono-material wood joint (Fig. 1, left) forms the premise for the exploration of the industrial robots' design space and the construction of a full-scale prototype (Fig. 1, right).

Specifically, the following objectives are defined as part of this research: (i) finger-jointing is suggested as a performative way of connecting plywood plates along their edges in a form and force-fitting manner at varying angles; (ii) the implementation of a robotic fabrication process greatly enhances the producibility of finger joints by facilitating their geometrically complex fabrication; (iii) the transfer of biological principles of structural morphology in living systems to architectural morphology provides a methodology to meaningfully populate the expanding design space of robotic fabrication, i.e. in areas that yield high structural and architectural performance; (iv) the established fabricational and biomimetic principles can be translated into geometric principles and synthesized in a generative computational design tool that implements architectural demands; (v) the geometrically intricate nature of finger-joints can be abstracted in a custom digital fabrication model that forms the basis for the production of a full-scale prototype (Fig. 1).

Related Work

The following chapter outlines the context of the research which includes (i) the historical development of jointing wood panels and current limitations in CNC fabrication, (ii) applications of CNC milling in the architectural context, (iii) the use of parametric design software in the generation of robotic milling paths, as well as (iv) existing research into the characteristics of plate structures.

Historical Developments in Wood Jointing

As wood has been one of the most predominant building materials throughout the preindustrial era it can serve as an indicator for advances in production technology (Schindler 2007 and Hoadley 2000). Furthermore, limited material supply and the exigencies of manual labor have led to the development of performative and robust yet simple connections such as finger joints. Finger jointing, an ancient and commonly used corner joint for over 3500 years (Kirby 1999), results in a form and force-fitting connection with high structural capacity as it withstands normal and in particular shear forces without the use of additional fasteners. Today, finger joints are mainly used in furniture design due to their aesthetic qualities and, in their wedge-shaped variation, for linearly extending timber slats within the industrial fabrication process of other timber products such as cross-

Figure 1 Close-up of the robotically fabricated finger joints connection (left); Interior view of the finished full-scale pavilion (right)

laminated timber or glue-laminated beams (Moro et al. 2009).

While manual manufacturing (Fig. 2, left) as well as current CNC technology (Fig. 2, right) for trimming and milling machines is limited to producing finger joints for rectangular or planar plate connections [1], connections for beam structures are already highly evolved in industrial timber fabrication [2]. Research publications by the National Aeronautics and Space Administration in the 1980s already demonstrated the finger joint's structural capacity by efficiently jointing plywood panels for continuous surfaces on wind turbines and airplane wings (NASA 1984, Spera et al. 1990). In the building industry, however, the development of mass production machinery during industrialization and their limited design space led to a preference for metal fasteners over geometrically complex mono-material connections (Schindler 2009).

The robotic fabrication technique, which was developed as part of this research, enables the efficient fabrication of performative finger-joint plate connections that lie outside the typical kinetic range of process-specific CNC-machinery.

Robotic Milling in Architecture

In the architectural context, CNC- and Robotic Milling are often used for mold making. Milling, a subtractive fabrication process, is sometimes regarded as inefficient due to its inherent material waste and the time consumed by the rough cut and fine cut fabrication steps. Alternative milling strategies that limit the material waste have been explored, such as flank milling, where a finished surface can be produced while omitting the rough cut resulting in ruled-surface geometry (Brell-Çokcan et al. 2009; Schindler and Scheurer 2007). However, when milling elements from planar sheets of material such as plywood, the overall efficiency of material usage is largely determined by the nesting efficiency of the elements on the stock sheet.

Current research into custom milling strategies questions the traditional design-to-production workflow and suggests a "production immanent design tool" for robotic milling that allows the user to explore design variations throughout the whole design-to-fabrication process. The result is a digital tool that automatically

Figure 2 Manual fabrication of dovetail joints (Baumgartner, left); Tools for machine-based fabrication of different finger joints restricted to a 90° connection (Super Carbide Tools and Porter Cable, right)

generates robot code inside the design environment and bypasses the typical post-processing step in the fabrication workflow where ISO-standard NC-Code, or G-Code, is converted into robot-specific control code (Brell-Çokcan and Braumann 2010).

Whereas streamlining the fabrication process is an obvious goal, writing machine specific control code limits the execution of the code to a particular brand of robots. Interoperability is traded for the specificity of a particular robotic fabrication setup. It can therefore be argued that the disadvantages of working with G-Code, i.e. the need for additional data-processing steps, and its advantages, such as interoperability and simulation of the complex 6- and more axis robot kinematics by robust post-processors, should be weighed for each particular milling project.

Generative Approach to Robotic Programming

The need for a generative approach to robotic programming has been identified largely as a consequence of the large amount of unique building components in non-standard fabrication projects. Since the planning effort usually scales disproportionally with increasing number of fabricated elements, the robot control data has to be generated directly from the design data (Donwetsch 2007). In a similar vein, Bechtold (2010) argues that even for small production volumes in customized construction, automated robotic programming strategies become necessary. To address the complexity of non-standard parts, he suggests automating the generation of robotic code directly from the parametric design model, thereby eliminating intermediate software environments.

The typical fragmentation of the data processing chain, aka "Digital Chain", from CAD and CAE to CAM and robotic fabrication into different software environments is commonly identified as a nuisance to the architect/designer. Therefore alternative approaches are suggested such as the "production immanent design tool" discussed above, where the end-effector is simulated in the design environment and specific robot code is generated directly from the CAD model (Brell-Çokcan and Braumann 2010).

In a fully integrative design process incorporating fabrication, material behavior, architectural and structural parameters in one generative design approach, it could be argued that it is insufficient to only retroactively test for collision avoidance, out-of-reach positions, singularities, etc. by merely simulating the geometry of the milling tool and spindle, and the robot kinematics, within the design environment. Instead, the reciprocity of design and fabrication should form the premise by which the CAD model is generated at the outset (Menges and Schwinn 2012).

Plate structures in Architecture

While typical triangulated lattice grid shells concentrate tension and compression forces in their edges and vertices, their trivalent geometrical dual carries load by distributed in-plane forces (Bagger 2010). Just as every facet in a triangulated system is triangular, every vertex in a plate structure connects three facet corners. Consequently, by following this topological rule no structural components other than the plates themselves are needed, ensuring much higher structural efficiency. In contrast to foldable or origami-type patterns, this principle provides that no bending forces occur along the plate's margins which ensures the system's kinematic stability and maximum structural efficiency. Rigid spatial plate configurations therefore always have trivalent vertices. This important geometric principle also proves to be a general characteristic found in many biological plate structures in nature at different scales (Wester 2002; Nachtigall 2004). Since stability and adaptability are a main factor in the animal's survival, biological plate skeletons have developed towards highly efficient load-bearing structures as in the case of sand dollars, a sub-species of the sea urchins (*Echinoidea*) (Seilacher 1979).

In the following research, the morphology of natural plate structure systems serve as role models from which architecturally and structurally performative principles can be synthesized. In the context of the expanding possibilities of robotic fabrication processes and the geometric differentiation of building elements now possible, biomimetics is therefore suggested as a filter in the development of a highly performative and novel architectural material system.

Methodology

Initiated by the robotic implementation of a newly developed finger-joint fabrication process, the case study presented investigates the transfer of morphological principles of a biological role model (*Clypeasteroida*) to a technological system. In the subsequent biomimetic design process the principles are translated into a generative computational design tool incorporating structural constraints as well as those of robotic fabrication leading to a full-scale built prototype (Fig. 1).

Robotic Fabrication of Finger Joints

The robotic fabrication process developed as part of the research presented opens up the design space significantly through the ability to efficiently join differentiated plate structures by means of three-dimensional finger joints (Fig. 3). This is facilitated by the development of a customized tool for combined flank milling and tip cutting by

Figure 3 Robotically fabricated finger joints. Connecting two plates with different material thickness at a specific angle (left); Prototype with differentiated finger joints (right)

one of the industrial partners of this project (Fig. 4, left).

A 7-axis robotic set-up enables the finger jointing of geometrically differentiated plywood plates with varying plate thickness and connection angles; empirical tests using the standard 6-axis robotic fabrication setup showed that it proved necessary to either manually reposition the work piece or introduce a 7th external revolute axis in order to manufacture convex as well as concave connection angles within the same fabrication process. This range of kinetic freedom is not offered by typical CNC woodworking or jointing machines and thus unique to the robotic fabrication setup (Fig. 4, right). The possibilities offered by the design space of this fabrication setup raises the question about a methodology which allows for the meaningful application of the new fabrication process and material system with respect to architectural and structural performance. (Krieg et al. 2011)

Biomimetic Design Strategy

In biology many examples show how morphological differentiation on several hierarchical levels allows for adjustments and adaptations to system-internal and system-external constraints while employing as little material and energy as possible (Knippers and Speck 2012). In order to filter the vast possibilities in geometric differentiation now emerging, biomimetics is suggested as a methodology to develop finger-joined plate structures with regards to a range of performance criteria. While many natural systems exhibit general design principles such as heterogeneity, anisotropy and hierarchy, the research focused on the morphology of the sand dollar (*Clypeasteroida*, Fig. 5, left). As a subspecies of the sea urchin (*Echinoidea*), it became of particular interest and subsequently provided the critical design principles, such as plate morphology (Fig. 5, middle) and plate connections (Fig. 5, right), which were translated into a generative computational design tool (Krieg et al. 2012). Structural analyses and physical tests (Fig. 6) confirmed that following the researched topological rules of rigid plate structures, mainly shear forces appear along the plate's edges (La Magna et al. 2009). Not only does this prove the finger joint's particular structural capacity in such a system, but it also explains the microscopic calcite projections along the sea urchin's plate edges.

Generative Design Tool

Following the analysis of the biological system's performative capacity, the aim was to integrate its characteristics into architectural design and to test the resulting spatial and structural material-system through the construction of a full-scale prototype. The focus was set on the development of a modular system that allows a high degree of adaptability and performance due to the geometric differentiation of its plate components. The algorithmic definition of the module's and its component's global arrangement applies structural principles from the biological model by transferring the rule of three elements per node to the connections between the modular units, while implementing biomimetically informed organizational strategies for the module's and component's arrangement during the design process. Through a physi-

Figure 4 Customized milling cutter for shaft and end milling utilized in the developed robotic fabrication process (left); machine setup: 6-axis robot connected with a separate turntable as an external axis (right)

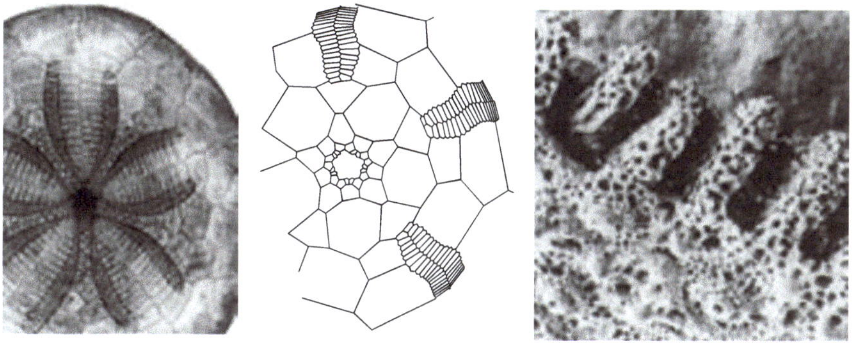

Figure 5 Close-up of sea urchin test (left); schematic top view of sea urchin test showing the outlines and arrangement of the plates (middle); microscopic view of a plate margin showing the calcite projections similar to finger joints (Seilacher 1979, right)

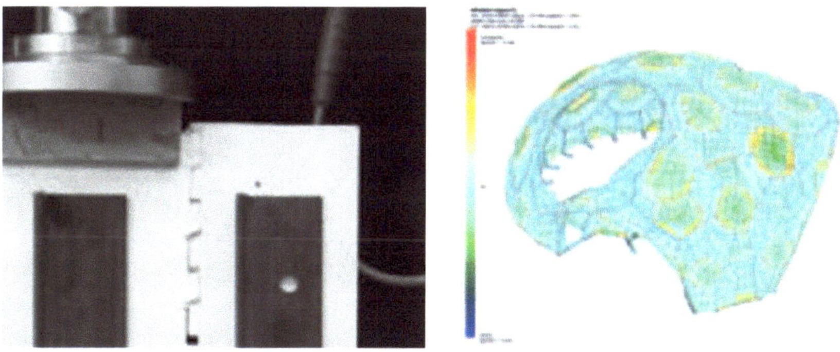

Figure 6. Setup for physically testing the load bearing capacity of robotically fabricated finger joints under shear stresses (left); FE plot of the structure (La Magna et al. 2012, right)

cally simulated form-finding process convenient interaction with the design tool is possible in order to control the spatial outcome while remaining in the context of the system's morpho-spatial capacity (Fig. 7) (Krieg et al. 2012).

Robotic Fabrication Programming

In addition to the biomimetic principles, the specific parameters of the robotic fabrication setup are translated as part of the generative rule set into the computational design tool (Schwinn et al. 2012). One of the main aspects of this translation is the mathematical description of the spatial relation between work piece and milling effector through trigonometry and linear algebra.

The different tool paths for the fabrication of a plate are a function of the angles between the plate and its neighboring plates (Fig. 8, above), which can yield different structural and geometric properties. For example the contact surface between plates decreases for angles close to 90 degrees providing less contact area. However, the length of the indentation of the finger joints increases towards 0 and 180 degrees resulting in extremely sharp finger joints that compromise structural stability and accuracy of fabrication.

Ultimately, the geometric plate relations, finger joint geometric properties, and the specific end effector geometry, consisting of milling tool, chuck and spindle bounding box, confine the preferred range of the joint angles to approximately 15 to 165 degrees (Fig. 8, below). As embedded parameters in the computational design tool these fabricational constraints directly inform the design process. The fabrication data model is parametrically driven by the geometry and design model that was generated with respect to the fabricational constraints outlined above. The programming of the robotic fabrication consists of a custom process including the topological analysis of the plate connectivity, which is the basis for automated tool path generation in form of an ordered point cloud as well as the automated extraction of machine code into an ISO-based CNC format (ISO 6983) (Fig. 9).

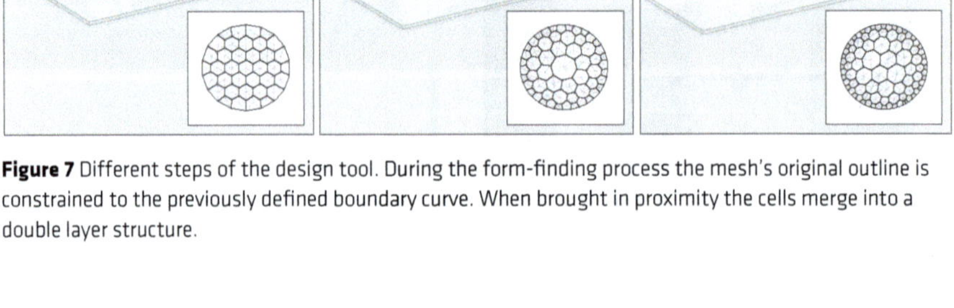

Figure 7 Different steps of the design tool. During the form-finding process the mesh's original outline is constrained to the previously defined boundary curve. When brought in proximity the cells merge into a double layer structure.

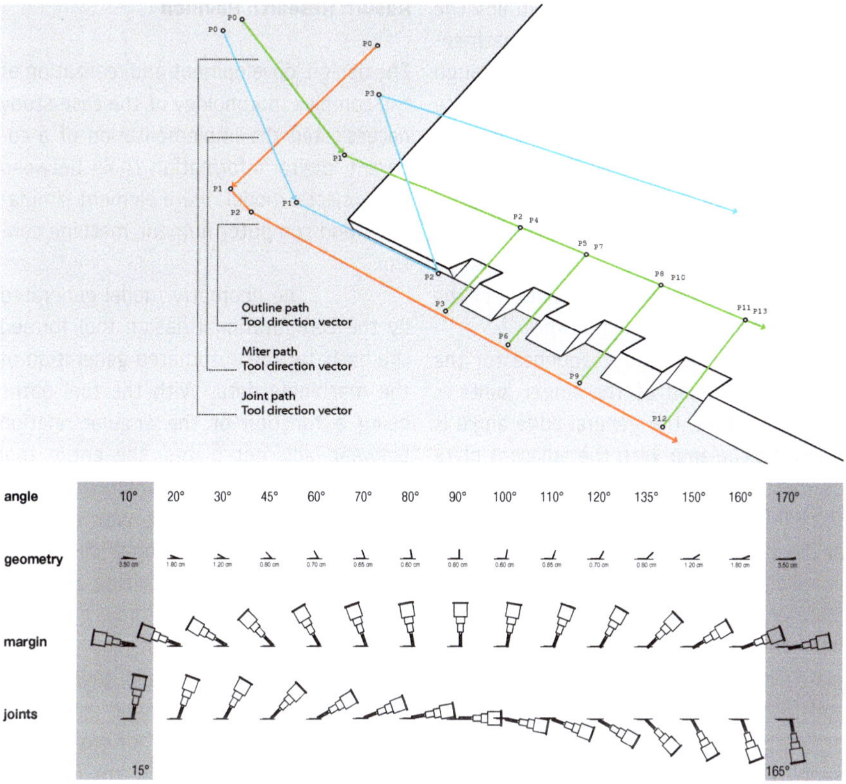

Figure 8 Geometric representation of the different tool paths (above); The finger joint fabrication is geometrically constrained due to possible collisions between the machine and the stock piece (below)

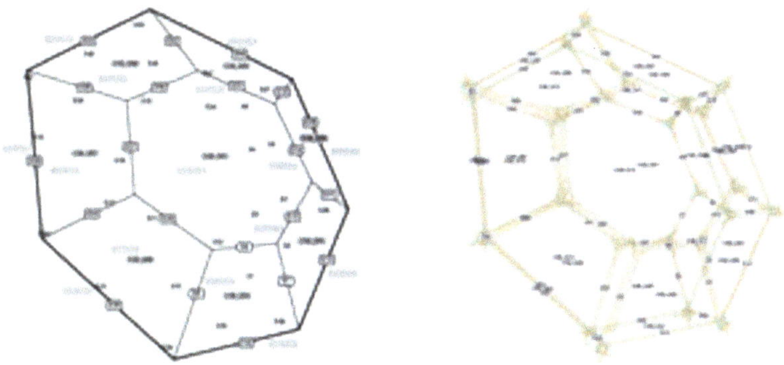

Figure 9 Tool path generation: Topological map of a module (left); Geometric representation of the generated tool paths (right)

The machine code contains the Cartesian coordinates of the tool path sequence which has to be translated through reverse transformation into the joint space of the machine. Due to the inherent complexity of the 7-axis inverse kinematics and to ensure a reliable and repeatable fabrication process, this step is implemented in a dedicated post-processor as opposed to relying on the automatic calculation of the robot control unit during runtime.

The resulting sequence for the robotic fabrication of the finger joints is as follows: first, the general edge angle is set to be coplanar with the adjacent plate by milling the plate's edge with the tool shaft (Fig 10, left); second, a mitered corner at the start and end segment of each edge is milled by aligning the tool axis with the plate angle's bi-sector (Fig 10, middle); finally, the finger joints are indented into the plate's edge normal to the adjacent plate's construction plane (Fig 10, right), producing accurately shaped force- and form-fitting joints as opposed to the rounded corners usually resulting from contour cutting with the tool shaft.

Result: Research Pavilion

The design, development and realization of the complex morphology of the case study necessitated the implementation of a coherent, digital information chain between the project's model, finite element simulations and computer numeric machine control.

The geometry model generated by the computational design tool formed the basis for the automated generation of the machining data. With the tool paths being a function of the angular relation between adjacent plates, the entire tool path information for the production of all 855 unique plate elements with its more than 100,000 individual finger joints were automatically generated utilizing a custom process developed for this project. Following the robotic production, the plywood plates were manually joined together to form individual plate cells and assembled on site. Ultimately, 200 m^3 of gross building volume were enclosed by 2m^3 of wood. Due to the structural capacity of the material system, the entire pavilion could be built

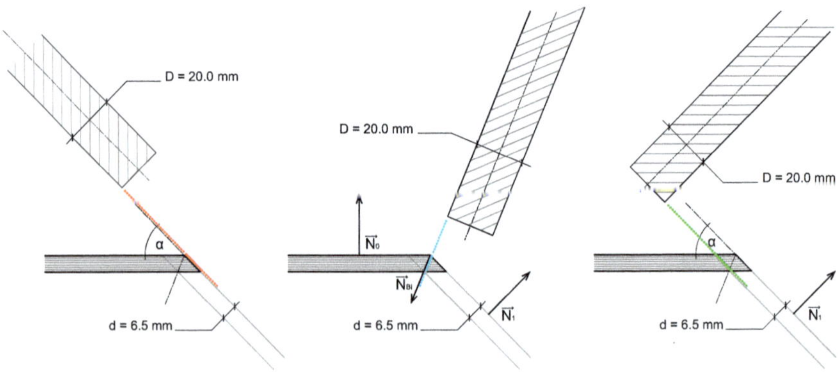

Figure 10 Three steps of the finger joint fabrication process: Milling the plate's outline (left); Milling the edge's miters (middle); Spot facing the finger joints (right)

Figure 11 Each module and its plate components respond to structural and architectural requirements (above); The spatial experience changes as the interior lighting emphasizes the double-layer's depth (below)

out of 6.5mm sheets of plywood. The two interior spaces illustrate the constructional logic of the pavilion: where the double-layer structure separates into two individual layers an interstitial space is framed that emphasizes the double-layered nature of the material system (Fig. 10). The main space of the pavilion is characterized by its prominent orientation towards the park and by a gradual change in the size of the openings in the inner layer that facilitate assembly and disassembly of the entire structure.

Discussion and Conclusion

The case study demonstrates the feasibility of a methodology for exploring and filtering emerging machinic morphospaces (Menges and Schwinn 2012) of a newly developed multi-axis robotic fabrication process through biological principles that act as role models for performative architectural morphologies. The performance capacity of the structural system is demonstrated not only by the extremely efficient material-to-built volume ratio; it is most evident in the fact that the entire pavilion, despite its considerable size, could be built exclusively out of extremely thin (6.5mm) sheets of plywood.

Thus the research pavilion demonstrates how an expanded machine design space offering a high degree of morphological differentiation enabled by robotic fabrication combined with biomimetic strategies provides for both the development of materially efficient structures and the exploration of novel architectural possibilities.

The presented methodology of biomimetically informing the computational design process also suggests alternative biological form-finding principles which will be explored in further research. For example, instead of translating morphological rules of a specific biological role model, a biomimetic formation process such as agent based form-finding can yield alternative spatial configurations that still maintain the topological and fabricational requirements of finger-joint plate structures.

Acknowledgements

This research project was conducted in cooperation with the Institute of Building Structures and Structural Design, University of Stuttgart (Professor Jan Knippers, Markus Gabler, Riccardo La Magna, and Frederic Waimer) as well has the Plant Biomechanics Group in Freiburg (Professor Thomas Speck and Dr. Olga Speck). We are particularly grateful for the extraordinary dedication of our students Oliver David Krieg, Boyan Mihaylov, Peter Brachat, Benjamin Busch, Solmaz Fahimian, Christin Gegenheimer, Nicola Haberbosch, Elias Kästle, Yong Sung Kwon, and Hongmei Zhai.

References

Bagger A 2010, *Plate shell structures of glass. Studies leading to guidelines for structural design*. Technical University of Denmark.

Bechthold M 2010, "The Return of the Future: A Second Go at Robotic Construction", *Architectural Design*, 80(4), pp. 116-121.

Bonwetsch T, Gramazio F and Kohler M 2007, "Digitally Fabricating Non-standardised Brick Walls", *ManuBuild Conference Proceedings*, Rotterdam, pp. 191-196.

Brell-Çokcan S, Reis M, Schmiedhofer H, and Braumann J 2009, "Digital Design to Digital Production" *Proceedings of the 27th eCAADe Conference*, Istanbul, Turkey, pp. 323-329.

Brell-Çokcan S and Braumann J 2010, "A New Parametric Design Tool for Robot Milling", *Proceedings of the 30th ACADIA Conference 2010*, New York, USA, pp. 357-363.

Brell-Çokcan S and Braumann J 2011, "Parametric Robot Control", *Proceedings of the 31st ACADIA Conference 2011*, Calgary/Banff, Canada, pp. 242-251.

Hoadley RB 2000, *Understanding wood: A craftsman's guide to wood technology*, Taunton Press, Newtown.

Kirby I 1999, *The Complete Dovetail: Handmade Furniture's Signature Joint*, Cambium Press, Plano.

Knippers, J and Speck, T 2012, "Design and construction principles in nature and architecture", *Bioinspiration & Biomimetics*, vol. 7, no. 1.

Krieg OD, Dierichs K, Reichert S, Schwinn T and Menges A 2011, "Performative Architectural Morphology", *Proceedings of the 29th eCAADe Conference 2011*, Ljubljana, Slovenia, pp. 573-580.

La Magna R, Waimer F and Knippers J 2012, "Nature-inspired generation sheme for shell structures", *IASS Conference 2012 Proceedings*.

Menges A and Schwinn T 2012, "Manufacturing Reciprocities", *Architectural Design*, 82(2), pp. 118-125.

Menges A 2011, "Integrative Design Computation: Integrating Material Behaviour and Robotic Manufacturing Processes in Computational Design for Performative Wood Constructions", *Proceedings of the 31th ACADIA Conference 2011*, Calgary/Banff, Canada, pp. 72-81.

Moro JL, Rottner M, Alihodzic B, and Weißbach M 2009, *Holzprodukte, Baukonstruktion - vom Prinzip zum Detail : Band 1 Grundlagen*, Spinger, Berlin, Heidelberg, pp. 267-284.

Nachtigall W 2004, *Bau-Bionik: Natur, Analogien, Technik*, Springer-Verlag, Berlin.

NASA 1984, *MOD-5A Wind Turbine Generator Program Design Report*, General Electric Company, U.S. Department of Energy, Washington, D.C.

Spera A, Esgar J, Gougeon M and Zuteck M 1990, *Structural Properties of Laminated Douglas Fir/ Epoxy Composite Material*. U.S. Department of Energy, Washington, D.C.

Scheer D, Feil A and Zerwer C 2006, *Nachhaltigkeit im Bereich Bauen und Wohnen – ökologische Bewertung der Bauholz-Kette*, Institut für ökologische Wirtschaftsforschung, Heidelberg.

Schindler C and Scheurer F 2007, „Architektonische Anwendungen des fünfachsigen Flankenfräsens", *Milling Technology in Architecture, Art and Design*. http://www.designtoproduction.ch/images/stories/publication/2007_Schindler_Flankenfraesen.pdf, 16.7.2012.

Schindler C 2007, "Information-Tool-Technology: Contemporary digital fabrication as part of a continuous development of process technology as illustrated with the example of timber construction", *Proceedings of the 27th ACADIA Conference 2007*, Halifax, Canada.

Schindler C 2009, *Ein architektonisches Periodisierungsmodell anhand fertigungstechnischer Kriterien, dargestellt am Beispiel des Holzbaus*, Eidgenössische Technische Hochschule, Zürich.

Schindler C 2010, "Die Standards des Nonstandards", *GAM - Graz Architecture Magazine 06*, Graz, pp. 181 - 193.

Schwinn T, Krieg O, Mihaylov B, Reichert S and Menges A 2012, "Machinic Morphospaces: Biomimetic Design Strategies for the Computational Exploration of Robot Constraint Spaces for Wood Fabrication", *Proceedings of 32nd ACADIA Conference 2012*, San Francisco, USA.

Wester, T 2002, "Nature Teaching Structures", *International Journal of Space Structures*, 17(2), pp. 135-147.

[1] http://www.getwoodworking.com/news/article.asp?a=859

[2] http://www.hundegger.de/index.php?id=57&L=1

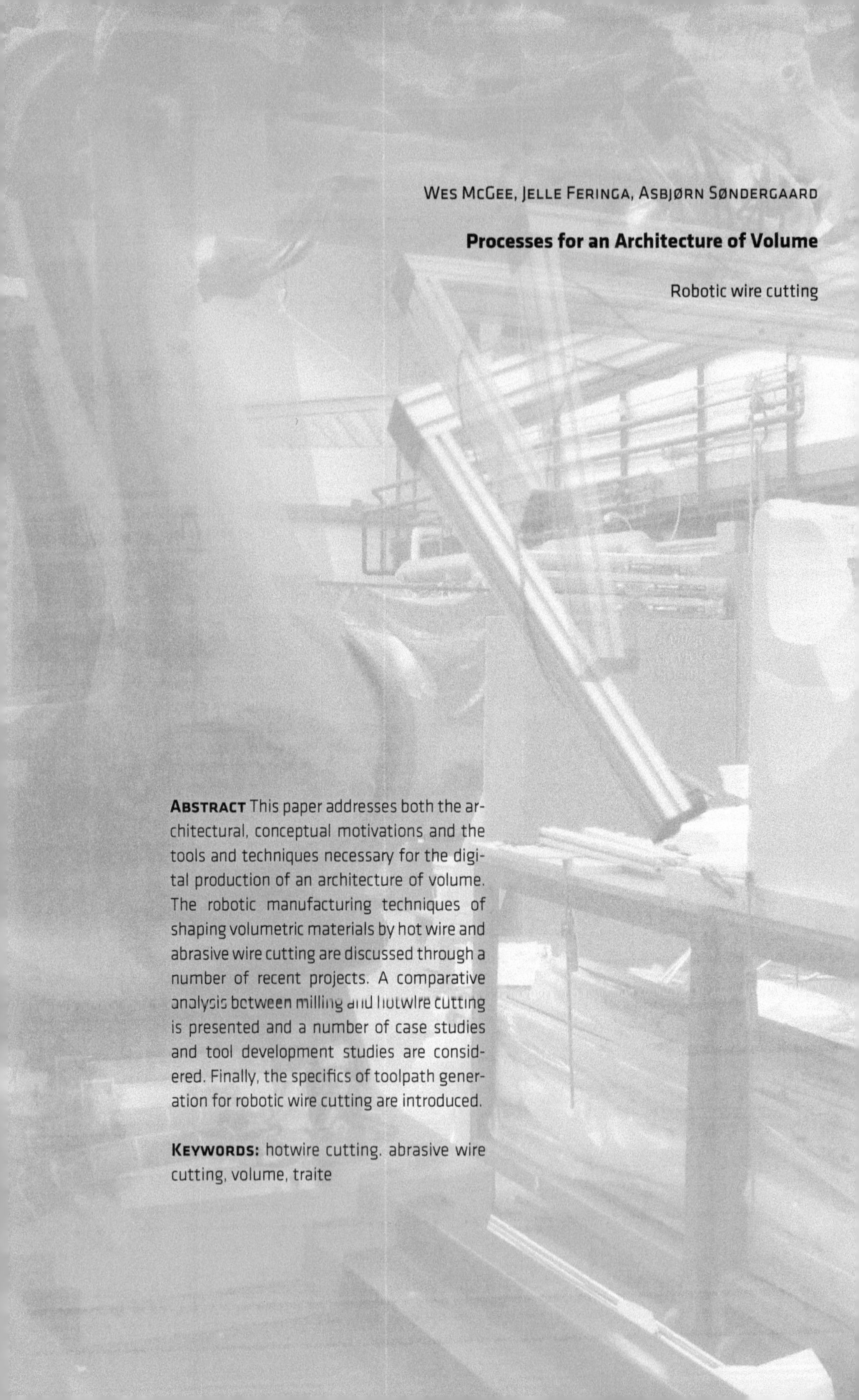

Wes McGee, Jelle Feringa, Asbjørn Søndergaard

Processes for an Architecture of Volume

Robotic wire cutting

Abstract This paper addresses both the architectural, conceptual motivations and the tools and techniques necessary for the digital production of an architecture of volume. The robotic manufacturing techniques of shaping volumetric materials by hot wire and abrasive wire cutting are discussed through a number of recent projects. A comparative analysis between milling and hotwire cutting is presented and a number of case studies and tool development studies are considered. Finally, the specifics of toolpath generation for robotic wire cutting are introduced.

Keywords: hotwire cutting. abrasive wire cutting, volume, traite

Introduction

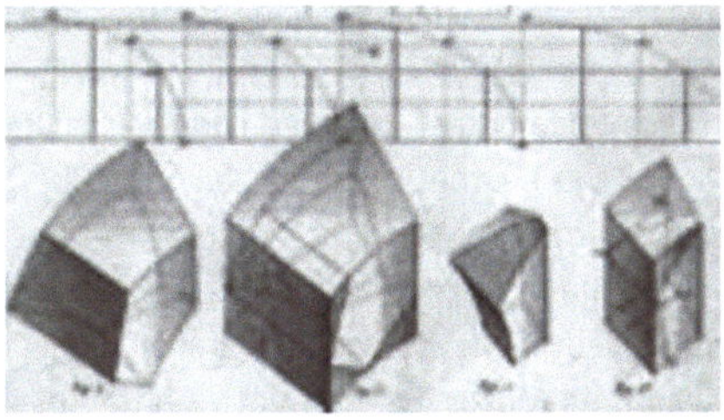

Figure 1 de La Rue, Traite de la Coupe des Pierres

There has been a growing interest in material processes that can support an architecture of volume, investigating materials which are unconstrained by the limitations of sheet based materials. Our initial investigations in processes for an architecture of volume explored the lightest and least expensive volumetric material available, EPS foam. This material has seen many applications in the mold making, highway and construction industry, as it is cheap, recyclable, extremely light and easy to shape. This material is typically carved using large CNC routers, and for double curved geometries this is still a requirement. The material can also be cut with a hotwire, which provides a method whose historical precedent can be associated with stereotomy and the developed surface of traditional stone masonry (de la Rue, 1782) (Fig. 1).

Architectural production has been systematically compressed into ever thinner layers by the constraints of industrially processed materials. CNC fabrication processes, initially heralded as liberating the designer from the disconnect between drawing and making, have accelerated the process, packaging components into the discrete 4' x 8' work envelope of the typical 3 axis router. In addition, streamlining the workflow from design software to fabrication processes (while this has many positive benefits), has in some ways allowed this "flattening", slicing, slivering and wafering of building construction methods to go unquestioned. Openly available scripts allow 3D surfaces to be ribbed, unrolled and nested into common sheet sizes, ready for production. Contemporary digital fabrication techniques continue to proliferate this limitation, producing a stream of contoured, folded, notched and otherwise surface-driven projects.

This scope - that of a representational model - is sufficiently nebulous in terms of scale to abstract architecture from its realization. The irony of the proliferation of CNC methods such as 3D printing, 2D laser cutting, and routing is that its obscures the industrial potential of the

Figure 2 Subdivided columns, M. Hansmeyer (left); Metropol Parasol, J. Mayer (right)

close coupling of design and manufacturing methods. This becomes problematic when these false or self-imposed constraints become the aesthetic of the building, where the approach of building a representational model has been projected to its full size, as observed in the "Metropol Parasol" project by Jürgen Mayer (Fig. 2, right). An interesting example that simultaneously illustrates the merits and limits of this approach is the work of Michael Hansmeyer. His "subdivided columns", a series of 2.7m high columns, built from 1mm layer grey cardboard (Fig. 2, left). While the intricacy and elegance of the work is not questioned, the project is antithetical in terms of fabrication; columns are load bearing structures that aren't made of cardboard, and the intricacy can be subscribed to the strong will of architectural students, rather than architectural engineering. A consequence of exploring methods of construction that have little or no manifestation in building practice can be construed as a form of technological self-censorship. With the mechanics, tooling and technology available, it is paramount to focus research on those modes of production that do scale, hence are of value to the construction industry.

Figure 3 Production of the RDM Vault at Hyperbody's robotics workshop in Rotterdam

While advanced manufacturing methods have traditionally been associated with costly manufacturing methods, robotic hotwire cutting (RHWC) breaks with this trend given that complex formwork can be delivered for the approximate cost of normative formwork. As such RHWC is both an enabler, technically, in terms of forms that can be produced, and economically since this can be achieved at little or no additional expense. With the many ongoing predicaments in the construction industry, and the modest cost of delving into robotics, this is an important aspect that is open to further exploration.

Hyperbody's robotics lab is equipped with two second hand ABB S4 robots, that were both acquired for less than what a makerbot costs (Fig. 3). Brand new robotic manipulators typically cost less than half the price of a typically capable dedicated CNC machine. Robotic fabrication presents a development platform for such considerations, given the trade-off of precision, ease of integration and programming, robustness, and market availability. As the technology has begun to gain acceptance in the building fabrication industry (admittedly it remains a very small fraction), these methods have started to challenge what type of construction can be delivered within a given budget.

RHWC

Hotwire cutting holds a number of advantages when used to create formwork for casting. At an architectural scale, traditional approaches such as CNC milling become prohibitively time consuming. At the sheer volume demanded for full scale architectural in situ casts, such as bridges and commercial buildings, the incremental removal of material offered by the milling technology necessitates considerable machining time and results in production fees unacceptable to most building budgets (McGee 2011, Feringa 2011). Machining hours may be reduced by tolerating a rougher surface, however, production times remains prohibitively high, and the rough tooling paths simultaneously frustrate the demoulding process. This limits the application prospect for CNC-milling technology primarily to detailing tasks, exclusive high-end building budgets and repetitive casts, where formwork may be reused. RHWC offers a number of advantages. The removal of material in this process is essentially volumic; the cutting process processes a surface in a single sweeping motion, whereas in milling the volume is removed layer-by-layer, constrained by the limited depth of the milling bit. Per surface, the length of the tooling path is parameterized over the radius of the milling bit, where often a roughening milling bit is used with a large diameter to approximate the shape quickly, while a milling bit of a smaller diameter is required to achieve a smooth surface. In addition, the RHWC leaves a surface considerable smoother than that of the milling process, producing a better surface finish on the cast product, while reducing demoulding adhesion. For extremely finished surfaces the mold can still be coated with polyurea, requiring less coats than a typical milled finish. The difference in production speed is easily understood geometrically; whereas milling essentially removes a sphere, RHWC removes a cylinder of material at an instance in time (Fig. 4) That amounts to a difference of 1 to 2 orders of magnitude, as the following comparative study shows, approximating the differences in production time for either production technique (Table 1).

It is important to mention that, while the increase in production speed is dramatic, the additional effort of rationaliz-

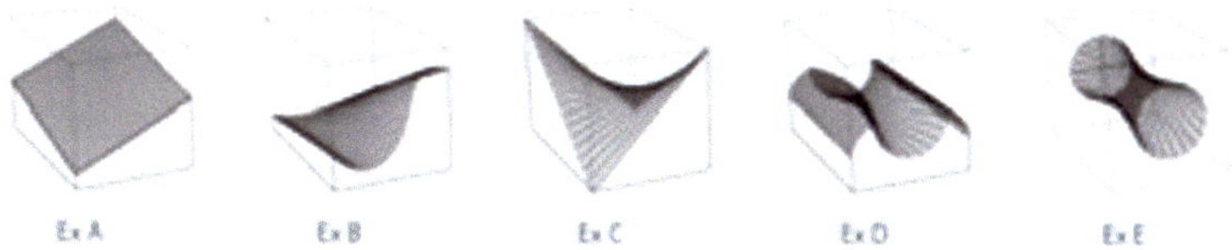

Figure 4 Comparative scheme of sample geometries

	ex. a	ex. b	ex. c	ex. d	ex. E
CNC rough.	3h 34m	5h 5m	3h 44m	4h 22m	6h 31m
CNC finishing	6h44m	7h 42m	7h 01m	h 31m	12h 14m
RHWC	0h 01,8m	0h 02,4m	0h 02,3m	0h 02,7m	0h 03,1m
Area cut	2,66 m2	2.95 m2	2.86 m2	4,01 m2	3,49 m2
Removed vol	1,44 m3	2.06 m3	1,44 m3	1,72 m3	2,41 m3

Table 1 Machining metrics comparing the CNC milling with RHWC

ing geometry to ruled surfaces – a key topic of architectural geometry – is not factored into this comparison. While RHWC is remarkably efficient, the geometric grammar that can be produced is a subset of what can be produced by milling. However, it is important to realize that architectural scale works in favor of RHWC. First of all, in the sense that forms which traditionally would not be manufactured by CNC methods can now be produced. Secondly, due to scale, the limitation to ruled surfaces becomes less of an issue, since a greater surface area makes constructing a satisfactory approximation less problematic.

Projects

A number of case study projects have been performed to validate the capabilities of RHWC. In "Periscope", by Matter Design Studio, the hotwire process provided the means for the rapid production of a 50 foot tower of foam (Fig. 5). The economy of time and material was of paramount importance, due to the pressures of a two week construction window and limited budget. The RHWC process was used to generate a large array of mass customized masonry units, which were assembled in a running bond to approximate the original, doubly curved column. In an effort to establish a characterization of the various approaches to working with volumetric materials, one could consider this a "slab based" process, whereby components are cut from a slab of material, preserving some portion of the slab surface on the top and bottom of the part (Fig. 6). In this case the preservation of the parallel top and bottom surface is important to support the assembly technique.

A more recent project by students at the University of Michigan uses the slab cutting process to shape AAC sheets into voussoir units to form a thick-shell compressive vault. In this case the prototype uses abrasive waterjet cutting to cut the 4" thick AAC block. Previous work at the University of Michigan saw the application of this technique to process 2" thick sandstone (in slab form), uniquely cut to form thin-shell vault components. Wire cutting becomes more efficient and precise at larger material thicknesses, and opens

Figure 5 Periscope completed

Figure 6 Slab cutting EPS

up the possibility of structural systems that respond to additional factors beyond the material efficiency of the thin shell vault (Clifford, 2012). Clifford describes this as a shift from form-finding to form-responding, and uses it as an approach to develop structurally viable forms requiring relatively thick sections. The ability to work with volumetric materials is critical to the success of the process (Fig. 7).

An alternative approach to "slab cutting", which adds an additional level of geometric freedom, is the "solid" based cutting process as explored in recent projects by Hyperbody (Fig. 8). A recent collaboration between Hyperbody and ROK-Rippmann Oesterle Knauss / ETH Zurich, the RDM Vault, explores a joint approach to the design and fabrication of vaulting structures, as evoked in Rippmann and P. Block (Rippman and Block, 2011). RhinoVault (Rippman, Lachauer and Block, 2012) provided intuitive tools for the design of a vaulting structure, while PyRAPID enabled the transliteration of the resulting geometry to robotic motion, cutting the "traites" out of EPS foam. The pavilion was erected at Hyperbody's robot lab which will host the RobArch workshop in Rotterdam (Fig. 9).

In this case the components are nested completely within a volumetric block of material. All faces of a component are wire cut, as opposed to the slab cutting process, which relies on a parallel top and bottom face. While the cut surfaces are still limited to ruled geometries, by shaping the entire exterior of the component the aggregation can more accurately approximate a freeform surface, while producing joint faces which are normal to the thrust vec-

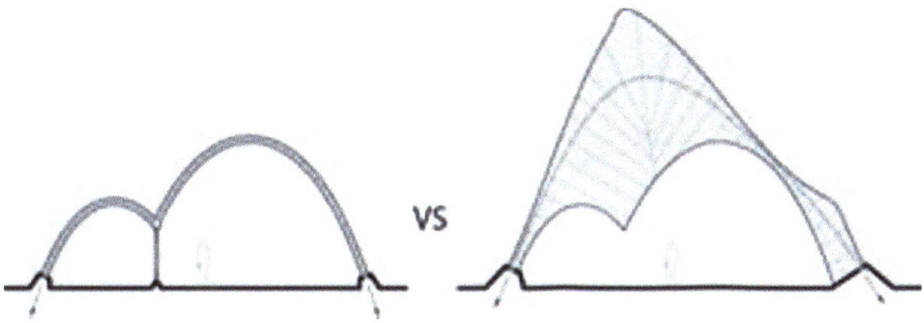

Figure 7 Thick funicular solver

Figure 8 Solid based cutting process, as explored in the RDM Vault

tors. Component sizes and shapes will still be governed by their ability to fit within an available volume of material.

From Hotwire to Abrasive wire

Building upon the work with EPS, there have been a number of investigations using more permanent volumetric materials, such as AAC (autoclave aerated concrete) and natural stone. It is against this background that several projects were undertaken using robotically manipulated abrasive wire cutting equipment. While numerous projects have investigated the geometric potential of hotwire cutting EPS, considerably fewer have dealt with developing the end of arm tooling to mount abrasive wire saws to robotic equipment for the purpose of cutting more rigid materials. While wire cutting harder stone materials remains a very slow process, there are several advantages when compared with CNC milling or multi-axis bridge saw cutting. The capital cost of the equipment is considerably less (one third to one half), with the generic robotic manipulator costing far less than a stone capable CNC, even after factoring in the cost of integration and tooling. In addition, the implication of the process as a semi-finishing operation makes it more appropriate to the tolerances possible using robotic manipulators, as opposed to more precise CNC equipment. There are also potential material efficiencies that can develop, given the much smaller kerf of the segmented wire compared to milling and sawing, although these will be highly geometry dependant.

Segmented diamond wires are well known for their ability to cut harder

Figure 9 RDM Vault

materials like natural stone (marble, granite) and reinforced concrete. Typically, the wire sawing process is used for either semi-finished flat slab cutting applications, or large scale demolition. There are exceptions, of course, such as this large 6 axis CNC wire profiling system by Pellegrini Meccanica Spa (Fig. 10 left). The machine clearly illustrates the possibilities for multi-axis wire cutting, but it also presents opportunities for a more flexible, portable approach to fabrication.

Dedicated CNC approaches are likely to always possess an advantage in terms of accuracy and overall capacity, but there are potential applications where the flexibility and portability offered by industrial robotic manipulators can fill a unique role in fabrication. Several researchers have tested applications for robotic wire sawing, but the capabilities of a robotically guided wire cutting operation to yield complex units in a finished/semi-finished state has not been studied extensively. It is worth pointing out that just over a decade ago "nearly eight hundred full-size DIN A0 templates were required to guide the stonemason's hand" in completing the translation from model to workshop (Burry, 2001). Shutao Li, et al. developed a proof of concept production line to machine AAC slabs directly from BIM data. The tooling developed utilized a segmented diamond wire circulating in a rectangular frame (Li, 2007). This approach has also been used in combination with a spiral cutting steel wire, cutting ruled geometries out of cured plaster (Bard, 2012) (Fig. 10, right).

The authors are currently engaged in developing end of arm tooling for robotic diamond wire cutting (RDWC), with a number of areas targeted for study. Robotic applications will require the tooling to be considerably lighter than the CNC applications highlighted above. This is not a trivial task, as even a typical manual profiling diamond wire saw can weigh 500 lbs [3]. In a typical wire cutting operation, as in band sawing, there are guides which support the blade opposite the travel direction. In the case of 3D wire cutting, the wire is capable of moving in any direction. The guide system needs to support this, and potentially will require a servo driven solution for positioning relative to the cut direction, similar to the CNC equipment described previously.

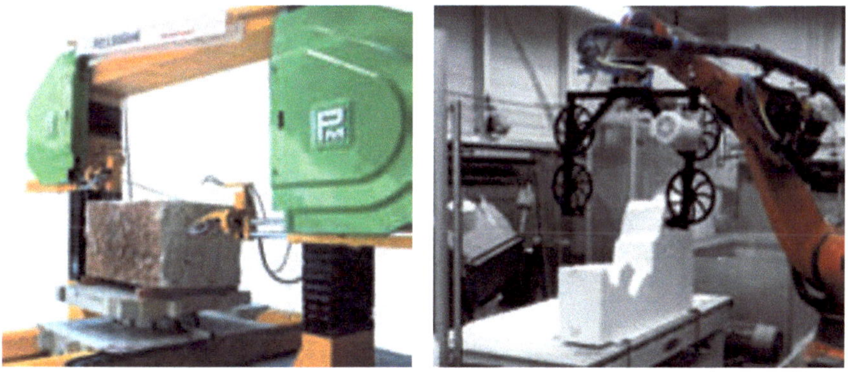

Figure 10 6 axis CNC diamond wire saw (left); Wire sawing end effector developed at the University of Michigan Taubman College (right)

Software

It can be argued the generic robotic manipulator utilized in this research is only incrementally different to it ancestors which were in continuous use in mass production for decades. Without a doubt, one of the driving factors behind its growing adoption in the architectural fabrication industry is the use of open source and bespoken software applications. While robotic manipulators provide incredible flexibility, this comes at the price of developing tools which suit both the designer and fabricator. Compared to CNC equipment, which has clearly also benefited from the developing culture around scripting and algorithmic design methodologies, robotics has the added benefit of compatibility with an open framework for fabrication. Closed-loop process feedback and the "ease" of integration into a multifaceted production process are relatively complex to perform using traditional CNC equipment; with robotic manipulators these capabilities are integral to the design. Such open frameworks are superMatterTools, developed by Wes McGee and Dave Pigram, Daniel Piker's lobster, Robots-in-Architecture's KUKA|prc, PyRAPID by Jelle Feringa and HAL by Thibault Schwartz.

A key motivation for the development of open frameworks is how existing approaches such as contour cutting can be adapted to the solid cutting process (Fig. 9), which are considerable more demanding in terms of toolpath generation, motion planning and collision avoidance. An interesting aspect is how the software developed for the exploration of RHWC maps with modest adaption to RDWC. Hotwire cutting is a comparatively safe production method compared to the brutality of the diamond wire sawing process, so the evolution of RHWC naturally paved the way for RDWC. Whereas the development of RHWC was demanding in terms of software develop-

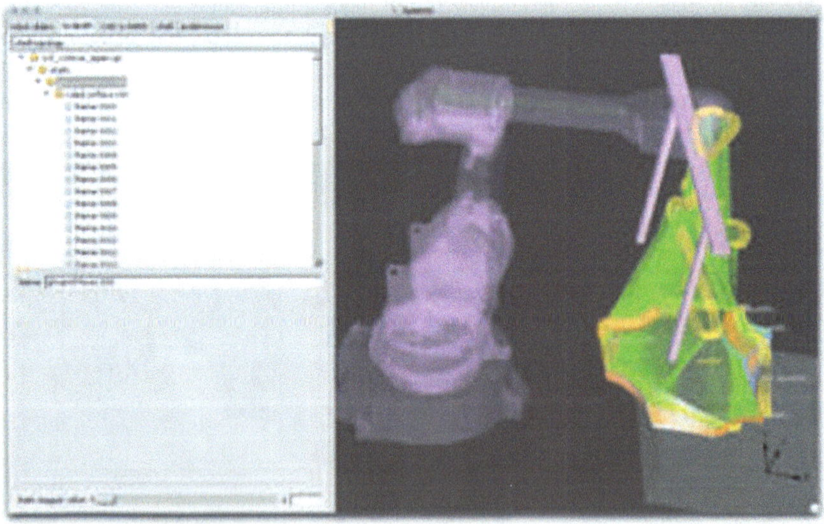

Figure 11 PyRAPID, custom RHWC software develop on top of PythonOCC

ment and trivial in terms of the required tooling, these roles are reversed in the continued development of RDWC, where building a practical wiresaw is demanding.

The potential of RHWC and RHWD was explored with PyRAPID a software application was developed in Python with PythonOCC, a wrapper of the OpenCasCade CAD kernel as it main dependency (Fig. 11). The application automatically clusters the faces so that they can be cut in a single sweeping motion, and generates a toolpath optimized for extending the reachability of the end-effector, and computes the inverse kinematics from that pose. As the tool orientation has a degree of freedom over the axis of the wire, the key is to exploit this, as it allows for considerable optimization of the reachability of the robot.

After clustering the faces the software tests whether an additional roughening step is required. The roughening step is specific to robotic hotwire cutting, while with a traditional hotwire cutting machine no clashes between the tool and workpiece occur. The downside, however, is that to cut large blocks, a considerably larger machine is required, while a robot is a fairly compact machine, certainly in view of its reachability.

Ongoing efforts include logistics, such as the integration of picking and placing, in order to facilitate a production workflow that minimizes operator attendance. The added flexibility and process integration of this setup is a topic of further research. An important argument why robotic hotwire cutting has considerable merits over a classic CNC hotwire machine is specifically the issue of process integration.

References

Rippmann M, Lachauer L, and Block P 2012, "Interactive Vault Design", *International Journal of Space Structures*, vol. 27, no. 4.

Bard J, Mankouche S and Schulte M 2012, "Morphfaux: probing the proto-synthetic nature of plaster through robotic tooling", *Proceedings of the Acadia conference 2012*, San Francisco, US.

Burry M, Burry J and Faulí J 2001, "Sagrada Família Rosassa: Global Computer aided Dialogue between Designer and Craftsperson (Overcoming Differences in Age, Time and Distance), Reinventing the Discourse", *Proceedings of the Twenty First Annual Conference of the Association for Computer-Aided Design in Architecture 2001*, p. 07.

Clifford B 2012, "Thick Funicular: Particle-Spring Systems for Variable-Depth Form-Responding Compression-Only Structures", ACSA, forthcoming.

McGee W 2011, "Matter and making", *FABRICATE: Making Digital Architecture*, Waterloo: Riverside Architectural Press.

Feringa J 2011, "Investigations in design & fabrication at Hyperbody", *FABRICATE: Making Digital Architecture*. Waterloo: Riverside Architectural Press.

Rippmann M and Block P 2011, "Digital Stereotomy: Voussoir geometry for freeform masonry-like vaults informed by structural and fabrication constraints" *Proceedings of the IABSE-IASS Symposium 2011*, London, UK.

Li S, Isele J, Häfele KH, and Geiger A 2007, "CAD/CAM Integrated Building Prefabrication Based On A Product Data Model", *Proceedings of the Joint International Conference on Computing and Decision Making in Civil and Building Engineering.*

La Rue JB de 1782, *Traité de la coupe des pierres. Imprimerie royale.*

[1] http://block.arch.ethz.ch/tools/rhinovault

[2] http://www.pellegrini.net/cataloghi/PELLEGRINI_RobotWire.pdf

[3] http://www.md-dario.com/saws/band-saws-for-marble.html

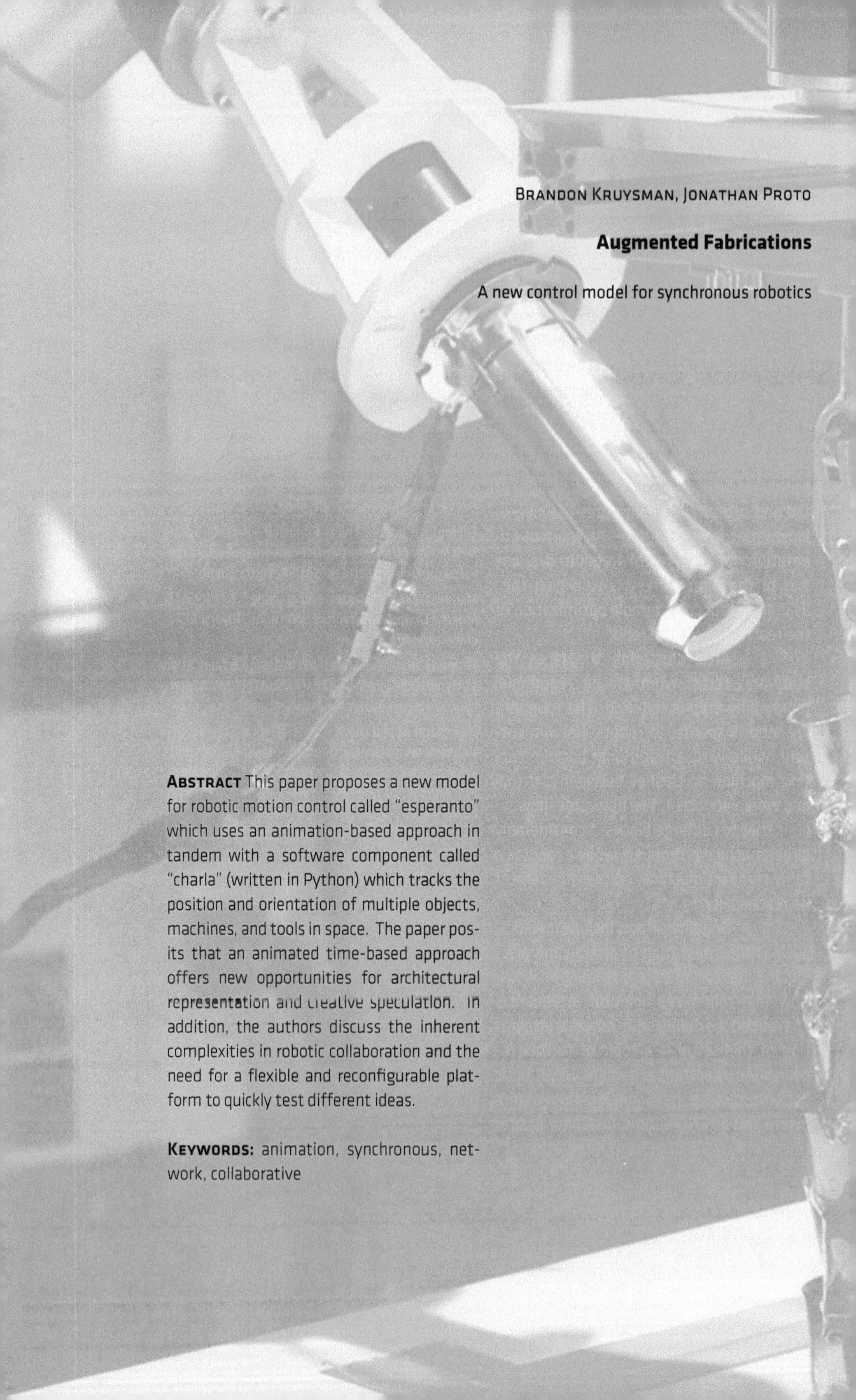

Brandon Kruysman, Jonathan Proto

Augmented Fabrications

A new control model for synchronous robotics

Abstract This paper proposes a new model for robotic motion control called "esperanto" which uses an animation-based approach in tandem with a software component called "charla" (written in Python) which tracks the position and orientation of multiple objects, machines, and tools in space. The paper posits that an animated time-based approach offers new opportunities for architectural representation and creative speculation. In addition, the authors discuss the inherent complexities in robotic collaboration and the need for a flexible and reconfigurable platform to quickly test different ideas.

Keywords: animation, synchronous, network, collaborative

Animated Machines

Robots and cinema have a long history. Cameras attached to robots have been used for years in modern cinematography. Mark Roberts and his company Mark Roberts Motion Control, have been providing motion control products for film and TV, winning an Academy Award in 1999 for their Milo motion control rig. These are custom robots developed specifically for filmmaking, with integrated software that allows designers to program all motion in Autodesk Maya. Since Maya is standard in the film industry, this can be easily integrated into the designer's workflow. The precise positioning of these robots makes it much easier in the construction of mixed-reality shots, or compositing of CGI onto real video footage. However, not until Jeff Linell introduced a company called Bot&Dolly, has this technology been accessible and more involved in the creative process. Bot&Dolly's new motion control system, called IRIS, is a platform that also uses Maya as the main software for programming motion, however, they are using standard industrial robot arms (Fig. 1). The animated approach to robotics, as well as the choice to use existing industrial robots rather than highly custom robots that require an immense effort in transporting, gives Bot&Dolly a unique approach to modern filmmaking. The flexible platform allows the designer to animate the physical world, and provides precise and expressive control over 6-axis industrial robot arms. The robot's tasks are not constrained to just holding a camera either, they could be used to set any object in motion. While our industries have many differences, Jeff Linell and Bot&Dolly share a desire to make robotics a medium that is open and inherently creative. Not constrained to just filmmaking, Bot&Dolly operates within a space that collides animation, cinematography, and automation through the design of mo-

Figure 1 Bot&Dolly's IRIS Motion Control System

tion. When synchronizing multiple mediums, robots, and/or objects, like Bot&Dolly so often has to do, the study of objects in motion becomes extremely crucial.

Character Rigging and Skeletal Control

Animation in cinema has over the years been an extremely powerful medium in the ability to give "life" to objects and in the construction of fantastic worlds that invites the audience to an alternative reality. The core technique of the custom platform, called "esperanto", is character rigging, typically found in the animation and film industry; the only difference is that we are not rigging virtual characters, but industrial robot arms. This platform opens up the potentials of animating the physical world, where designers have the unique ability to experiment with materials in motion, rather than only executing fabrication related tasks.

The basic concept behind the platform is to "rig" a skeleton for a physical machine. Conceptually, this is not limited to a 6-axis industrial robot arm. Rigs could be applied to any machine that can be driven by a series of motors that can be then controlled by G-Code. These rigs serve as the virtual driver for robotic motion control. Using Python scripting, the animated rigs are then translated to the language for industrial robot arms, with its corresponding speed, i/o for tooling (pneumatic or digital), and synchronization information for multi-robot applications. This translation from digital model to physical motion is the core of the esperanto platform. While much of this information is easily translated, one variable, time (which in the case of collaborative robotics is the most important), has been the most challenging. In the following sections we will describe these issues and how they have been resolved in greater depth.

Using the specifications provided by the industrial robotic manufacturer, custom character rigs were constructed for each robot type using Autodesk Maya. Typical "joint" construction of skeletons was bypassed in favor of a more accurate and precise expression-based character rig using trigonometry functions. Once calibrated, a 3D model of all 5 industrial robot arms could be used as a way to simulate, visualize, and program the robot arms. Since visualization and representation has always been a driver in the development of the custom robotic platform, Autodesk Maya offered a number of appealing features that suggested its use as the main software: its ability to produce stunning simulations, ease of animation, and time-based approach to digital modeling. In any multi-robot configuration, computing power for the visualization of multiple robots was a primary concern, as well as a necessity to understand objects in motion over time, where the coordination of multiple robots in space is critical. Many platforms that were tested could not visualize multiple robots over time in the same 3D model. The expression-based character rigs were crucial in providing a lightweight solution to having 5 robots moving in one digital model.

The robots can be rigged a number of ways, depending upon the desired motion effect. For example, we have experimented with standard methods of control, such as an inverse kinematic or forward kinematic rig, but have also experimented with hybrid "rigs". These hybrids offer an experimental method of advanced rigging control that is very different from the approaches found in an engineering environment. Some robots could be driven by half of their axis by forward kinematics and the other half by inverse kinematics etc. One example is that we developed 1 skeleton that rigs 3 robots (conceptually) as one machine. We have also rigged tooling, which act as an additional rig attached to the robot that can be animated and translated to inputs/outputs for the robot. Although these hybrid rigs are still at the early stages of development, these unique methods of control could eventually provide designers with the opportunity to expand a range of motions and control that may previously have been regarded as inachievable using these machines.

New Representations in Architecture and Manufacturing

The difference between the animators in the film industry and the animators in the robot lab is that the worlds that the designers create in the lab have the potential to come into physical existence. What is typically constrained to the "space of the frame", now has the ability to exist in reality in 4 dimensions, interacting and affecting real materials in space. The way that we represent objects in space is not only directly related to how we see space, but rather how we envision new ways of making space (As, Schodek, 27).

Motion graphics beyond typical fly-throughs and rendered perspectives offer a very convincing and powerful tool that allows a design idea to be communicated to literally any type of audience, without any knowledge of traditional graphic conventions (As, Schodek, 116). For the designer the platform has the unique ability not only to design motion, but also to program, simulate, and speculate all at the same time (Fig. 2). This type of animation space suspends the distinction between representation and simulation, simulation and speculation, making it very hard for one to determine which is which. Are we representing the construction, or constructing the representation? Many times we are performing

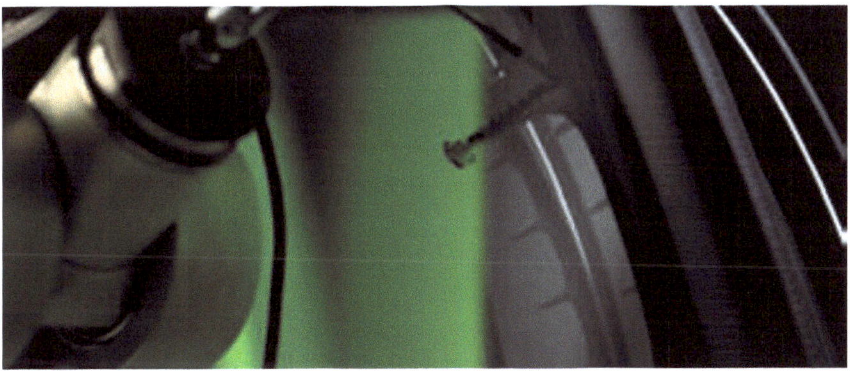

Figure 2 Composite Rendering used to combine animation techniques with physical testing as a new method for simulation and speculation in advanced manufacturing

both. Motion in architectural design is not reduced to previous notions of analogical representations in the generation of form, but instead taken literally, to use the design of motion to make new things for architecture. It is the direct interaction of designers with their built environment and is not mediated through reductive graphic representations that often reduce the richness inherent in design. The animation of a complex multi-robot scenario can be directly translated to robotic motion.

Although the platform has a multiplicity of functions regarding how it may be used, we must not forget that it exists in two very different spaces, with very different rules. It can manifest itself in complex ways and on many different levels. It can be used to create fantastic worlds, but also be used for production purposes. However, some of the most interesting results been in hybrid or mixed reality environments in robotics (or advanced manufacturing speculation). In this animation space, the composite (both virtual and real), can be both analytical and generative. It is the layering of tracings, and the ability to spatially synchronize different architectural representations, through which we see a new type of speculative medium in architecture and manufacturing.

This type of medium in robotics relies on the tracking and positioning of objects in motion. It is highly dependent on the synchronization of cameras (Fig. 3) (the double description of the virtual and physical camera) and the motion of multiple robots in space and a set of parameters: the robot's position and orientation in space over time in physical space and the virtual counterpart in the animated space. This combination of animations and real video gives the designer a unique toolset in constructing manufacturing environments that

Figure 3 Custom Camera Rig, with arduino controlled follow focus. This tool allows the designer to see "through the arm" from a location outside of the robotic cell. All features of a DSLR camera can be controlled from an external workstation.

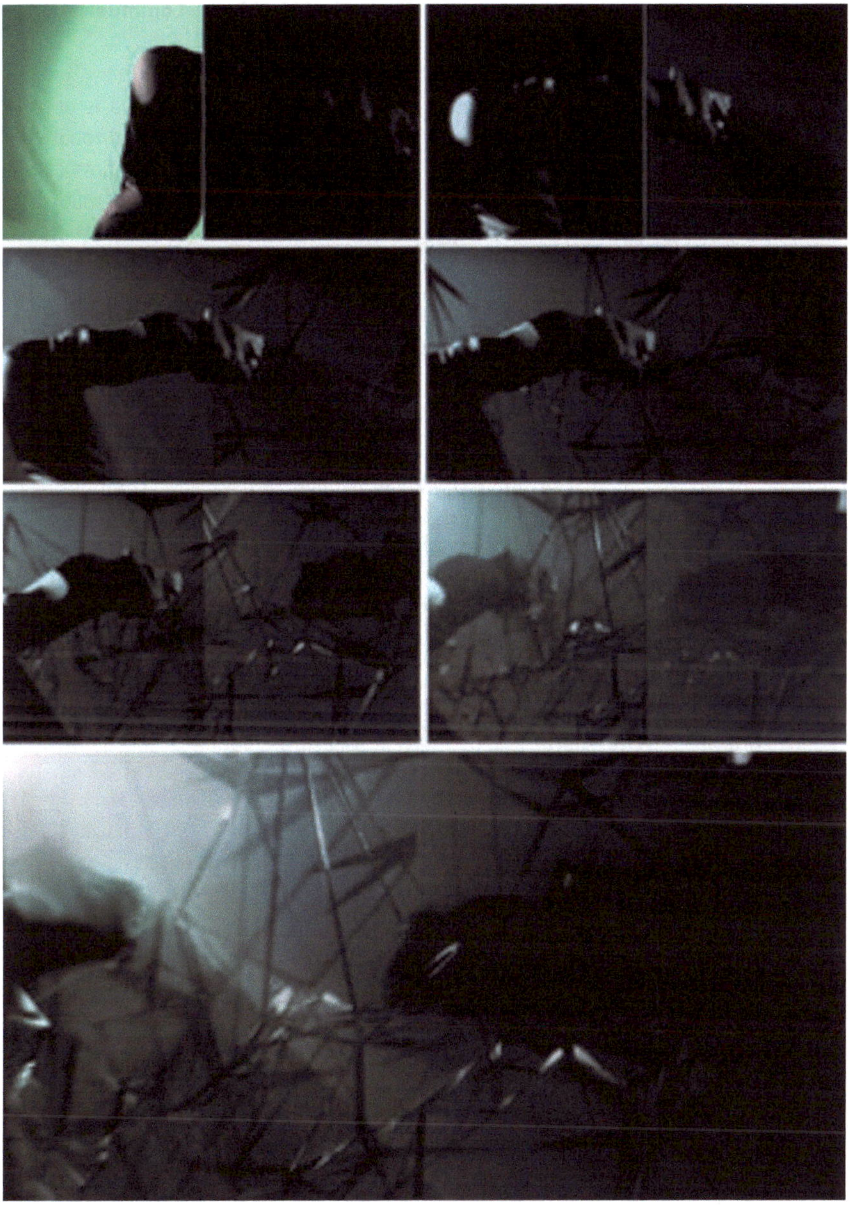

Figure 4 Sequence of images showing the breakdown of "the net". First, starting with filming robots using green screen, then adding layers of virtual environment (foreground, midground, background), then adding lighting, color correction, and additional visual effects.

may not yet want be constrained by real parameters such as material costs, etc. This has the potential to act as a virtual mock-up, without expensive investment in materials, etc. In the experiment entitled "the net" in collaboration with Curime Batliner, we explored the potential of hybrid mixed reality environments by precisely tracking 3 robots in space, and synchronizing both the virtual animation with real camera footage. The robots are positioned in a highly complex digital environment (a net), which is being manipulated by the robot's motions and simulated using the physics engine in Maya. The series of images shown captures the layers of imagery in order to obtain the shot, which is a dynamic camera path, with real robots composited in a complex virtual environment (Fig 4). The tracking, coordination, and overlaying of these different representational techniques has provided much feedback in understanding time, motion, and the translational discrepancies between digital and physical space.

Networked Machines – Communication Protocol

The global time slider in Maya is used to serve as the basis for physical robot motion. Animation of multiple robots can be performed and exported to the robot controllers for execution. However, there are discrepancies in the timing of translating the digital model to actual motion. The basic unit of time, 1/24th of a second, (or the "keyframe" in animation space), isn't always the case in physical space. Depending upon the complexity of the motion sequence, the conversion of speed from the digital model to the physical motion can be slightly varied. For example, a minute long (or 1440 keyframe) animation may actually take 60.5 seconds to complete. Although this may not be a problem in some cases, it is a cause of much concern for any task requiring coordination. A second software component was developed in order to provide a flexible solution to collaborative ro-

Figure 5 Custom ipad interface for charla designed in Touch OSC that allows the user to configure and control the robotic network from one device. The interface also has a feedback component, which allows the user to visualize each robot's status in their program in real-time.

bots called charla. The tracking and positioning of multiple objects, machines, and tools in space is made possible by this additional tool. This custom software component enables many objects to be in communication with one another, creating an animation timeline that adjusts in real-time, based on real motion control constraints. The first version of charla was constructed using a hardware approach, where robots were physically wired together using a series of relay switches. Although in many cases this system proved to be successful, we found many problems in the processing of information. For instance, mechanical delays and hardware malfunctions required us to rethink our approach. The second version, called eCharla, uses a much faster and reliable ethernet based approach that writes a TCP server with socket communication. eCharla is written in Python, using methods of synchronous programming. All objects in the network require their own open port that sends data to a server which relays all information to the other nodes in the network. All coordinated tasks require this higher level control model or "brain" which addresses many of the complexities of collaborative robotics; such as multiple robot arm scheduling and programming, as well as collision detection (Fig. 5).

This system is based around the notion that the network is susceptible to modification. The goal was to be able to virtually rig a machine, quickly locate its precise position in space in relation to other nodes, and integrate it into the robotic network quickly so that the machines could be programmed and in communication with one another. This level of adaptation is extremely important in our research, where the designer can quickly configure alternative network behaviors. This approach is not constrained to just industrial robot arms, but can also be objects in the environment or end arm tooling.

Networked Manufacturing

This next example tests the platform on multiple levels; the implementation of animated machines, study of materials in motion, synchronous communication protocol, and embedded representational strategies. "Hot-Networks" is an experiment in repetitive unpredictability. It aims to capture the transient transformation of a material in motion and uses the randomness and wildness inherent in materials as opportunities for design. By applying the precision and accuracy of robots with the reaction of prolonged exposure to heat, the objects express themselves in multiple ways; highly iterative and regular on one hand, and completely wild and irregular on the other. While typical manufacturing scenarios are obsessed with the production of large quantities of almost exactly the same objects, this duality produces a tension in each piece made, where no one object is ever the same and variation in material reactions is encouraged.

The robotic cell consisted of multiple robots with different tools and tasks, in communication. Everything is interconnected in the network: representation, material behavior, synchronous motion, and coordinated end of arm tooling. The project explores additive manufacturing using a variety of different fabrication techniques; combining iterative picking and placing, with spraying and heating.

The exercise used heat as a way to transform components in the form of piles and stacks, and also used paint as an additional process within the robotic sequence to apply various levels of transparency (Fig 6). The robotic cell consisted of 5 robots with overlapping work-spheres. Each robot performed a separate task; one filming the fabrication sequence, another picking and placing components, one air-brushing, one positioning the work-surface (allowing for neighboring robots to reach when needed), and one applying heat for the fusion of components. The programming of the robots was designed so that the robots' tasks are offset (meaning when one robot moves to get another piece of material, two of the other robots work together to paint the pieces that are placed). The sequence is also designed so that the work surface is dynamic, where it has the ability to move into neighboring robots' work-spheres for collaboration, while having an extremely accurate way of positioning the tube shaped components in space. Many calibration tests were performed with the heating sequence that associated timing with formal implications. This allowed the objects to have material and formal characteristics that ranged between control and wildness (Fig. 7). The objects could range from "pile" to "stack", depending upon the amount of heat applied and the relationships defined between robots. Variation in form was a result of the timing and coordination of robots.

Figure 6 Hot Networks Experiment 1, showing robot collaboration to fuse PETG plastic components using heat and applied pressure

Conclusion

The animated approach to robotics described above provides the designer with a unique and robust platform that exists on multiple levels. It satisfies the desire for speculation in manufacturing, blurs the boundaries between digital and physical spaces and projects ideas beyond the 2-dimensional computer screen. While also pushing ideas of architectural representations and communication, the platform offers a "scaling up" of control models that allows designers the ability to visualize, choreograph, and program complex motion sequences of multiple robots or machines at once. This approach to collaborative robotic manufacturing caters to the designer. The interface is open and flexible, and welcomes an experimental approach to robotics in architecture.

References

As, I and Schodek, DL (eds.) 2008, *Dynamic Digital Representations in Architecture: Visions in Motion*, Taylor & Francis, Abington, England.

Dagognet, F (ed.) 1992, *Etienne-Jules Marey: A Passion for the Trace*, Zone, New York.

Mahall, M and Asli, S (eds.) 2009, *How Architecture Learned to Speculate*, De Bruyn for Institut Grundlagen Moderner Architektur Und Entwerfen (IGMA), Stuttgart.

Mataric, MJ (ed.) 2007, *The Robotics Primer*, MIT, Cambridge, MA.

Menache, A (ed.) 2011, *Understanding Motion Capture for Computer Animation*, Morgan Kaufman, Burlington, MA.

Telotte, JP (ed.) 2010, *Animating Space: From Mickey to Wall-E*, University of Kentucky, Lexington, KY.

Figure 7 Hot Networks Experiment 1, stack/pile of plastic PETG components with gradated paint

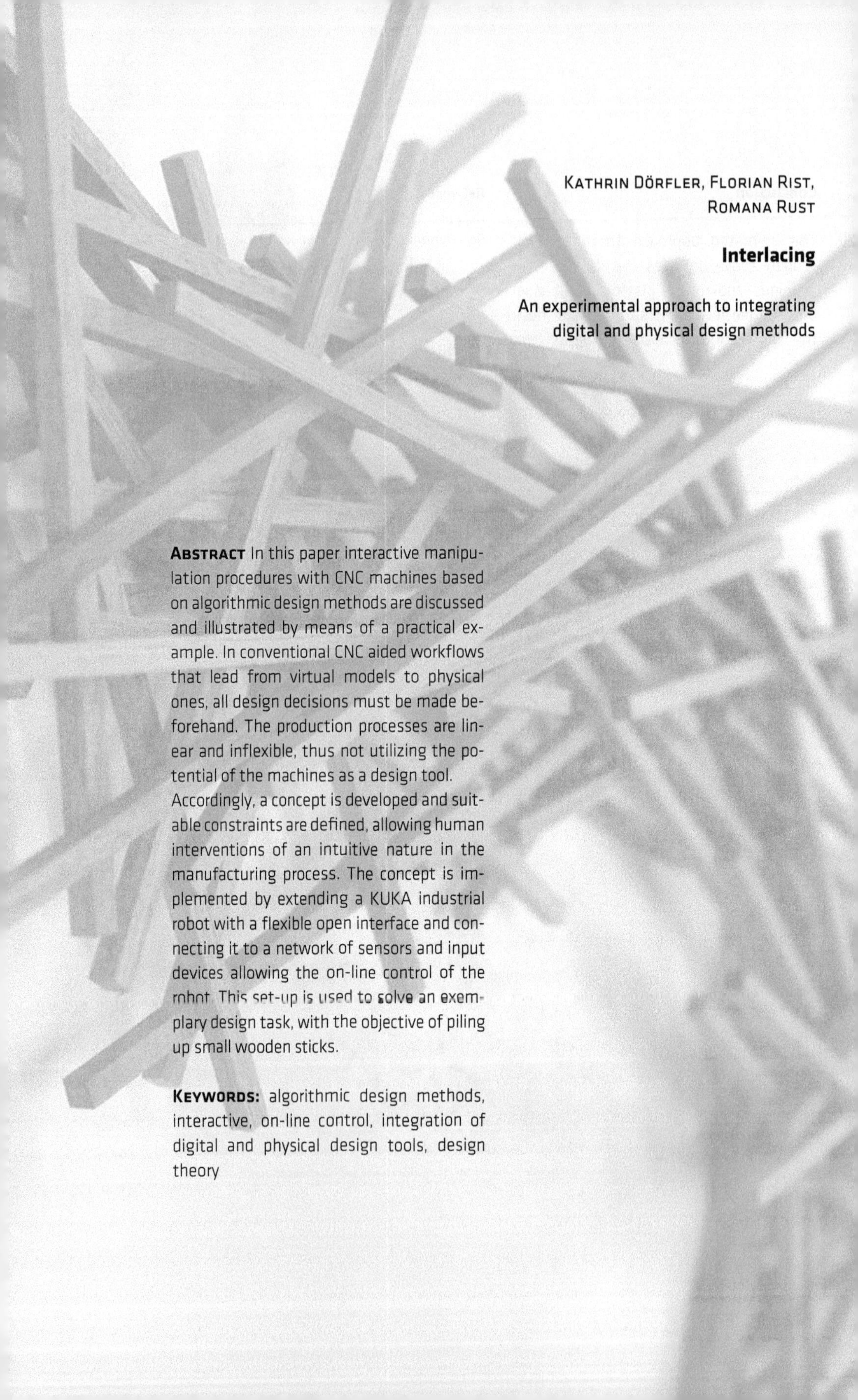

Kathrin Dörfler, Florian Rist,
Romana Rust

Interlacing

An experimental approach to integrating digital and physical design methods

Abstract In this paper interactive manipulation procedures with CNC machines based on algorithmic design methods are discussed and illustrated by means of a practical example. In conventional CNC aided workflows that lead from virtual models to physical ones, all design decisions must be made beforehand. The production processes are linear and inflexible, thus not utilizing the potential of the machines as a design tool.
Accordingly, a concept is developed and suitable constraints are defined, allowing human interventions of an intuitive nature in the manufacturing process. The concept is implemented by extending a KUKA industrial robot with a flexible open interface and connecting it to a network of sensors and input devices allowing the on-line control of the robot. This set-up is used to solve an exemplary design task, with the objective of piling up small wooden sticks.

Keywords: algorithmic design methods, interactive, on-line control, integration of digital and physical design tools, design theory

Introduction

Design processes in architecture are non-linear processes which are aided by physical and digital design tools. In this context the term "digital design tools" refers to software tools capable of mapping relations and structures of mutual dependencies (Hovestadt, 2010, p.16), rather than electronic drawing boards. The use of such advanced rule- and information-based digital design tools requires a rationalization and quantification of design problems in order to be operationalizable and processable by the computer (König, 2011, p. 275). Physical tools, on the other hand, allow a direct sensual dialogue and connection with materials and their properties. They allow intuitive and spontaneous actions, but they lack the ability to represent semantic relations or information other than the ones inscribed in their physical properties, and therefore hardly meet today's need to process all kind of information alongside the design process.

The two worlds could complement one another in advantageous ways: *'A sensible model for the use of algorithmic methods must attempt to automate the resolution of operational problems and facilitate the solving of non-operational problems through a combination of man and computer. [...], a combination of intuitive and rational design strategies.'* (König, 2011, p. 278) The established rationalized digital design processes, based on parametric and algorithmic shape generation, need to be opened-up to spontaneous design impulses, as well as to the physical world and its constraints.

In this examination we show a practical approach to bridging the gap between these digital and physical design tools and try to interlace them by establishing interfaces between the physical and digital realities. The basic principles of a complex design tool, complementing digital information processing with intuitive, physical interaction with real-world objects, are illustrated.

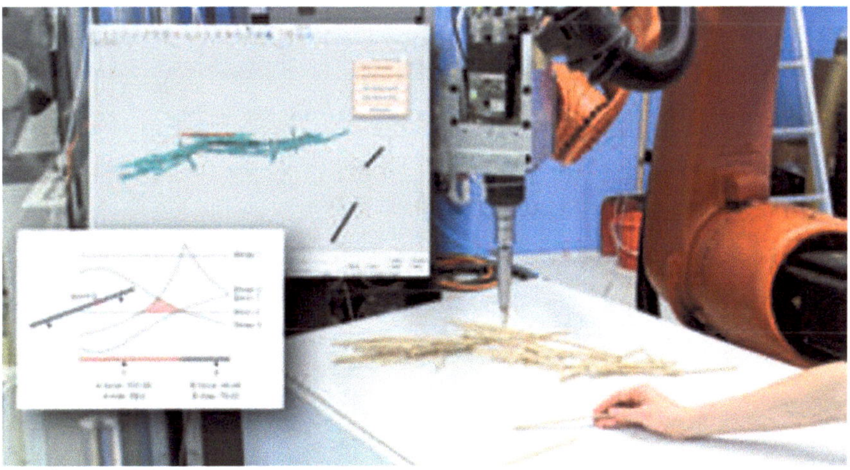

Figure 1 Case study: KUKA robot piling up wooden sticks

In order to link physical and virtual realities, intermediaries between these two worlds are necessary. In this case actuators, which can transport information from the virtual into the physical and sensors inversely from the physical to the virtual world, are applied.

In an environment where digital and physical realities are interwoven, physical objects become representatives of immaterial information, and are at the same time turned into sources of information for their virtual pendant.

Materials and Methods

Requirements

In the following experiment, CNC-machines are not merely understood as manufacturing automatons, but as universal actuators linking the digital to the physical world. They are part of a design tool that can be controlled interactively by the designer, during the fabrication process. The design and the fabrication process are no longer treated separately, they are rather unified in an open system, where design decisions can be made while the physical manufacturing process is in progress. Ultimately, the designer is not required to pre-describe the finished form anymore, only the process of fabrication needs to be designed. (Gramazio & Kohler, 2008, p. 9).

Design Goals

The implementation of such a design/fabrication system requires certain things (Fig. 2): first of all, actuators, manipulating in the physical world (in this specific case, an articulated arm industrial robot), need to be

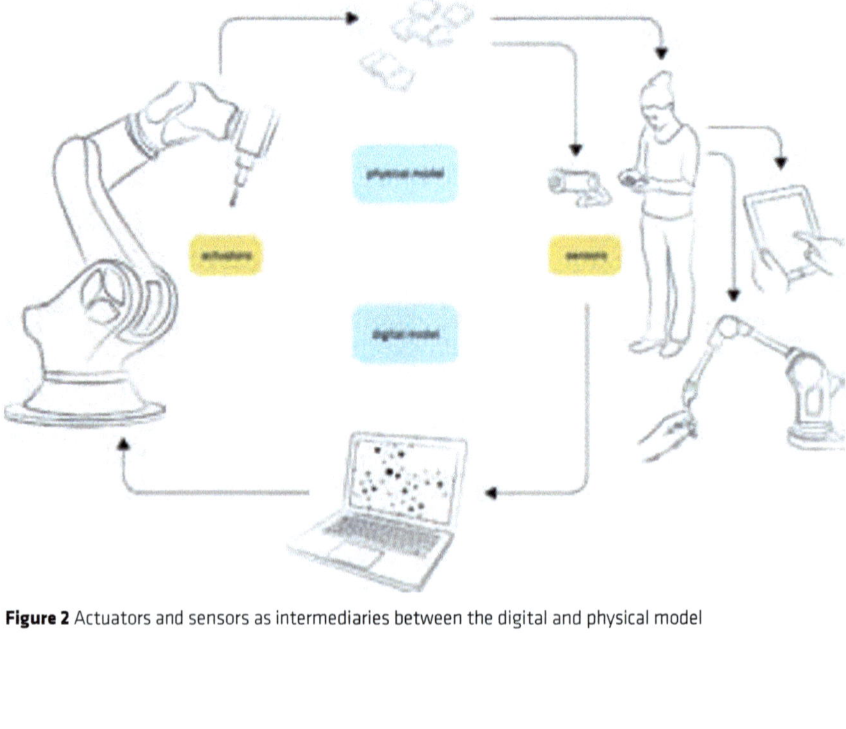

Figure 2 Actuators and sensors as intermediaries between the digital and physical model

able to receive and execute commands at any time in real time, instead of executing a set of predefined orders. On the other hand, sensors (video camera, iPad, 3D coordinate measurement arm) have to deliver a stream of information from the physical world to the digital, for the possibility to influence and moderate the fabrication process.

System

The whole set-up has to be controlled by software which routes and processes the exchanged information to all the different components of the system.

One primary requirement was to build the system as simply, cheaply and openly as possible, to give students the ability to experiment with it, expand or transverse it to new platforms. Consequently, the existing hardware was used, but proprietary interfaces were replaced by simple open (network) interfaces, enabling a communication based on XML telegrams, transmitted via TCP/IP on Ethernet.

The network of sensor and actuator clients is linked by a server application (Fig. 3), written in Python. This server handles the communication and the data exchange between the clients and it also includes a module containing the implementation of specific fabrication planning algorithms. All software was developed on Mac OS X, but with portability and platform independence in mind, meaning it should also run on Linux or Windows.

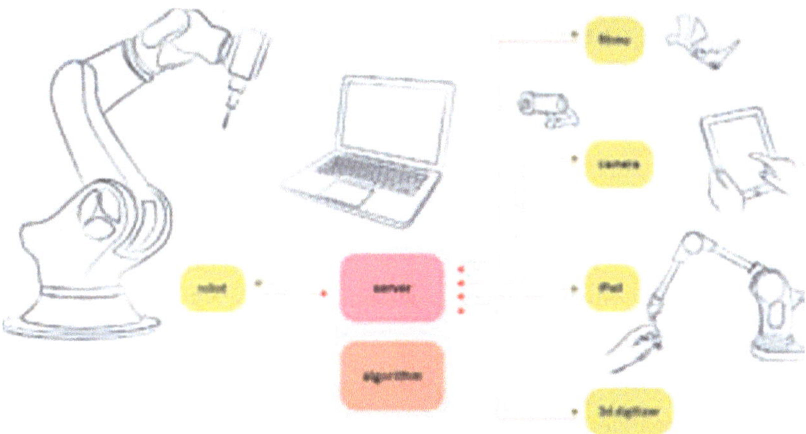

Figure 3 Server-client application

Figure 4 Communication between an iPad and the KUKA robot via the server-application: simultaneous movement of the end effector of the robot to the orientation of the iPad

For the control of the KUKA KR 60 HA robot, the KUKA.Ethernet KRL-XML module is used, which provides rudimentary functions to exchange data as XML telegrams via TCP/IP during a running KRL program.

In the simple example above (Fig. 4), the server receives the accelerometer data from the iPad over WLAN, evaluates and converts it into frames (E6POS), and sends the coordinates to the KRL program running on the robot controller, which then executes the commands and rotates the end effector according to the orientation of the iPad. Although the KUKA Ethernet KRL-XML module is rather slow and induces significant lag in the communication, the motion of the robot still follows the movement of the iPad at a decent pace.

Algorithms

The data from the sensors or other input devices is not only used for directly controlling the movement of the robot, but also as an input for advanced algorithms to fabricate complex objects, influencing high level calculations through feedback from the physical to the digital domain.

To design the shape-generating algorithms for this specific system, one has to consider certain aspects. The algorithms are not only based on a model of reality, but they are directly linked to the physical reality. They have to be built upon the choice of materials, tools, sensors and input devices, and their respective properties.

There are essential differences in designing algorithms for shape generation in a solely virtual environment: in the physical world, for example, structures can only be assembled and disassembled in a specific order. While this and similar problems are easy to describe and comprehend, a rather difficult "soft" problem has to be kept in mind while designing these algorithms: As user interaction is desired, it is necessary to ensure that the interaction is direct and transparent, in the sense that the user is able to understand the causality between his action and the reaction of the system, at least to a certain degree. One of the keys to achieving this is to keep the system highly responsive, by ensuring that cycle times are as low as possible. Therefore, recursive algorithms solving a problem step by step, while each iteration requires only a short time to calculate and execute, are generally preferred, even though more efficient approaches may exist.

Through the definition of rules, the algorithm describes not only one possible solution, but contains a solution space. "*The result depends on the particular circumstances found in the individual formation processes. The algorithm, the code, can be interpreted as the genotype, and is the basis of any development, whereas the distinct generated shapes are the phenotypes.*" (Alexander, 1964)

Case Study

General

The equipment/setup for the following experiment consists of a KUKA KR 60 HA robot (as actuator) with a customized vacuum lifter as end effector (Fig. 5, left), a standard webcam for optical tracking, mounted underneath a light transmitting translucent tabletop (Fig. 5, right), and wooden sticks,

5×5 mm, each with a different length between 150 and 300 mm (see also Fig. 1).

The rather simple problem of piling wooden sticks is well suited to exploring certain advantages of the proposed system. The design task: to create a pile of straight wooden sticks with equal square cross sections but with different and unknown lengths, features both intuitive and complex rational aspects.

As the pile of sticks grows, it becomes increasingly difficult for a human being to maintain structural integrity without limiting the overall shape to some simple, easy to understand structural system. So this task is handled by rationalizing the problem of stability and implementing an algorithm that ensures that the pile is stable, while not limiting its shape in an unnecessary way. The algorithm keeps track of all sticks placed and calculates stable positions for placing new ones.

While the algorithm only creates stable piles it does not create a specifically shaped pile. There is no explicit formal definition of the final shape inscribed, which opens up the design/fabrication process to intuitive and spontaneous formal decisions. The designer gives an initial impulse for the shape generation by positioning several sticks, the base layer, in a distinct manner (Fig. 6). As the pile grows, it is up to the designer to offer one or several sticks of choice to the robot. The stick's orientation, center point and length are extracted from a camera image and fed to the piling algorithm.

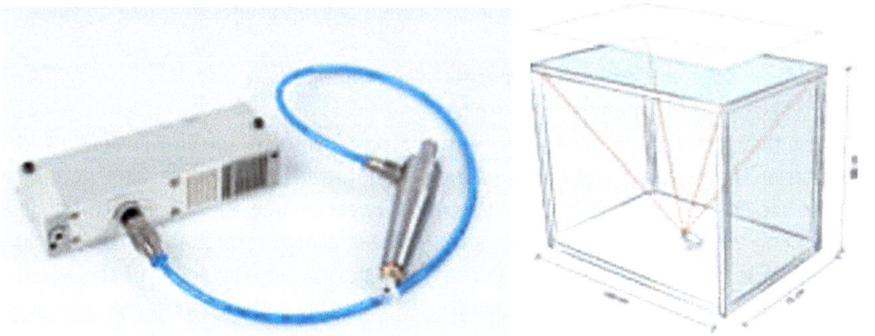

Figure 5 Customized vacuum lifter (left); table for camera-tracking (right)

Figure 6 Initial position, final state, and a rendering of the simultaneously generated 3D model in McNeel Rhinoceros 5

Thus, the overall shape can be controlled spontaneously by the designer. The final outcome derives from the initial configuration of sticks in the base layer, and depends on the different sticks placed on the table for the robot to pick them up and integrate them into the structure. In the ongoing process the designer can also take elements away that don't fit certain conditions, or let the robot rebuild parts of the already built structure.

Piling Sticks

The main problem addressed by the piling algorithm is stability. Once the dimensions of new beams are determined by analyzing the camera image, it is necessary to find a suitable place on the pile where the new beam fits and would not cause the pile to collapse.

To solve this problem it is necessary to keep track of all members of the pile, and how they interact with each other. This information is stored in a graph and updated with each beam added or removed. Each beam along with its properties like length and weight is represented by a node in the graph, the edges of the graph correspond to the points at which the beams rest on each other. For every beam the actual reaction at the bearing points, as well as the maximum bearing load under which the structure remains static, is stored.

With this information and the dimensions of a new beam at hand, it is possible to determine a domain of stability, within the parameter space of all possible positions and orientations of the new beam. From this domain any point can be chosen, and the beam can now safely be placed without compromising stability (Fig. 7).

If the position of the first bearing on one stick is defined, there are various maximum and minimum conditions for limiting the solution space for the position of a second bearing on another stick. The intersection area of all maximum and minimum conditions represents this solution space.

Recognizing Sticks

In the experiment, the small wooden sticks are detected by a simple high definition USB webcam (Fig. 8). Although the camera delivers a high resolution image, this image is distorted. The edges of the wooden elements are detected by applying a Hough

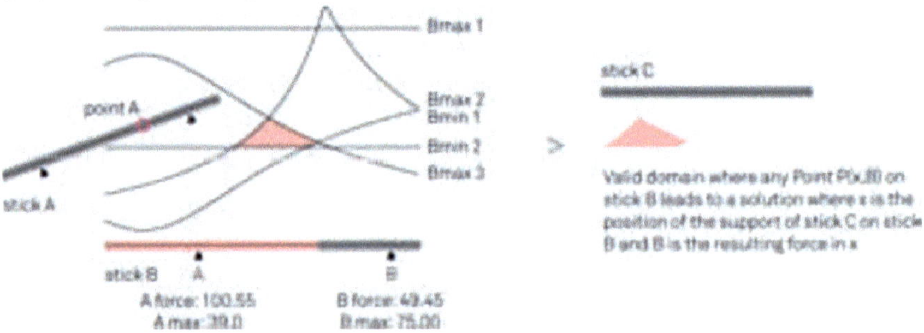

Figure 7 Intersection (colored area) represents possible positions for a stick C on stick A and B

transformation to the image. From the result the length, orientation and center point of all visible members are calculated. The coordinates of the camera need to be transformed into the base coordinate system of the robot and, due to the distortion referred to,these results also need to be linearized before being used. The linearization is achieved by performing a spline interpolation on a 15 by 10 points reference grid. The achieved overall accuracy turned out to be +/- 0.5 mm.

Visualizing Sticks

Along with the graph, a 3D model of the pile is created in McNeel Rhinoceros 3D (Fig. 8, 9). Rhinoceros is integrated into the system as just another client, linked to the server application via Ethernet exchanging data by XML.

On one hand, this enriches the experiment with the possibility of contrasting the physical with the virtual model. In contrast to the common chronological order, the virtual model evolves synchronously with the fabrication of the physical object, and does not exist beforehand. On the other hand, it is possible to utilize Rhinoceros' functions to perform geometry processing, and it also allows the use of its powerful graphical user interface. This enables high level user interaction with the digital model. In the experiment a function was implemented to remove sticks from the digital model in Rhinoceros which in turn can be automatically removed from the physical object by the robot. Besides that, if the user removes a stick from the physical structure, the digital model needs to be updated manually to ensure consistency. This can be also accomplished within the Rhinoceros graphical user interface.

Discussion

Each element in the physical structure has its representative in the digital structure, and is therefore directly linked to an immaterial digital logic, while on the other hand each computed position of a stick derives from its respective physical properties (length and weight). In this case study the algorithm is limited to rules that refer only to static conditions, but it is extendable, and one could integrate rules/conditions of any kind.

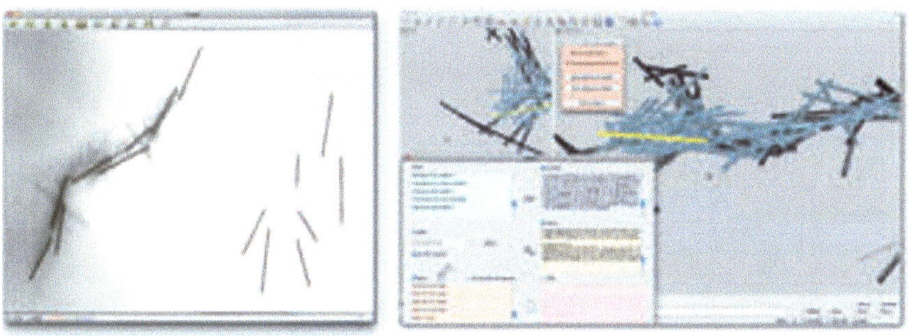

Figure 8 User Interface of the server application and digital model of the stick pile in McNeel Rhinoceros 3D. / Camera tracking: finding the positions of the sticks in the base layer

Feedback

The experimental setup was designed to achieve user interaction, without using a dedicated external input device, but through direct manipulation during the production of the physical object. As an important consequence, a feedback loop is created (Fig. 10). As the input parameters for the shape generation algorithm are extracted from sensors monitoring the physical object, the actions planned by this algorithm and carried out by the actuator are monitored as well and affect the input data. This feedback loop incorporates the virtual along with the physical domain. The feedback can be utilized in various ways. Together with a properly designed algorithm it can help to create robust and fault tolerant systems, capable of recovering from errors or improper user inputs. But feedback is also a key ingredient in the creation of systems that show complexity even though they are based on very simple rules. While this kind of emergence can already be observed while studying purely digital feedback loops, incorporating the physical world opens up a lot of new possibilities. In many cases it can even help to simplify the algorithms, as the direct incorporation of physical properties makes their digital simulation obsolete.

Responsiveness

As observed in the described experiments and studied in depth in "Nach vor und zurück" (Dörfler, Rust, 2012), responsiveness is a very important issue when designing the system to link digital and physical design tools. The user interaction has to be direct and transparent, in the sense that the user is able to understand the causality between his action and the reaction of the system, at least to a certain degree.

The time lag has to be low, while within this context, specifying "low" depends on the set-up. In one situation, a 10 second delay may be fast enough to give the user a feeling of interaction, whereas in another case, even a fraction of this would be too long.

Future Development

The combination of automated processes with intuitive actions in an interlaced physi-

Figure 9 Piled up sticks, walnut

cal/digital setup, as approached in the described case study, can offer the designer a playful environment within which the solution of rationalized problems is taken over by the machine, thus freeing the designer to follow direct impulses.

With the robot's individual customization, the development of the external control via network, and its extension with different sensors, the tool is raised above its predetermined functionality and can be used within this new context. To expand the functionality further, the next steps would be to take full control of the tool path interpolation and establish a hard real time system for the data exchange between the clients.

Acknowledgement

This project uses a number of open source software libraries and tools. Among others these are: numpy and scipy, openCV, roslib, pyqt, Qt, Python, Eclipse; We thank the community for providing and maintaining these useful tools.

References

Hovestadt L, 2010, *Jenseits des Rasters – Architektur und Informationstechnologie*. Birkhäuser Verlag AG. Basel . Boston . Berlin.

König R, 2011, "Generative Planning Methods from a Structuralist Perspective", in: Valena T, (ed.), Avermaete T (ed.), Vrachliotis G (ed.) *Structuralism Reloaded*. Edition Axel Menges.

Gramazio F, and Kohler M, 2008. *Digital Materiality in Architecture*. Lars Müller Publishers. Baden, Switzerland.

Herresthal K, 2008. "Von fließender Systematik und generativen Prozessen – Christopher Alexander im Gespräch mit Rem Koolhaas und Hans Ulrich Obrist", in: *ARCH+ Zeitschrift für Architektur und Städtebau*. no. 189, October 2008.

Alexander C, 1964. *Notes on the Synthesis of Form*. Harvard University Press. Cambridge, Massachusetts, London.

Dörfler K, Rust R, 2012. *Nach vor und zurück - Verschränkung digitaler und physischer Gestaltungsprozesse*. Master Thesis TU Vienna / TU Graz.

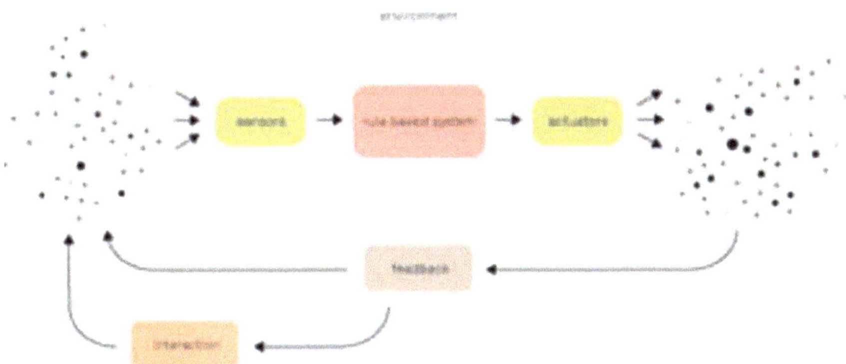

Figure 10 Feedback process, diagram of one iteration.

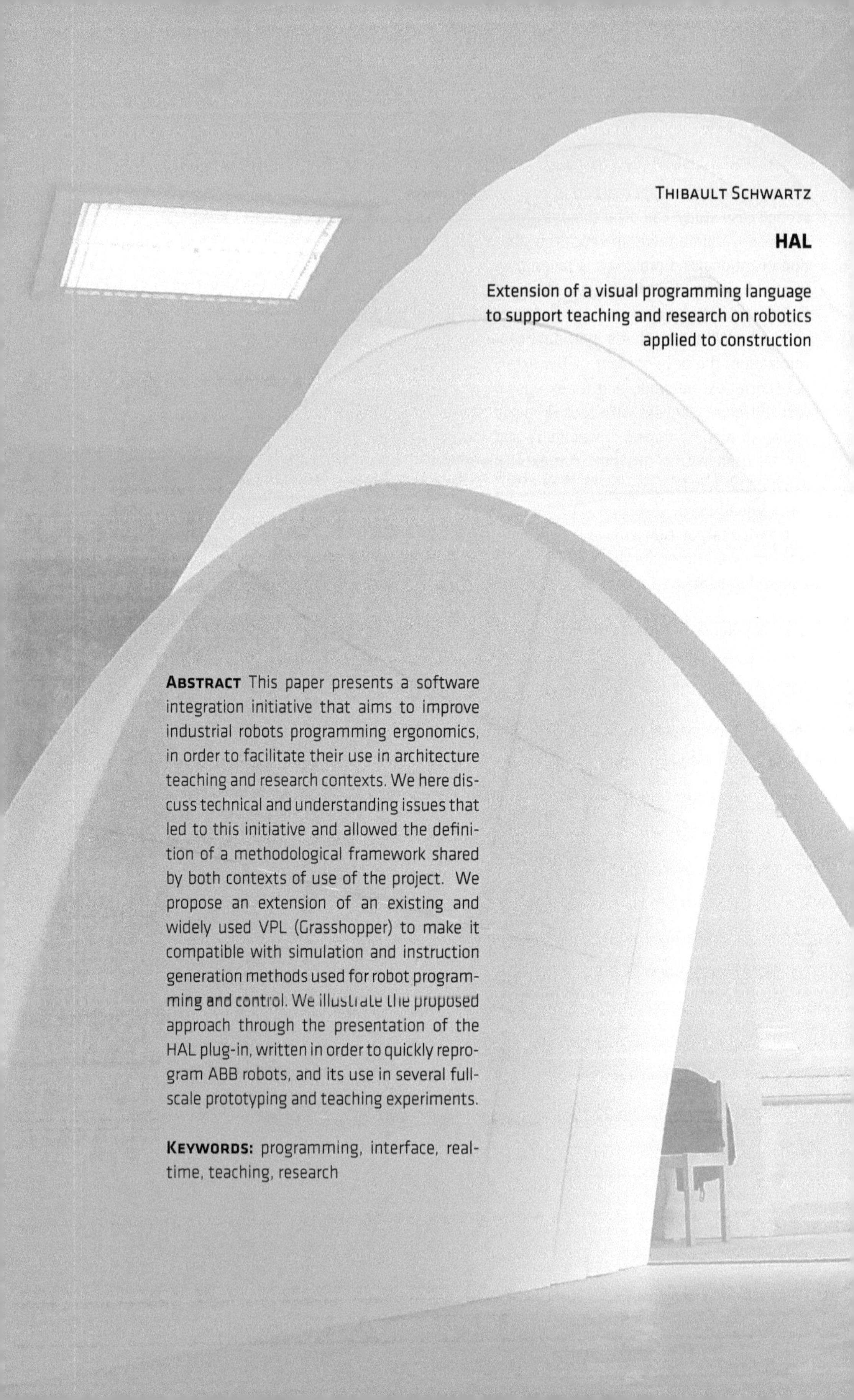

Thibault Schwartz

HAL

Extension of a visual programming language to support teaching and research on robotics applied to construction

Abstract This paper presents a software integration initiative that aims to improve industrial robots programming ergonomics, in order to facilitate their use in architecture teaching and research contexts. We here discuss technical and understanding issues that led to this initiative and allowed the definition of a methodological framework shared by both contexts of use of the project. We propose an extension of an existing and widely used VPL (Grasshopper) to make it compatible with simulation and instruction generation methods used for robot programming and control. We illustrate the proposed approach through the presentation of the HAL plug-in, written in order to quickly reprogram ABB robots, and its use in several full-scale prototyping and teaching experiments.

Keywords: programming, interface, real-time, teaching, research

Introduction

Democratization of CAD technologies, perceptible in architecture schools as well as in the construction community, progressively led during the last decade to the creation of groups combining students and professionals seeking to extend their morphological research undertaken at a "virtual" level, towards a systematic experimentation practice of manufacturing methods of these geometrical abstractions. As a result, and taking advantage of lower cost CNC machines (routers, laser cutters, 3D printers, etc.), rapid prototyping workshops of numerous universities and R&D offices have become genuine micro-factories. Even so, the limited size of this equipment – whose primary function is to produce models – makes it difficult to shift to the production scale required to experiment with building systems and their industrial manufacturing processes. This context of educational and research infrastructure development participates nowadays in the gradual integration of robotic arms – industrial machinery if ever there was one – in universities and construction sciences-related offices.

However, the whole range of difficulties involved in this integration is often underestimated: if most of the traditional workshop equipment is almost immediate and intuitive to handle, the use of a robot requires a significant learning time. In an educational context, this period of apprenticeship and equipment mastering is highly problematic, given the short project cycles (semesters) that punctuate students' education and during which they only have a few weeks to discover, master and use extensively a new type of machinery they barely knew anything about. In a research and development (R&D) environment, learning is often easier, depending on the professional activity of the participants. In such a case, the integration problems lie at the general workflow level, where the implementation of a robust and flexible interface in line with existing hardware and software infrastructures can prove to be onerous.

Therefore, this paper proposes to address several hypotheses concerning the development of robot control interfaces, in order to pool and meet the needs faced by these environments of experimentation.

Constraints: Programming Interfaces Requirements and Limitations.

Intimately linked to the invention of CNC

Figure 1 Pre-punched program tape, control panels and general view of the MIT's Cincinnati Hydrotel 3-axis milling machine, early working prototype of NC (Numerical Control) technology (Pease, 1952)

machines, programming strategies and their relative – software and hardware – interfaces have been a controversial subject since the early 40's. Historically, as Noble (1984) exposed, two strategies were – and still are – mainly discussed: programming by demonstration (PbD) also known as "teaching" and promoted by Sponaugle (1944), de Neergaard (1945), Leaver (1949), G. Devol (1952, 1956); and offline-programming, favored by Parsons (1958) and by Pease, McDonough and Forrester (1962) and around which CNC-routers technologies have been initially developed at the MIT (Fig.1). The first strategy is an attempt to ease the access and control of the machines for operators and on-site workers but is very time-consuming and can be relatively imprecise; while the second strategy is to restrict control to engineers and programmers but allows the full use of the machine's ability to process language(s) and to integrate abstract – and absolute – data to a program for interaction purposes.

Regarding robot technology, the requested computing capacities and the complexity of simulation systems have long been obstacles to offline programming: unlike 3-axis routers that have each axis correlated with one vector of a three-dimensional Cartesian coordinate system (e.g., axis1=X, axis2=Y, axis3=Z), robotic arms operate in space with – usually – at least 6 degrees of freedom. For this reason, the position solving of a robot is not instinctively "predictable" and necessitates much more computing time. If we now have far more sophisticated computers that are able to achieve these calculations within a few hundredths of seconds, the gains of the evolution of offline programming interfaces for robots are still very limited. As a matter of fact, the programming approach largely remains "manual" and time consuming: these digital interfaces have long persisted in reproducing analogue teaching procedures, inducing significant programming time and constituting an ergonomic contradiction with computers solving capacities, which are theoretically able to exploit complex geometrical models to generate tool paths with minimal human intervention.

During a few years, robotics software solutions have integrated – as has been the case so far for 3 to 5 axis CNC machines – a certain level of compatibility with 3D models within their products (Lee & ElMaraghy, 1990; Bottazzi & Fonseca, 2006). This recent generation of programming interfaces, mixing a lot of robot control technology and very little control of the geometry, are primarily intended for industrial manufacturers in order to produce parts designed by third party developers, and perfectly fulfill their task.

However, architects, building scientists and students are in an opposite situation: as designers, they need to be able to integrate robotic manufacturing constraints at the very outset of their 3D sketching or geometrical programming practice. Furthermore, and as mentioned in our introduction, their interest in robot programming is primarily based on the need to rapidly produce prototypes of components whose geometry is, in most cases, unique. Therefore, this constraint of non-standard production requires the user to remain at a certain level of abstraction allowing both geometry and robot motion control, in other words a programming language level, in order to minimize manual operations that are repeated through each component to be produced, such as drawing extraction or tool configuration.

Nonetheless, the traditional CAD/CAM programming approach (model import, objects extraction and manipulation, tool path generation from objects) remains viable under the obvious condition that such operations are fully automated: the main interest in CAD contexts is that constituent elements of the final geometry to be manufactured are already isolated within the model, which allows a direct link between the geometric or algorithmic model and the tool path generation and simulation to be considered.

On the basis of the precedent hypotheses, we can then draft the list of constraints and limitations of an optimal programming interface for designers:

1. Advanced geometry modeling functionalities for sketching and model manipulation.
2. Compatibility with 3D files formats, for both meshes and NURBS geometry to optimize data-exchange between CAD platforms and preserve high-precision models.
3. Real time processing of geometry in order to refresh robot positions and tool path modifications preview.
4. Compatibility with a full geometry-based programming and configuration approach, allowing basic users to almost ignore text-based robot programming knowledge requirements.
5. For research and advanced use purposes, compatibility with common programming languages (C++, VB/C#, Python, Java, etc.) in order to be able to link specific peripheral applications (e.g. for structure optimization and form-finding, stock management, database synchronization, product documentation, etc.) as well as robot-specific applications.

Integration Strategy

To partly solve the problems related to building such an interface, we considered different scenarios of software plug-in creation for CAD and CAM tools presenting interesting features, and finally chose to use Grasshopper (Fig.2), a visual programming language (VPL) developed by David Rutten and running within the Rhinoceros modeler, as a main support of development. Since this selection allowed many of the previously mentioned prerequisites to be fulfilled, but had no specific functionality to control or program robots, we have been able to freely interpret the accessibility and prioritization of these additional functions.

The simultaneous and posterior emergence of similar approaches using different robot brands - KUKA|prc for KUKA robots by Johannes Braumann & Sigrid Brell Çockcan (2011)[1] and [S]GSC for Stäubli robots by Brian Harms (2012)[2] – convinced us from the earliest stages of development of the absolute necessity of full compatibil-

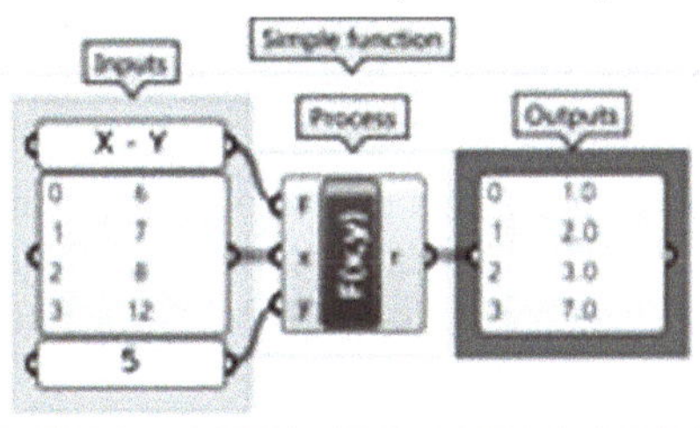

Figure 2 Screenshot of an annotated expression-based operation defined using Grasshopper

ity between this new interface and the existing Rhinoceros and Grasshopper function libraries, as well as with other third-party programs and plug-ins that enrich this language and designers' sets of tools over the months. In this perspective of systematic genericization, which we think is the main quality of any design tool, we developed our interface depending on two main integration objectives.

As a first step, we considered that the use of specific data types with which other components would not be compatible had to be avoided, as the partitioning of data types contradicts their possible manipulation with other tools, which would lead to some significant limitations for both use in an experimental teaching, or broader R&D workflows. On the other hand, the use of custom data types eases the wrapping of multiple primary data into complex – and conceptually more accurate – entities, thus reducing the amount of elements to declare in highly specific algorithms, and minimizing the amount of code involved to rebuild properties between those primary data. Taking kinematics solvers as an example, the reduction of robot configuration-related data (articulations locations, rotation domains, joints geometry, etc.) into a single "Robot" entity facilitates the simultaneous use of several solving algorithms while maintaining a common data structure throughout the computation process. In order to maintain cross-compatibility with other robot-related tools while making use of these new data types, we then decided to expose the generic "Robot" wrapper – and a lot of other custom types – in a separate public class library that we initiated (GenericCAM.dll) in order to centralize these CAM-related data types and methods. This particular measure has a major impact in terms of use. It is thus possible to develop tool correction algorithms, tool paths or reference axis systems modifications synchronized with sensors (e.g. to create real-time robot teaching applications using external devices), or even various mechanical simulation integration, by simply inserting complementary (eventually developed in-house) components between the basic robot interface functions. "OSC to HAL" and "HAL to Controller" plugins – which enable the use of smartphones, midi devices and grasshopper buttons for real time control of ABB robots – are a good illustration of this cross-compatibility aptitude. To make such interactions possible, a second point had to be carefully studied: the subdivision of simulation, instruction generation and calibration processes, as well as their adequate representation in order to preserve their generic and neutral status, separated from any construction or tooling-specific logic. By extension, this subdivision is also supposed to ensure the compatibility of most of these functions with different brands of robots, although each manufacturer uses different programming languages for their machines. Specific utilities and components can then be sorted according to:

1. Data or objects (as generic data types) they are designed to manipulate: targets, end-effectors presets, robot-specific instructions, list of attributes, data structures syncing, etc.
2. Specific equipment and/or machining methods they can control: milling, hot wire cutting, pick and place, 3D printing...
3. The overall programming chronology in

which they fit: collision simulation, data streaming, kinematics solving, etc.

While the other previously mentioned interfaces available for various robot brands usually focus on the provision of simplified simulation functionalities encapsulating many variables and thereby emphasizing their ease of use, the approach introduced in the development of HAL has always been to provide highly "neutral" components in order to expose most of the robot programming language capabilities, and to facilitate the simulation of complex automated systems involving different robots and third-party devices. However, just like KUKA|prc and [S]GSC, HAL also offers simplified functionalities to enable fast learning of the basic industrial robotics-related concepts, and to enable a quick transition to the most advanced features involving RAPID instructions programming and the development of complex manufacturing strategies.

Implementation

Although it required several months of successive trials and improvements (v0.01: June-October 2011; v0.02: October-December 2011; v0.03: January-February 2012; v0.04: August-September 2012), the implementation of this library occurred relatively naturally, and broadly follows the theoretical prioritization mentioned above. An ABB IRB120 robot (580mm) was used as a main testing machine, while some debug operations were also carried on an IRB140 (810mm), an IRB1400 (1440mm) and an IRB6400 (2400mm) robot.

The software operates with five core components:

1. An OpenSoundControl (OSC) messages translator that allows real-time robot control and teaching using smartphones, tablets, midi devices, etc.
2. A multi-robot compatible forward kinematics solver, allowing to link real-time

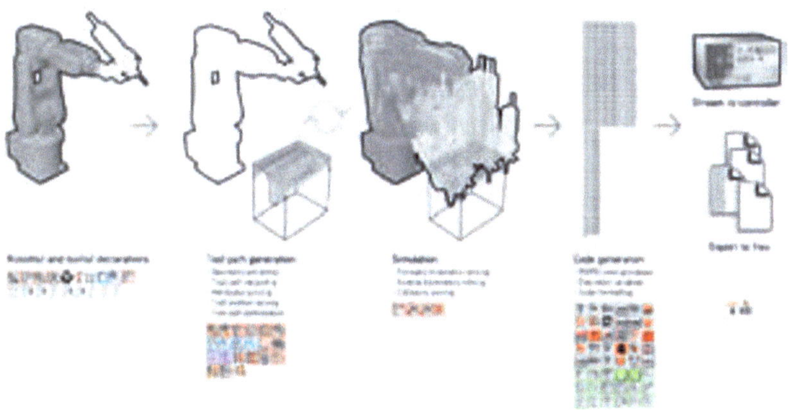

Figure 3 Chronology of a basic robot program elaboration process and enumeration of some of the main algorithms that can be used for each steps. HAL provides default algorithms for each of these stages, as well as several utilities, easing their modification or substitution with custom-made components.

control/teaching processes to the main simulation.

3. A multi-robot compatible inverse kinematics solver, around which simulation components can be used to program tool paths from geometrical data.
4. A post-processor that automatically generates the robot instructions from the solving data using the ABB-RAPID programming language, which results can be formatted and supplemented with additional programming functions libraries.
5. A bridge to virtual and physical robot controllers for data streaming and program execution control.

As evocated previously, we took advantage of this implementation exercise to try to facilitate some advanced approaches, especially in terms of multi-robots simulation and advanced tasks programming. The inverse kinematics solver is thus computing differently depending on the data structures it receives: with a robot base axis systems data tree structure as input, it will automatically create as many robots as axis systems. If one target is provided, it will attempt to compute all robots kinematics in order to reach this specific target. If a data tree of targets is provided, it will try to match robot entities with targets of the equivalent allocation level. This mechanism, also present in many other components (e.g. to automatically generate nesting of RAPID loops or procedures), allows an easy and quick simulation of the behavior of multiple similar robots and their possible interactions with a minimal amount of code and programming effort.

R&D Applications

As a research interface, HAL has been tested in different project prototyping situations. For EZCT's u-Cube construction system project (EZCT, 2012), the interface has been used to realize EPS molds for casting Ultra High Performance Fiber Concrete (UHPFC) lattices. HAL was performing in conjunction with experimental surface discretization algorithms and other procedures being developed in Mathematica and Grasshopper, and allowed several mold solutions to be quickly prototyped (Fig.4).

In order to test the interface robustness in a production context, three full-scale EPS shell prototypes were also produced. The first two "Automated Foam-Domes", were both composed of 152 parts, with a height ranging from 2.4m (#2) (Fig.5)(Schwartz, 2011) to 2.7m (#1). These pieces were entirely prefabricated using 5 to 6 standard 80mm insulation foam panels, and machined using a 300mm wide hot wire cutter mounted on an IRB120. HAL was used in order to automatically generate the 76 different tool paths allowing the prefabrication of the 152 constructive components of which all 6 faces were cut (computing time of both simulation and code generation without collision solving: ±200ms by part, ±15s for the entire prototype). The link between geometrical generation and real-time robot simulation allowed the optimization of material needs and fabrication process and reduced fabrication costs and duration (only 16 hours of prefabrication for the second prototype: ±7min by part).

A similar approach was used for the third full-scale EPS prototype, a 2x5x3.5m parabolic vault (Schwartz, 2012), in which the shell discretization has been specifically studied in relation with the IRB120 dimensions and kinematics in order to obtain 500x650x90mm panels (slightly bigger than the robot itself), therefore multiplying the average panel size used for precedent prototypes by 4. Temporary parts were extracted from 1200x1200x600mm EPS blocs directly by the robot using a 670mm wide hot wire cutter, before being individually machined properly using the same end-effector. HAL facilitated the very delicate programming of both extraction and machining of the 160 different panels, operations which were difficult to manage due to the large size of the cutter and required very precise calibration and tool path configurations because of the use of the robot maximum kinematic capabilities and the related multiplication of imprecisions. Assembly slots were also cut in the panel during the final part machining. The tool path extraction and instruction generation including complete collisions solving (robot-robot, tool-robot, tool-part, robot-part, tool-context, robot-context) were computed in ±900ms for each part (±2min 30s for the whole prototype).

Figure 4 Structural UHPC mesh sample from EZCT's u-Cube construction system

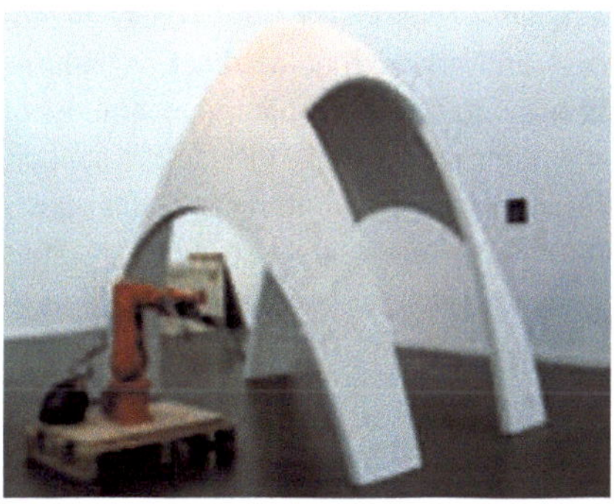

Figure 5 IRB120 robot running a demonstration program next to the Automated FoamDome #2 during the Synthetic 2012 exhibition

Teaching Applications

As a pedagogical support, HAL has been used in different structures, both as part of architecture student general courses and as part of research-oriented workshops. For instance, several seminars have been proposed during the early phase of the plugin development by Felix Agid (EZCT) at the Ecole des Beaux-Arts du Mans (ESBA TALM) in partnership with the Ecole d'Art et de Design de Valencienne (ESADV) and the TU-Delft Hyperbody Research Group; and at the Paris-Malaquais school of architecture (ENSAPM). The first 2 week seminar, held at Le Mans and Rotterdam, was part of a new larger pedagogical laboratory (Synthetic-lab) mixing robotics, material sciences and neurosciences within art and engineering. During these series of workshops (one week of initiation in generative programming and robot simulation, and one week of production), a 3-meter-high full-scale foam structure was realized by 8 art students. The second seminar, at Paris-Malaquais, allowed 14 undergraduate architecture students to experiment with the robotized hot wire cutting process linked with Kinect sensors. In only one week the students implemented simple fabrication strategies linked with capture-based 3D models, and produced a series of EPS prototypes using Grasshopper with Firefly and HAL plugins (Fig.6)(Agid & Nguyen, 2012). Since its first release on the internet in November 2011, HAL has also been used and tested by several architecture schools, such as the Harvard Graduate School of Design where a design team used it to shape tectonic ceramic elements with CNC robotic wire cutting (Andreani & Garcia del Castillo & Jyoti, 2011). This experimentation has been extended to an educational workshop, Ceramics 2.0, during the Smart Geometry 2012 conference (Fig.7) during which 10 students and professionals developed and produced a series of various ceramics elements (ornamental and structural components).

Conclusion

This paper introduces a VPL-based robot programming interface implementation based on pedagogical and research requirements in the area of architecture and building sciences. It is shown that needs for an easy to use, yet, robust integration of robot programming in 3D CAD software leads to several adjustments of traditional robot task modeling and planning in order to meet architectural and constructive purposes. The observable multiplication of robotized fabrication solutions and their related control technologies - leading to an inevitable proliferation of similar software development experiments – invites to a broader debate on the structurization of such programs, on the type of knowledge they allow to acquire, and on the industrial production strategies they help to shape. In light of the increasingly common teaching experiments conducted in architecture schools involving CNC machine programming, it also seems desirable to maintain a relative rigor in the teaching of such technologies in order to preserve a strong link between this growing automation-oriented knowledge, and the geometrical, structural, economic and social valuation of their resulting (digital and material) production. As a guaranteed intellectual and conceptual benefit for architecture practice from robotics still remains uncertain (unlike the very predictable savings resulting from the usage of programmable devices in construction sites), we believe that the systematic exposure of control and simulation technologies mentioned in Chapter 3 is necessary to preserve at least a minimal consciousness of the design process limitations that automation induces or helps to shape. Future improvements of the proposed interface include the support for external axis, interchangeable post-processors, and extended functionalities

Figure 6 "Gestes et Trajectoires" workshop at ENSA Paris-Malaquais

for human machine interfaces (HMI) experiments. Since September 2012, we have been undertaking additional pedagogical and research applications involving HAL at the Bartlett (London), TU Innsbruck (Innsbruck), the Paris-Malaquais school of Architecture (Paris) and ENSCI's FabLab (Paris), which will assuredly continue to shape and mature this new, but yet robust, tool.

References

Agid F and Nguyen MM 2012, "Gestes et Trajectoires" workshop, ENSA Paris-Malaquais, Paris.

Andreani S, Garcia Del Castillo JL and Jyoti, A 2011, "Flowing Matter, Generation of complex tectonic ceramic assemblies with CNC (robotic) wire-cut ruled shaped components", Material Systems and Processes: Ceramic Lab, Harvard Graduate School of Design.

Bottazzi V and Fonseca J 2006, "Off-line programming industrial robots based in the information extracted from neutral files generated by the commercial CAD tools", *Industrial Robotics: Programming, Simulation and Applications*, Huat LK (Ed.), InTech (http://intechopen.com), p. 349.

Devol GC 1952, "Magnetic process control", U.S. Patent 2 590 091.

Devol GC 1956, "Magnetic storage and sensing device", U.S. Patent 2 741 757.

EZCT Architecture & Design Research, 2012, "U-Cube, Universal Cube for Discrete Construction'" *Multiversité Créatives (exhibition)*, Centre Pompidou, Paris.

Forrester JW et al. 1962, "Numerical control servo-system", U.S. Patent 3 069 608.

Leaver EW and Mounce GR 1949, "Method and apparatus for the automatic control of machinery", U.S. Patent 2 475 245.

Lee DMA and ElMaraghy WH 1990, "ROBOSIM: a CAD-based off-line programming and analysis system for robotic manipulators", *Computer-Aided Engineering Journal*, Vol.7, No.5, October 1990, pp. 141-148.

Neergaard (de) LE 1945, "Method and means for recording and reproducing displacements", U.S. Patent 2 628 539.

Noble DF 1984, *Forces of Production: A social history of industrial automation*, Knopf, New York, pp. 152-192.

Parsons JT et al. 1958, "Motor controlled apparatus for positioning machine tool", U.S. Patent 2 820 187.

Pease W 1952, "An Automatic Machine Tool", *Scientific American*, Automatic Control (special issue), September 1952, pp. 101-115.

Schwartz T 2011, "Automated Foamdome #2", *Synthetic 2012: Foam* (exhibition), ESBA TALM, Le Mans.

Schwartz T 2012, "Habitat Prospecteur 2012", Graduation project, ENSA Paris-Malaquais, Paris.

Sponaugle LB 1944, "Method of operating machine tools and apparatus therefor", U.S. Patent 2 484 968.

[1] http://www.robotsinarchitecture.org/

[2] http://www.nstrmnt.com/

Figure 7 "Ceramics 2.0" workshop cluster during Smart Geometry 2012

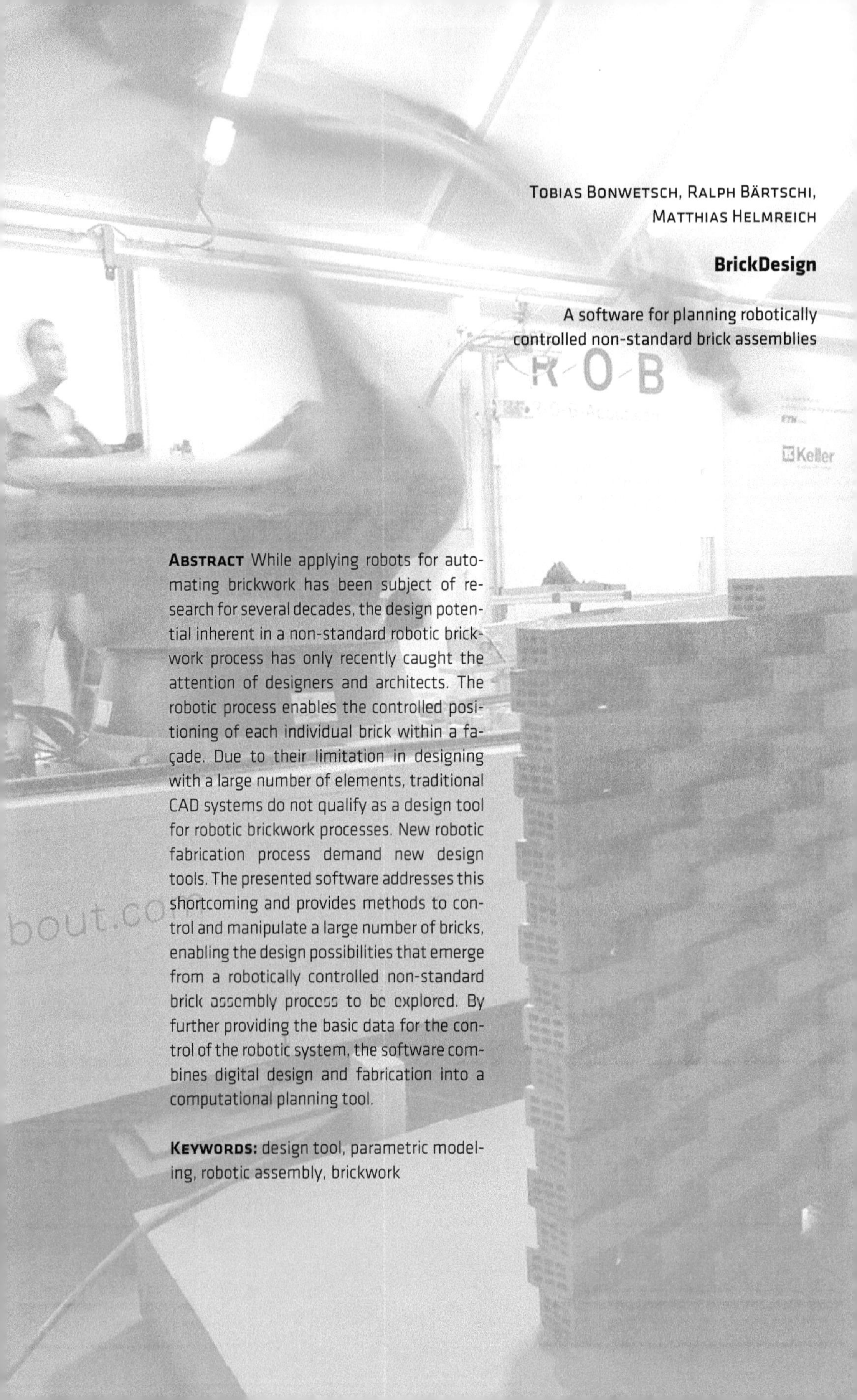

Tobias Bonwetsch, Ralph Bärtschi,
Matthias Helmreich

BrickDesign

A software for planning robotically controlled non-standard brick assemblies

Abstract While applying robots for automating brickwork has been subject of research for several decades, the design potential inherent in a non-standard robotic brickwork process has only recently caught the attention of designers and architects. The robotic process enables the controlled positioning of each individual brick within a façade. Due to their limitation in designing with a large number of elements, traditional CAD systems do not qualify as a design tool for robotic brickwork processes. New robotic fabrication process demand new design tools. The presented software addresses this shortcoming and provides methods to control and manipulate a large number of bricks, enabling the design possibilities that emerge from a robotically controlled non-standard brick assembly process to be explored. By further providing the basic data for the control of the robotic system, the software combines digital design and fabrication into a computational planning tool.

Keywords: design tool, parametric modeling, robotic assembly, brickwork

Introduction

The key to activating the flexibility inherent to a robotic system is its programmability: As illustrated by several recent brickwork projects, digital design data can directly be used to control the robotic system, thereby enabling the architect to explicitly control the building process and realize novel constructive systems and non-standard design solutions (Bonwetsch et al., 2007; Gramazio and Kohler, 2008; Baertschi et al., 2010). Given the potential to individually position each brick, the question arises about how to design a non-standard brick façade consisting of several ten thousand bricks. The conventional way for an architect to represent a brick wall in plan is to define its outer boundaries, without explicitly defining each single brick. Traditional CAD systems entail this representation but offer no tools to efficiently compute and represent a non-standard brick wall from its constituent elements.

For standard brickwork this information is sufficient. Given the bond type and the dimension of the brick unit a mason with his implicit knowledge and experience can easily erect such a wall (Lynch, 1994). Moreover, in order to explore the full design space spanned by the non-standard robotic fabrication process, a computational model depicting each single brick is necessary. In a conventional CAD-system this would imply modeling every single brick and placing it 'manually' at the desired position in space. Such an operation would be unjustifiably time-consuming and in terms of impracticality resembles the manual construction of such a brick façade. Therefore, the possibilities opened up by digitally controlled robotic fabrication processes clearly require new design tools.

In this paper we discuss a comprehensive new approach to this field of architectural research: the design tool *BrickDesign,* which is conceptually based on the creative control of a large number of units in order to foster a systemic, unifying planning process. In this, the software allows the design of a façade from its constituent elements – the bricks – rather than through an overall geometry. Thereby *BrickDesign* enables to explore the full design space spanned by the possibilities of robotic fabrication processes and implements these in a wider architectural planning process.

Design Tools for Robotically Assembled Brickwork

The extensive research on robotically fabricated brickwork in the 1980s and 1990s focused solely on standard brickwork (Kodama et al., 1988; Slocum and Schena, 1988; Lehtinen et al., 1989; Malinovsky et al., 1990; Chamberlain et al., 1991; Altobelli et al., 1993; Rosenfeld et al., 1993; Pritschow et al., 1994; Andres et al., 1994). Specific software tools were not necessary for the initial design stage, but only became important at the stage of generating the control code for the robotic system. In a top down process a given wall defined by its boundaries is broken down into the position of each individual brick (Herkommer and Bley, 1996). More recently, Cavieres et al. (2011) have emphasized the importance of integrating knowledge-based parametric tools already at a conceptual design stage. For a very similar problem of concrete masonry walls they propose parametric templates that embed construction and structural design knowledge, while omitting fabrication constraints. Although applicable at an early design stage, the initial brick distribution relies on an input surface similar to the top down process of Herkommer and Bley (1996). Due to the lack of appropriate software tools our own robotic brickwork projects mentioned above generally relied on custom scripting (Baertschi et al., 2010). The developed scripts are highly project specific and cannot be readily generalized on a more abstract level in order to address a broader scope of design exploration. A first attempt for a design tool for robotic brickwork was the web-based software *Creator*, which allows the user to map images on a brick wall element. However, the design possibilities are very limited. The user can merely choose the height and length of the wall, as well as an image to project onto it [1].

ROB-Walls

The limitations of the design tools contrasts with the possibilities opened up by the robotic fabrication process. In a previous project,already, a non-standard brick façade system was realized. In collaboration with *Keller AG Ziegeleien* a material and construction technology was developed, as well as generic computational driver for generating control code for the specific robotic setup [2].

The driver software enables an individual design of a brick assembly from a CAD program to be automatically exported to the robotic system. The driver converts the data into control data for the fabrication process, it generates the necessary gluing paths and handles collision problems specific to the robotic setup (Fig. 1).

Figure 1 Robotic setup of automated brickwork process

BrickDesign

The applied research project *BrickDesign* addresses the above mentioned limits on the design software and complements the preliminary work of the robotically fabricated brick façade system *ROB-Walls* by providing the essential starting point in a digital chain that directly links the design with the fabrication process. The goal is to provide a design tool that integrates the parameters of a robotic brickwork process, while offering a high degree of freedom at an early design stage [3].

Foremost, it enables the controlled design and manipulation of a large amount of discrete elements, the individual bricks. It offers a parametric design environment, which allows for a fast buildup of façade geometries, which can easily be adapted to changing design intentions and requirements. Further, different methods that manipulate the local position of individual bricks can be applied to map patterns on a façade. These methods can be extended ad libitum by the user through an open script interface that gives access to a number of brick parameters. *BrickDesign* is designed as a plugin to a common CAD system, so that the design of a brick façade can be integrated in the standard design and planning process. In order to meet practical demands the software was developed in close collaboration with architects. Following the spiral model for software development defined by Boehm (1986), in an iterative process several software prototypes were built and tested on real world

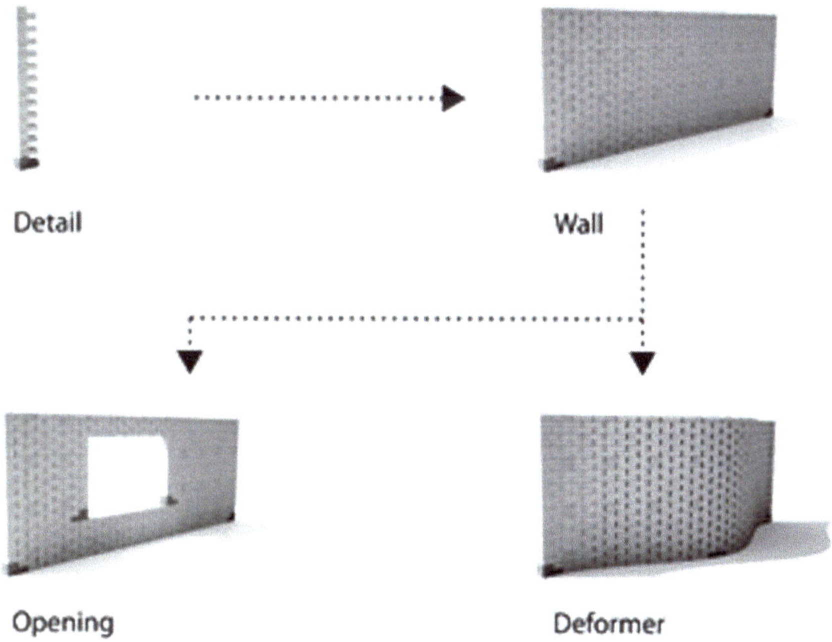

Figure 2 BrickDesign objects, the dark tinted bricks highlight the defining bricks. All other bricks are generated automatically by the software.

projects. For the development,*Rhinoceros* [4] is used as a host system, but the core of the software is independent and can be ported to other CAD systems in the future. Below we highlight some of the concepts of the software.

Designing with bricks

The core concept of the software is that the brick unit is the basis for every action performed. Basically, a design is generated through drawing, placing and manipulating individual bricks. This bottom-up approach distinguishes itself from the software solutions mentioned above, which follow a top-down approach by allowing the user to draw a surface which gets discretized into individual brick units in a secondary step. On the one hand, the bottom-up approach is closer to the reality of building since it forces the user to think in a constructive logic distinctive to brickwork. On the other hand, populating an arbitrary surface with brick units is not a trivial problem and automating this process can lead to a severe limitation to the user's design freedom.

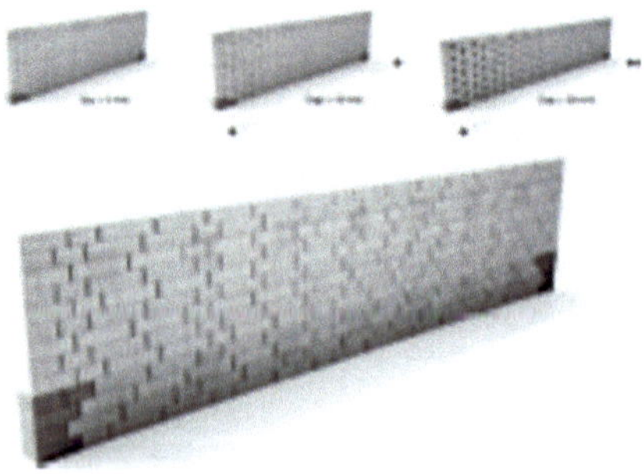

Figure 3 Example of a running bond pattern with varying butt joint width (top), a so called 'wild bond' requires the definition of a more sophisticated brick sequence and a corresponding corner detail (bottom).

These characteristics are reflected in the data model of the software. *BrickDesign* offers four classes of CAD objects: *details*, *walls*, *openings* and *deformers*. The starting point for a design is always the *detail*. *Details* can be endings of a wall, corners or reveals. The user can pick a predefined *detail* out of a library, or manually draw its defining bricks. Repeating rows are then automatically generated. The predefined library *details* represent complete parametric objects, thus offering the advantage of automatically adapting to changing geometries and brick formats. A *wall* object spans between two *details*. The bricks in a *wall* are automatically generated based on a user defined bond pattern. A *wall* can have an arbitrary number of *openings*. These are defined by selecting the brick at the lower left and the upper right corner of the projected opening. Also an arbitrary number of *deformers* can be defined within a *wall*. Through selecting a brick in a *wall*, it can be transformed into a *deformer* brick. A deformer allows additionally manipulating the course of the footprint of the *wall* (Fig. 2).

Unlike traditional brick work, the software does not define a bond pattern by a fixed grid, but through the definition of a sequence of specific brick formats. A traditional brick bond is based on the relation between the brick module's width and length, and a predefined butt joint. In a non-standard robotic assembly process the width of the butt joint can constantly change due to the position of the individual bricks and is subject to design decisions. Hence, *BrickDesign* does not apply a vertical raster when placing bricks. A bond pattern in a façade is specified by the interlocking defined through the *detail* and the definition of a sequence of brick formats: e.g. a simple running bond is created by a *detail* where the connecting bricks are offset half a brick every second layer and the brick sequence is defined by a single full brick (Fig. 3).

In order for the designer to have a constant visual feedback of his decisions, the software always displays the complete assembly with each individual brick. With such numbers of bricks involved, most CAD systems run into performance issues. Therefore, the four object classes are displayed as joined meshes, enabling to work with up to one million bricks in the chosen host CAD system.

Parametric and Static Mode

BrickDesign differentiates between a parametric mode and a static mode. Generally, all *BrickDesign* objects establish a parametric relationship to each other. This means, for instance, that when moving a *detail* all objects connected to that *detail* automatically get updated, i.e. the geometry of a connected *wall* adapts, as well as the number of bricks. These adaptations are performed in real time. The immediate visual feedback makes the parametric mode an ideal environment for design exploration and fine tuning the overall geometry (Fig. 4).

Other actions that are computationally too complex to be performed in real time are executed on a static instance of the parametric model. Mainly because they do not allow for a real time visual feedback and would thus disrupt the intuitive and interactive experience of the parametric mode.These actions include the application of user defined scripts, ornamentation of the façade, optimization routines, feasibility checks and solvers for constructability. Nevertheless, in *BrickDesign* it is possible to switch back and forth between the parametric and the static mode at any time. In order to increase productivity, actions performed in the static mode can be recorded in an associative chain and played back on other instances of the parametric model in the same sequence and with the same parameters.

Ornamentation of a façade is realized through small variable deflections of the individual bricks. A brick can be shifted along its X- and Y-axis, as well as rotated around its vertical axis. Additionally, a pattern can be created through using a palette of different colored bricks. Besides the possibility for the user to script an own rule set

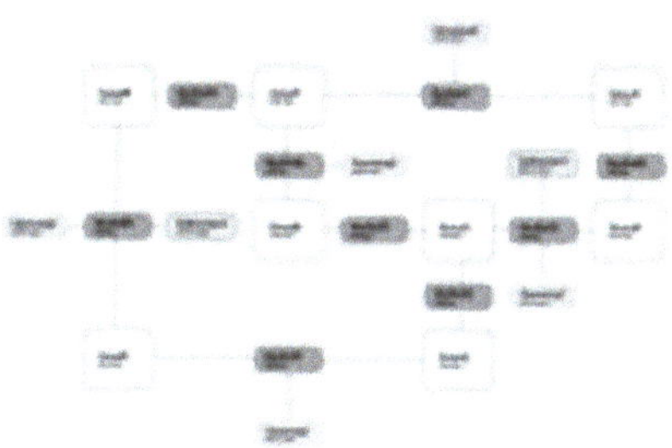

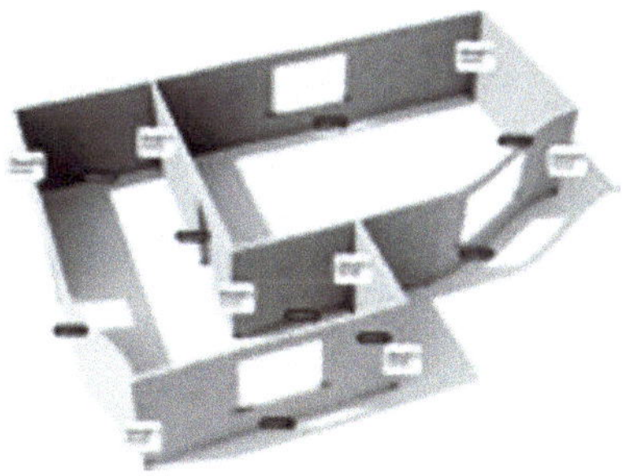

Figure 4 Example of a parametric model's data structure (left) and the corresponding design (right)

for patterning the facade, two predefined methods are already implemented in the software. Either, the user can place force fields in the model that act upon the bricks, or an image can be used as data input for a pattern. Each of these methods offers several parameters for the user to define how the input data should be transformed onto the façade and which brick parameters will get affected (Fig. 5).

Conclusion

New robotic fabrication processes demand new design tools. In order to fully exploit the potentials inherent to a fabrication process, its parameters have to be made available at an early design stage. Thereby, parameters of fabrication can inform the process of design exploration.

Digital control over robotically assembled brickwork empowers the designer to inform each individual brick within a façade. Traditional CAD systems, which were more or less conceived to simulate manual drafting, do not offer the appropriate tools to creatively control and manipulate a large number of parts for designing non-standard brick facades.

In this paper we present a software tool that extends the spectrum of architectural planning and manufacturing methods and therefore creates a new level of robotic use in architecture. *BrickDesign* fosters information penetration across the whole process of making, from design to fabrication, opening up new ways of thinking about architectural design and materialization. The synchronization of design and making is essential to leveraging new architectural potentials. By being able to control each individual brick and assess brick assemblies consisting of a large number of interdependent parts – both in terms of their visual appearance and their feasibility –the software allows the exploration of design solutions outside the commonly known standards of brickwork.

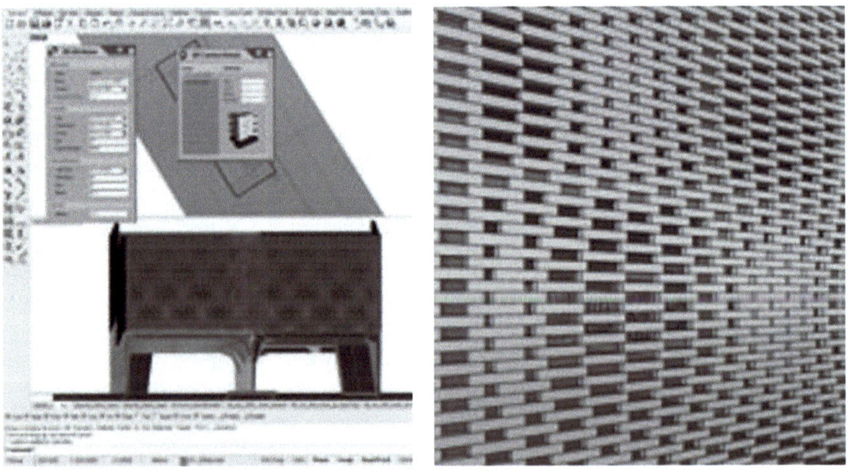

Figure 5 Screenshot of BrickDesign software. The design of a brick façade is integrated in the standard design and planning process.

Acknowledgements

This research was partly funded by the Swiss Commission for Technology and Innovation. The joined research project involves the group of Gramazio & Kohler, Architecture and Digital Fabrication at ETH Zurich and the industry partners Keller Ziegeleien and ROB Technologies.

References

Altobelli F, Taylor HF and Bernold LE 1993, "Prototype Robotic Masonry System", *Journal of Aerospace Engineering*, 6, 19-33.

Andres J, Bock T and Gebhart F 1994, "First results of the development of the masonry robot system ROCCO", *11th ISARC International Symposium on Automation and Robotics in Construction.*Brighton, England.

Baertschi R, Knauss M, Bonwetsch T, Gramazio F and Kohler M, "Wiggled Brick Bond", in: Ceccato C, Hesselgren L, Pauly M, Pottmann H, and Wallner J, eds. *Advances in Architectural Geometry*, 2010 Wien. Springer, pp. 137-148.

Boehm B 1986, "A spiral model of software development and enhancement", *SIGSOFT Softw. Eng. Notes*, 11, pp. 14-24.

Bonwetsch T, Gramazio F, and Kohler M, "Digitally Fabricating Non-Standardised Brick Walls", in SHARP, D. M., ed. *ManuBuild, 1st International Conference*, 2007 Rotterdam. pp. 191-196.

Cavieres A, Gentry R, and Al-Haddad T 2011, "Knowledge-based parametric tools for concrete masonry walls: Conceptual design and preliminary structural analysis", *Automation in Construction*, 20, pp. 716-728.

Chamberlain D, Speare P, and Ala S 1991, "Progress in a masonry tasking robot", *8th Int. Symposium on Automation and Robotics in Construction.* Stuttgart, Germany.

Gramazio F and Kohler M 2008, "Structural Oscillations", *In:* GEISER, R. (ed.) *Explorations in Architecture: Teaching, Design, Research.* Basel: Birkhäuser Verlag AG.

Herkommer F and Bley B 1996, CAD/CAM for the prefabrication of brickwork. *Automation in Construction*, 4, 321-329.

Kodama Y, Yamazaki Y, Kato H, Iguchi Y, and Naoi H 1988, "A robotized wall erection system with solid components", *5th International Symposium on Robotics in Construction* Tokyo, Japan: Japan Industrial Robot Association (Jira), cop.

Lehtinen H, Salo E and Aalto H 1989, "Outlines of two masonry robot system" *Proceedings of the 6th International Symposium on Automation and Robotics in Construction.* San Francisco, CA.

Lynch G 1994, *Brickwork history, technology and practice*, London, Donhead.

Malinovsky E, Borschewski EA, Eler EA, and Pogodin VM 1990, "A robotic complex for brick-laying applications" *7th International Symposium on Automation and Robotics in Construction* Bristol, England.

Pritschow G, Dalacker M, Kurz J, and Zeiher J "A mobile robot for on-site construction of masonry. Intelligent Robots and Systems '94. 'Advanced Robotic Systems and the Real World', IROS '94" *Proceedings of the IEEE/RSJ/GI International Conference* 12-16 Sept. 1994. 1701-1707 vol.3.

Rosenfeld Y, Warszawski A, and Zajicek U 1993, "Full-scale building with interior finishing robot", *Automation in Construction*, 2, pp.229-240.

Slocum AH and Schena B 1988, "Blockbot: A robot to automate construction of cement block walls" *Robotics and Autonomous Systems*, 4, pp. 111-129.

[1] http://www.rob-technologies.com/media/creator/creator.jnlp

[2] http://www.keller-ziegeleien.ch/de/fassaden/robmade_fassaden.

[4] http://brickdesign.rob-technologies.com.

[5] http://www.rhino3d.com.

Andreas Trummer, Felix Amtsberg,
Stefan Peters

Mill to Fit

The Robarch

Abstract Digital handcraft sets new standards for traditional building techniques. Linking the topics as geometry, the work piece, tooling and joining, entirely new meanings are produced. Stonecutting in particular has a long tradition of using machines and industrial robots. The "Roboterdesignlabor" established at TU Graz offers a process chain from CAD data, cam data for 3- and 5 axis tooling translated to the machine controller language for milling processes with an ABB IRB 6660. The hardware allows milling of hard materials like stone or concrete. During the workshop the focus is directed towards pushing through digital information, illustrating interfaces and carrying out work pieces. An arch made of foam-glass as a placeholder for brick, composed of several work pieces, acts as a case study. Thereby the focus is placed on the gap between the artificial bricks. Mechanical principles and restrictions of tooling influence the modification of the gap. The aim is to visualize the flow of forces and make the gap as transparent as possible.

Keywords: milling, contact face, foam-glass arch

Industrial and Individual

On opening the discussion about modelling and fabrication of structural elements it is worth looking at the work of K. Wachsmann (1901-1980). His lifelong thinking about industrial processes in the building industry anticipates topics of the current discussion about the individualisation of industrial processes. Two aspects are noteworthy. One is his search for the universal joint for timber constructions, the general panel system. The other is the stance he took on the industrial building process. In Wachsmann's opinion a complete building production chain needs a kinematic positioner for building elements in space. As a result of this research he presented the Location Orientation Manipulator (LOM) as a universal machine for positioning in 1969 (Nerdinger 2010). This kind of machine is still not in use but the building industry has profound expert knowledge in milling, drilling and cutting prefabricated building elements for finishing. Milling is especially time intensive but, depending on the material, offers a wide scope for shaping a structural element. The question therefore is: how can we use these techniques to bring complex structures close to an efficient but flexible building process?

Two different cases should be taken in consideration. On the one hand there is the shape of an element, such as a beam. If you need many different beams, it makes sense to divide similar elements into different classes and produce them in an industrial way. To give the single element the right shape you can use a milling machine in a second step to finish them. Using this strategy it was possible to realize projects like the wooden curtain facade of the Kilden Performing Art Center in Kristiansand, N (Stehlig, H and Scheurer, F, 2012) or the wooden roof of the Centre Pompidou in Metz. Both examples show that timber is an easy-to-use material that offers a high level of forming potential.

Figure 1 Simulation of robotic fabrication

The other case is structures that demand a high level of precision because of the joining details. The Mansueto Library Chicago (Sobek 2011) is a prefabricated steel grid shell. The geometry of almost every joint is different. Precise milling of the contact faces was required for the shell geometry and the glass cladding. This example shows that the knowledge about materials and fabrication processes can be the key to implementing geometrically complex structures on the construction site.

In discussing questions of complex structures and form finding in architecture it is mainly aspects like geometry that are taken in consideration. Parametric design is based in many cases on CAD programs that makes it possible to vary and handle this 3D data. In special cases like shell structures the aspect of force should determine the geometry. There is the question about the extent to which the design process could be informed as regards statics and structural details. On the other hand there is the field of material, jointing and fabrication that influence the construction.

In researching these topics architects and engineers of the TU Graz established the Roboterdesignlabor to focus on questions of fabrication and finishing. The decision to use a robot for milling is motivated by the flexibility of this machine as a part of different worksteps. As a part of the ultra high performance concrete (UHPC) group there is the goal of verifying whether a machine center set up of this kind can be used successfully in the process chain of double curved, thin walled prefabricated concrete elements employed for shells and domes (Trummer, A, Peters, S and Amtsberg, F, 2012). The work steps are milling the contact faces of compression stressed joints as well as operating a flexible formwork by using only one machine.

The Robarch - Visualizing the Contact Face

To advance the topics discussed towards reality the Robarch is investigated. Arches, like shells, are lightweight load-bearing structures which because of their special geometry have a favorably load-bearing behavior that results in immanent low material consumption. In addition to this, they also fascinate us by their elegance and the

Material	specific weight	dimensions	weight/element	compression strength
	kN/m^3	mm	kg	N/mm^2
Foamglas F	1,65	600/450/140	3,78	1,80

Table 1 Material for the Robarch

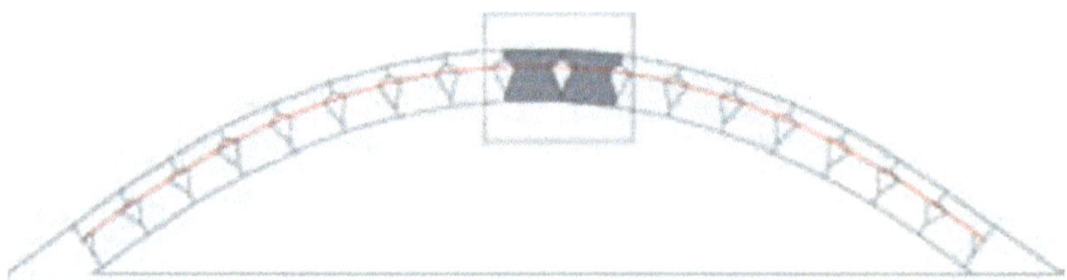

Figure 2 Side view of an "robarch"- variation influenced by the thrust line of uniformly distributed load

logical combination of form and distribution of forces. The focus is on the process chain from design to production. The correlation between form force and jointing is particularly important. The geometry of the structure shows the capacity of a compression stressed joint depending on the force flow. Using the industrial robot as a milling machine the accuracy of the machine set up is tested again. The machine center consists of a high precision robot (ABB IRB 6660) and a 6000 mm long traverse rail. In the workspace of 6000/1200/1200 mm the accuracy is limited by +/- 0.2 mm in each position.

The Robarch is made from foam glass. This material is commonly used as high-quality insulation, in this case it is employed as a structural mineral element. It can be easily treated, is light but brittle. The load carrying capacity is comparable to the strength of timber perpendicular to the grain. Depending on the compression strength of the material it is possible to minimize the area of the contact face to make this area as transparent and visible as possible. The material is shaped by milling. The aspects of force flow and joining are considered at the same time.

The milling process is used (in addition to the visualization of the contact face) to reduce the dead weight of the structure, too. The Robarch thus represents

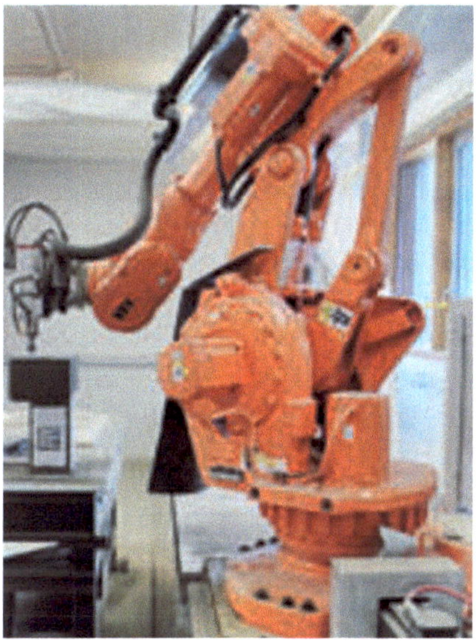

Figure 3 Milling an element

Figure 4 Rendering of the rob-arch bridge

a lightweight design because of its geometry and the shape of its elements. The open space between two elements varies along the arch and depends on the force flow. The shape of this opening results as a negative of two assembled elements (Fig. 2) and characterize the elevation.

Workflow

The material, the dimensions of the elements, the geometry of the arch and the loading determine the dimensions of the contact points. The aim is to find a geometry of the interfaces that satisfies the needs of load transfer and mounting. The use of the milling tools also limits the range of the possible shape. The basic geometry is given by a CAD file. As well as geometric control, the system is statically checked using a finite element program. The CAM software Hypermill creates the milling surface and gives a first control of collision of the tool. From that point it is necessary to combine data and real determining working object. The program pi-path translates the CAM code to a robot code and allows the movement of the robot to be tested (Fig. 3). Positioned on the fixed abutments the milled segments are assembled and the "robarch" can be walked across (Fig. 4).

Force Flow - Compression-stressed Lightweight Structures

As structures domes and arches are well known from both an empirical and analytical point of view. It is still fascinating to span large distances by force-fitting small modules (stones) together. The principles of physical models used by A. Gaudi as well as F. Otto and H. Isler still provide a background for the formfinding of force influenced models. Tools like "Rhino Vault" [2] combine up-to-date parametric designtools with the aspect of force flow.

There is an analogy between arch and cable structures (Fig. 5). A single cable changes its form if the load distribution varies. Thereby a funicular curve results which relates to the specific loading conditions. Mirroring this geometry and retaining the support conditions lead to an appropriate arch geometry that matches the given loading conditions, which means that the loads create normal internal forces only. Compared to cable structures arches cannot change their geometry when the loadcase varies. In addition to this, arches made of mineral materials like brick or glass have only a low bending strength. Therefore it must be considered that the funicular curves belonging to the different loadings

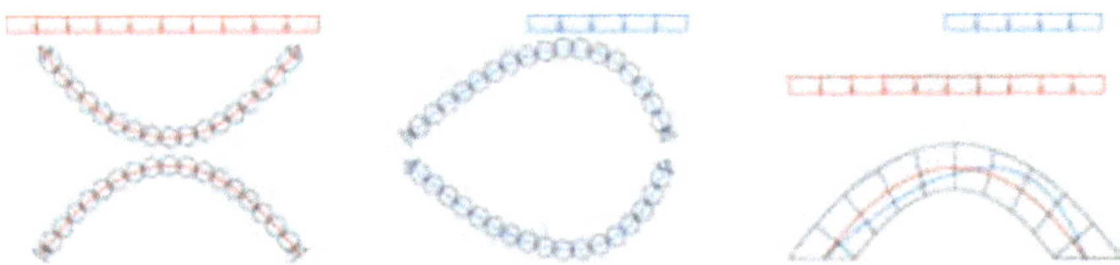

Figure 5 Functional curve: uniformly distributed load (left); nonuniformly distributed load (middle); inside an arch (right)

deflect inside the cross section of the arch with a certain distance to the edges. This circumstance can be influenced, e.g. by increasing the cross section or by putting extra load on the arch as a kind of prestressing. The Roman arch and the cable stressed arches by V. Suchow are comparable examples of this (Nerdinger 2010).

However, it is also possible to enable a brick arch to take bending moments. Here it is necessary to make sure that the compression forces, compared to the bending forces, are high enough to prevent tensile stress occurring at the edge of the cross section. Assuming that the sum of the compression forces "Fc" is transferred at only three points of the cross section, the bending moment capacity increases with the distance of the contact points.

The "Grasshopper" script used needs 3 curves as basic information: the thrust line and the upper and the lower template (Fig. 6a). To guarantee an advantageous load transmission in the contact area it creates a normal of the thrust line at every specified section (by the desired number of segments) and subdivides the arch (Fig. 6b).

Weight reduction and architectural appearance are controlled by the cutting tool (Fig. 6c). The opening at the sides is controlled by the forces that appear,

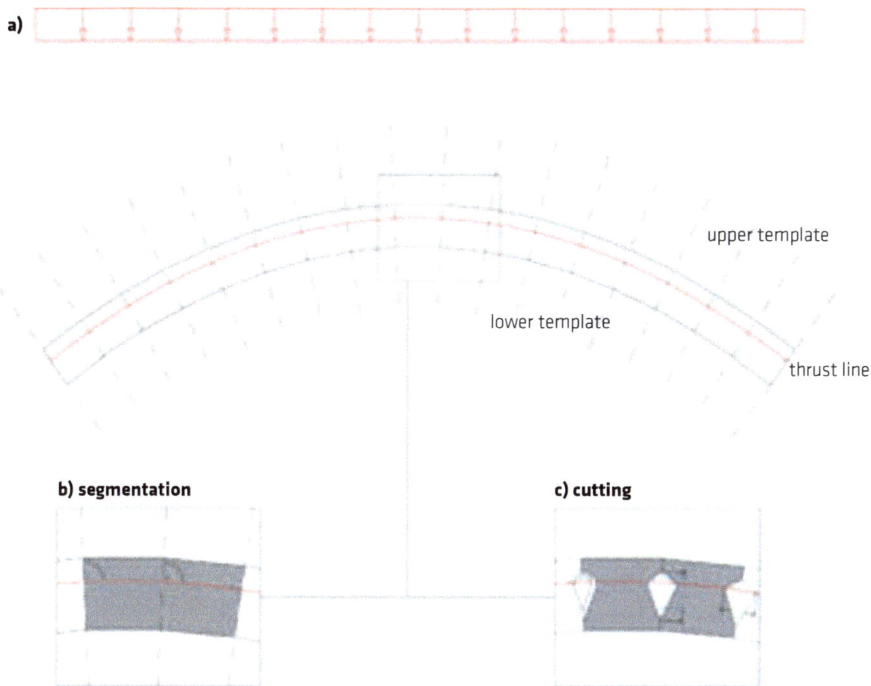

Figure 6 Formfinding. segmentation and cutting process

Figure 7 The contact zone (red) and the negative diamond (green) form one segment (left), assembly (right)

while the top surface is opened constantly to allow access by foot. The script offers an adjustment of material properties like the compression strength of the material or the internal forces. As a factor in our equation this value helps to determine the required contact face. The calculated area is divided in 3 sub-areas, which are positioned at the edges of the top area and at the bottom, where they act as a contact zone (Fig. 7). As seen in Fig. 4 a parabolic curve results because of uniformly distributed load. This curve is analyzed and can be divided into the required number of segments. To avoid shear forces in the joint the contact section always deflects as a normal of the pressure line. The area not needed to carry loads is milled out by the robot (Fig. 3). As demonstrated, the triangle is a material-efficient cross-sectional shape.

Conclusion

The starting point for this case study was one of the most intelligent and efficient structures in building history. Taking a close look at the mechanical behavior of arch structures and at the same time taking the aspects of parametric design and manufacturing into account allows a new way of thinking about this construction method. An industrial robot is used as a milling machine to shape the contact faces. Milling is carried out with the maximum possible precision. This case study combines traditional building methods with current tools and technologies and is intended as input for the increased use of robots in structural engineering and the building material industry.

References

Nerdinger W, 2010, *Wendepunkt des Bauens*, Edition Detail, München.

Sobek W, Blandini L, Krtscha A, 2011, "Die Sondertragwerke der Mansueto Library in Chicago - vom Entwurf zum Bau". *Stahlbau 80* pp. 3-6.

Stehlig H and Scheure F, 2012, "Waved Wooden Wall", *Proceedings of the Advanced Building Skin Conference*, Graz, A, pp. 25-25.

Trummer A, Peters S and Amtsberg F 2012, "Beyond parametric design", *Proceedings of the Advanced Building Skin Conference*, Graz, A, pp. 51-51.

[1] www.block.arch.ethz.ch/tools

Martin Bechthold, Nathan King

Design Robotics

Towards strategic design experiments

Abstract The use of industrial robotics in architecture is characterized by the dominance of two distinct approaches. The first attempts to solve practical problems using engineering methods without affecting design scope. The second is dominated by creative and artistic design experimentation, primarily seeks to inspire, and consciously leaves the practicalities and constraints of the construction industry out of the investigation. "Design Robotics" as a third, more strategic approach links design innovation to the reality of industrial production. The paper articulates its associated research methods and approaches by reviewing recent examples of research conducted by the Design Robotics Group (DRG) at Harvard University. The work, focused on robotically enabled ceramic systems, is a highly systematic form of research that bridges the gap between primarily artistic endeavors and the construction automation research of the building industry.

Keywords: design robotics, automation, ceramics, fabrication, computation

Introduction

To define when design becomes research and vice versa remains a difficult task. The term "design research" today refers to a variety of approaches that deploy design methods to solve a broad range of research problems (Laurel 2003). These methods, however, are insufficient for work in the area of architectural robotics, a field that began as *construction automation* in the 1980s using methods borrowed from engineering disciplines. Related contemporary academic research has veered toward the other extreme, the production of remarkable, often artistic installations that foreground design. Design Robotics, articulated in this paper, is an alternative approach to "design research" that combines analytic research with open-ended discovery and iterative feedback from material experimentation. This approach is illustrated here using projects on architectural ceramic systems conducted by the Design Robotics Group (DRG) at the Harvard Graduate School of Design.

Established Paradigms for Robotic Technology in Architecture

The short history of robotic technology in architecture is dominated by two opposing trajectories. The first, a pragmatic approach, is focused on resolving the shortcomings of manual labor – inefficiency, low-productivity, unavailability – through on-site construction automation. This effort originated in the massive research and development efforts of many large Japanese construction firms beginning in the 1980s. The second, and currently prevailing approach, is focused on broadening the scope of design by realizing one-off, often highly complex experimental aggregations that seek to understand unique design opportunities for robotically fabricated assemblies.

Construction Automation Approach of the 1980s and 1990s

Beginning in the 1980s several large Japanese construction firms developed automation strategies for the construction of tall buildings. The shared objective was to reduce the demand for construction workers, increase productivity, and improve site safety (Tanijiri 1997). By offering comfortable, almost factory-like, working conditions the industry hoped to attract young workers who could find less physically demanding and well-compensated jobs in other industries. Comprehensive construction automation systems were developed by Fujita Corp., Obayashi Corp., Kajima Corp., Shimizu Corp., Taisei Corp., Takenaka Corp., as well as by others. Different systems were conceived for pre-cast concrete and for steel construction, some "extruded" the building using an automated assembly floor at the top (Obayashi ABCS) or near the top (Fujita Corp.), others developed "push-up" systems that assembled all pre-configured modular components on the ground floor by incrementally jacking up the growing buildings with one complete floor at a time.

These systems, despite their unquestionable sophistication, depended on standardization of construction. Obayashi's ABCS system, for example, was initially designed to support the construction of high-rise buildings with rectangular plans and cores at opposite ends, and the first 5 applications over 10 years were limited to this building type. Only the last use of ABCS expanded the system's capability to a square plan with a central core (Ikeda and Harada 2006). Floor-to floor variations, typical for many contemporary high-rise designs, would have created inefficiencies or, if vertical material transport systems were affected, would be impossible to accommodate.

The initial R & D costs for each system were significant. Site productivity increased slightly compared to conventional construction, but the added value of these comprehensive automation systems was ultimately small. When the Japanese construction boom collapsed in response to the national recession, all automated construction systems were retired. Personal interviews by M. Bechthold in Japan in 2007 showed that none of the large corporations intended to reuse their automated construction technology. Along with economic conditions architectural preferences had changed, with demand for standardized buildings diminishing. Japan's automated construction systems were unable to effectively support the construction of non-standard, contemporary architecture.

Construction automation approaches today have shifted towards supporting pre-fabricated building systems (e.g. brick, steel, concrete), and industrial automation drives the high-volume production of ubiquitous building products such as ceramic tiles. Both approaches are geared towards improving productivity and replacing human labor with robots, as stated by Andres (1994) and Pritschow (1995) in their work on robotic bricklaying systems. Design-driven work on robotically placed non-standard brick patterns, by comparison, is a more recent phenomenon (Bonwetsch 2007) that continues to thrive in the academy today.

Robotically Enabled Artifacts and Installations

The introduction of industrial robots in academic laboratories triggered a wave of creative and complex installations that intellectually continued the experimental design work conducted with numerically controlled machine tools and routers in the early 2000s. Beginning with the ETH Zurich work cell (2005) and Harvard's robotic environment (2007), the popularity of industrial robots as more capable devices for experimental architectural fabrication continues to spread. Today's pursuit of industrial robotics at schools of architecture has clearly raised the cutting edge of digital fabrication to a new level. Throughout Europe and the United States robotic experimentation is geared towards furthering the understanding of new opportunities that robotic fabrication may bring to component, building, and product design. The current situation is largely driven by the academy, while fabricators, for the time being, remain spectators and are only beginning to invest in robotic fabrication technology. The situation, thus, is markedly different from the 1980s construction automation approach.

A common strategy employed in this type of robot-related work leads to the development of a project through a "bottom-up" approach that takes specific robotic process opportunities, including the ability to individualize or efficiently handle large numbers of units, as a starting point. This "discovery" phase remains intentionally loosely defined and open ended – an informed "play" that combines both digital and material experimentation. A second step is to systematize and rationalize, to some degree, the most appropriate experiment and to design a prototype that best illustrates the most novel design discoveries. In the third, and final step this piece is executed and evaluated.

Custom-automated code generation strategies have emerged, are now widely used, and have led to the creation of plug-ins and software components that automate robotic tooling within the digital design environment. DRG's automation tools link geometry data from a number of software platforms (e.g. Rhinocerous, DigitalProject, Catia, etc) to the robot control interface by bypassing proprietary manual robot programming and enabling the simplification and translation of many highly individualized model-driven movements with ease. The usually striking end results are often inspirational, but are difficult to connect to the reality of architectural production. The authors thus propose a strategic design research approach.

Design Robotics: A Strategic Research Method

A rigorous analysis of the chosen building or material system is the first step in more strategic research on architectural robotics. This analysis, while including obvious technical aspects relating to fabrication, must be broad enough to understand all relevant aspects of the given system, including production, distribution, economics, and end-use. This step is crucial when defining the problem or opportunity to be addressed such that new solutions can emerge. Deeper knowledge is acquired as the work proceeds, often allowing the definition of the research problem to be incrementally improved. Evaluative frameworks emerge incrementally and guide the research. The design of an experimental installation frequently serves as a proof of concept. Its features are strategically chosen such that its design to production yields generalizable knowledge that addresses the research problem within its broader industry context. Design Robotics thus represents a hybrid research method that combines bottom-up, technology driven design inquiry with traditional, problem-centered approaches. DRG is not the only group pursuing this type of research. Research on non-standard assembly of brick, wood slats or other materials can follow similar principles (Gramazio Kohler 2008). But DRG's approach has pursued the customization of the basic module itself as its core interest, thus potentially supplementing and enhancing robotic assembly procedures. Expanding the scope of robotic intervention forces the analysis of existing processes to penetrate deep beyond assembly procedures and embrace far broader industrial production issues. The following illustrates DRG's approach through a discussion of ongoing research projects on architectural ceramics.

Ceramic Industry: Contemporary Mass-Production

The DRG has been in collaboration with an industry association of Spanish tile producers (ASCER Tile of Spain) since 2009. The initial phase of research included a comprehensive analysis of the industry in terms of products, production processes, and research and development infrastructure. Spanish tile producers emphasize superior quality and innovative surface finishes as they compete with many other international brands. The industry primarily uses large hydraulic presses and steel molds to form flat tiles from dry clay bodies. Only a few companies form clay by extrusion through shaped steel dies. After the initial forming process tiles are post-processed on highly automated computer-controlled production lines designed to treat surfaces and edges. Surface finishing equipment includes numerically controlled ink-jet technology that prints patterns and images on pressed tiles. Computer-controlled techniques are widely used for packaging, storage, and logistics. Most companies sell their tiles through distributors; only one company has built a strong brand and a related distribution network. Tile production is based on demand predictions, and tiles that cannot be sold produce storage costs, and are eventually sold at discount prices abroad.

Several problems became clear. First, high-volume production techniques based on predicted market demands make customization of tiles – beyond what digital printing technologies could deliver – virtually impossible. Only one facility was able to produce customized, three-dimensionally formed ceramic elements for an ambitious architectural project. Considering current architectural trends towards complex form and individualized construction the need for custom building products is bound to increase. Growing demands on operational building performance may also reinforce that trend (Bechthold 2011). Product customization and responsiveness to more individualized market needs appeared to be a challenge for the industry, and thus was identified as the primary research agenda for DRG.

A second problem became evident while comparing automated tile production lines with downstream manual tile installation processes. Standard manual tile installation is archaic – slow, costly, and prone to error (King 2012). Interviews with tile producers revealed disconnects between the production industry and tile installers, even though installation is a significant cost factor of finished tile surfaces. It became clear that a potential for innovation might exist when considering tiles as a material system from production to installation, instead of merely looking at manufacturing aspects. Further research

Figure 1 Automated Spanish tile production facilities

showed that cutting waste and the amount of spare tiles purchased for future replacements exacerbate the cost disadvantage of tile finishes compared to other surface finishes. Cutting waste also presents an environmental burden. For the 2010 U.S. tile consumption of 23.2 million square meters the embodied energy of an assumed 5% cutting scrap is equivalent to a staggering 13.6 million liters of regular gasoline.

Technological innovation always requires investment. To understand potential opportunities for robotics in the production and installation of ceramic systems several cost analysis were conducted. First the cost impact of manufacturing equipment on the cost of a tile (as sold from the factory to the distributor) had to be determined. Data for Spanish production was unavailable, so published information from the Italian tile industry (with a similar product and process structures) was analyzed (Fiori 2007). The research showed that on average only 7% of the manufacturing cost of a pressed tile is spent on equipment amortization, a small yet not uncharacteristic amount for high-volume producers.

Next, the ratio of installation costs to material costs was analyzed. The typical US installation cost starts at 270 $/h for a square meter of mosaic tiles, which easily outweighs the cost of tile and grout. Prices for placing non-standard patterns (e.g. custom mosaics) are far beyond average construction budgets (King 2012). The cost analysis of manufacturing and installation assures that robotic interventions are fundamentally realistic from a cost standpoint.

Is the Spanish ceramic industry technologically ready for robotic systems? Factory visits showed that many Spanish producers already use 6-axis robotic manipulators for packaging. These companies possess the technical know-how needed to operate robotic work cells and in addition to proving viability based on cost, the initial research also concluded that robotic interventions are realistic from skill-level perspective. The potential customization of ceramic products offered through robotic intervention allows greater product differentiation that responds to dynamic architectural needs. New ceramic systems must also address existing problems of waste and inefficient installation procedures.

Material Systems: Ideation Stage

How can *design research* "invigorate" a material system as old and well established as ceramics? The broad research agenda outlined above leaves many different pathways open. To gather a number of ideas several experimental studies were conducted both by the research team and by students in as-

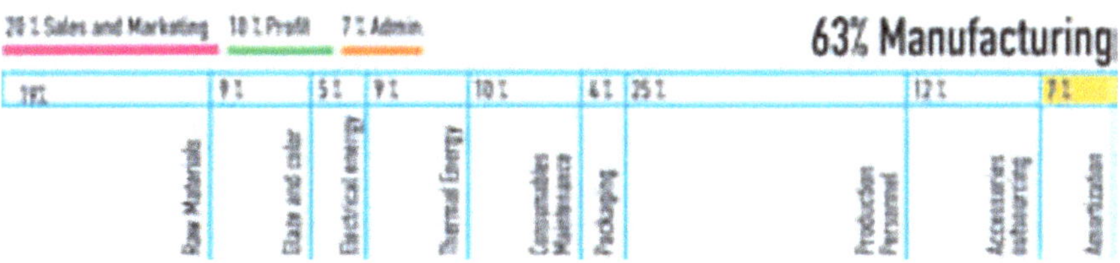

Figure 2 Financial breakdown of the ceramic industry showing only 7% amortization costs

sociated courses. These open ended, almost playful, studies used hands-on, computational and robotic explorations that were guided by the broadly framed agenda of "customizing ceramics" and "waste reduction". A bottom up approach largely characterizes this phase, albeit guided by the general theme of customization and waste reduction as defined at the outset. The bottom-up approach takes inspiration from the material system itself. Within the material system we refer to the following:

- General physical and aesthetic material properties: for ceramics this covers the properties of various clay bodies and their admixtures, such as moisture content, colors, dry-time, mechanical strength in green and fired state, porosity, as well as many other factors.
- Ability of the material to be shaped and formed: clay as a plastic material can be freely formed through processes such as slip casting, extrusion, and molding. Industrial methods work mostly with pressed tiles that can accommodate a minor degree of three-dimensional shaping. Extruded forms are linear and have a range of flat to complex profiles. To enable customization new processes must address these limitations.
- Opportunities and limitations of 6-axis robotic manipulator: Repeatability and precision can be a factor to be considered for the material system, but in the case of clay the material shrinkage is usually larger than process-inherent tolerances. Tooling for the robot is another question – in order to reduce process complexity it is often desirable to limit tool changes where possible.
- Characteristics of related industrial production processes: in the case of clay all production equipment is geared towards linear movement of ceramic pieces along conveyer belts and rollers, from initial forming to surface finishing, drying, firing, and dimensional rectification, Integration into industrial processes means at least one flat surface of the ceramic piece has to exist such that parts can rest on standard conveyer belts.

Many early ideas were discussed with industry experts. Clay extrusion was identified as the manufacturing process with the largest potential for part customization. The steel dies used in extruding linear clays forms are relatively costly, thus prohibiting small productions runs for custom parts. Several experiments explored robotic intervention geared towards supporting individualized production methods. The first design experiment developed a variable extrusion die that could change shape during extrusion. An industrial version would include numerically-controlled drive motors that alter the die geometry, thus enabling continuous product variation while maintaining a constant wall thickness.

Another extrusion-based experiment addressed customization through variable robotic cutting after the initial shaping process. During industrial processing the linear extrusion is cut to length or into its final shape using a variety of cutting mechanisms including wires and blades. Extruded hexagonal tiles, for example, are cut from flat slabs using hexagonal blades. The first attempt to adapt this cutting process and enhance its versatility through robotic intervention involved a fixed blade assembly that was manipulated to generate a family of façade components. A third customization project proposed a robotic wire-cutting process that shapes 5 sides of an extruded block into ruled surfaces with varying degrees of complexity. The envisioned industrial scenario was simulated by equipping a 6-axis industrial robotic work cell with a custom wire-cutting end effector designed for use with clay materials (Andreani 2012). Customization opportunities for pressed tiles exist primarily in the installation phase. Initial experiments and detailed precedent research confirmed the potential to rapidly and precisely place tiles using an industrial robotic work cell. Several experiments were conducted that involved the placement of dimensionally modular tiles at a digitally defined position using a pneumatic suction gripper. (King 2012) The system can also accommodate tiles of varying shapes and formats. In addition to developing technologies for robotic tile placement the entire tile installation workflow was reconsidered. Instead of installing robots on site, panels are robotically tiled off-site. Tile panels are then transported to site where they are installed. (King 2012)

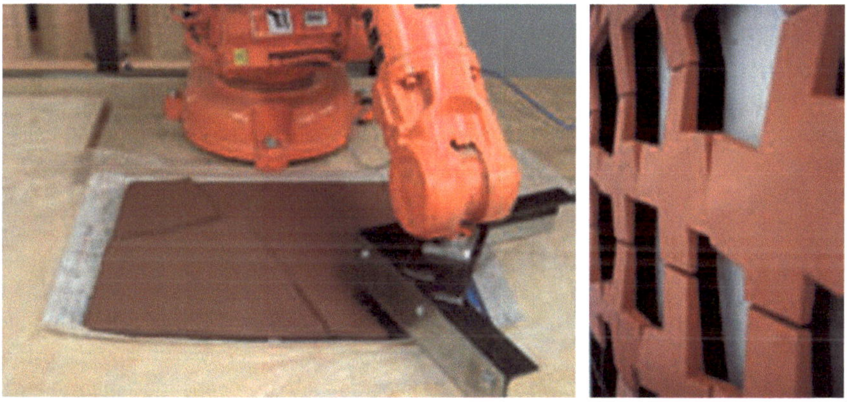

Figure 3 Post-extrusion robotic cutting and resulting prototypical facade assembly (students: Mauricio Loyola and Jeremy Keagy)

Figure 4 Prototypes produced during flowing matter research using a custom end effector (students: Stefano Andreani, Jose Luis Garcia del Castillo y Lopez, and Aurgho Jyoti)

From Ideas to Concepts: Evaluation and Development

During initial ideation, experimental "sketches" are evaluated and refined in pursuit of the larger research objective. Refined "sketches" are used to guide the ongoing creative experimentation, but without over-constraining it. Two research projects will illustrate the approach, first, the robotic extrusion of individualized ceramic façade elements, and second, the automated placement of non-standard tile patterns based on digital images or other algorithms.

Building on initial research into the possibility of variable extrusion a prototypical façade system was envisioned that enabled the creation of high-performance, custom components that can respond to specific environmental or aesthetic parameters. By strategically identifying the ceramic façade as a research platform several research trajectories emerged that led to the production of an Integrated Environmental Design to Robotic Fabrication Workflow (Bechthold 2011). Here a custom workflow linked a digital design model through several Grasshopper-based optimization modules that accounted for environmental performance optimization, material properties (shrinkage and deformation), design for robotic fabrication, machine code generation, and building integration. Parallel to the digital workflow was the development of a novel manufacturing process that utilizes a robotically actuated pin-mold and novel extrusion-based robotic material deposition system designed to create accurate individual façade elements and building components.

Figure 5 Image-based pattern generation and automated robotic tile placement processes

To evaluate the potential for customization using robotic tile placement a second workflow was established that incorporates both image-based and pattern-based algorithms into a design model that can be used to automate the programming of robotic movement during tile placement. (King 2012) A novel modular production strategy was proposed that enables the factory-based placement of tiles on modules that would be transported and installed onsite. This project combines the value of non-regular, non-standard tile patterns with a reduction in overall labor costs and shorter onsite installation time.

Typical evaluation criteria for DRG's ceramic research projects beyond purely technical questions include the following:

1. What kind of customization can be achieved with the process? Robotically manipulated extrusion processes typically result in geometry variations, while robotic tile placement can generate non-standard tile patterns.
2. What value does customization add to the material system and its applications? Robotic extrusions can achieve greater formal freedom for shading lamella and other elements while maintaining excellent building performance. Robotic tile placement produces non-standard tile patterns that are not economically possible with manual placement techniques.
3. Can waste be reduced? Both robotic extrusion and robotic tile installation are on-demand processes that can potentially reduce waste. The pattern generator can be configured such that tiles do not need to be cut – dimensional differences can be accommodated with varying grout line width.
4. How are installation procedures affected by part variation? Façade elements would normally be installed using custom connectors. Robotic tile placement requires a new approach of semi-prefabricated sheets, factory made, with on-site installation reduced to mounting pre-tiled sheets on prepared wall surfaces.
5. To what degree can the process be incorporated into state-of the art industrial production lines? Modular concepts are crucial when considering industrial integration. Robotic extrusions, for example, could be a stand-alone concept, but the configurable molds could easily be used for slumping flat extruded clay slabs. Robotic tile placement leaves current production methods for pressed tiles unchanged, but requires new business models for installers that move much of their activities to the factory floor.
6. How could parts be packaged and shipped? Flat tiles for robotic placement are shipped on pre-tiled sheets that can be efficiently stacked. Shipping costs for custom robotically extruded façade elements can be reduced through nesting algorithms already implemented to optimize kiln use.
7. Is there need for a new distribution model associated with the proposed processes? Robotic extrusions could easily integrate into existing supply chains of producers, installers, and façade companies. Robotic tile placement requires more direct links to be forged between the end-user or designer (whoever configures the pattern) and tile installers. Online pattern configuration would most readily provide this connection.

Work in this phase iteratively develops and test ideas for technical feasibility, design interest, and industrial integration, thus systematizing and rationalizing the initial experimentation.

Process Prototypes: Proof of Concept

The proof-of-concept involves the production of a prototype large enough to provide credible evidence of research agenda, and also allow for critical evaluation. Full production of large prototypes is unlikely to yield new insights in the academic setting because the work is geared towards the industrial or professional fabrication context, not towards the making of an artistic artifact. During the research described, several types of prototypes were developed to test ideas and provide iterative feed back during process development. These prototypes are critical to the work but are often specific to certain aspects of the research agenda, dry-placed tiles to tune accuracy or flat robotically extruded shapes to test material properties, for example. The proof of concept prototype is strategically defined to resolve certain speculative aspects of a proposed system as well as reconcile in-depth analysis of a given material system, novel process development, and design potential. In some cases the proof-of-concept prototype represents a piece of a larger system or an entire system in itself. During prototyping the robotic arm may be used to emulate a proposed process such as production line integrated wire cutting, or, in the case of robotic-tile placement and robotic extrusion, represent an actual proposed production process.

During the development of the previously described robotic extrusion process a design experiment was chosen that tested the workflow using an extreme scenario requiring shading and controlled views on the east, south, and west sides of a semi-circular glazed atrium space. The entire facade was used to calibrate the digital workflow but only a representative section of the shading system was ultimately fabricated. This section contained enough complexity and variation to both illustrate technical solutions and design potential of the novel manufacturing process (Bechthold 2011). In the case of robotic tile placement the entire workflow was demonstrated during the production of a single modular image-based mosaic (King 2012) This prototype used a custom pattern generation algorithm to reproduce a recognizable image using a series of modular tiles (see Fig. 6). The resulting digital image was used to generate robot code that in turn enabled robotic tile placement. In addition to presenting the technical feasibility of robotic placement the physical prototype also validated the proposed modular installation strategy.

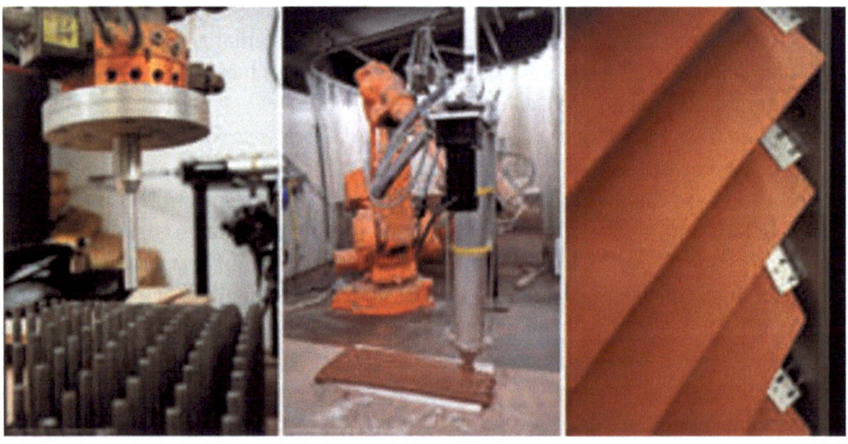

Figure 6 Prototypical manufacturing strategy including robotically actuated variable pin mold, robotic extrusion process, and finished proof-of-concept prototype

Conclusions

The use of robotics in the academy is entering a strategic mode of operation that differs markedly from both the traditional industrial automation approach to solving problems and from the digital crafting of one-off installations. DRG studies both part customization as well as the robotic assembly of modules. Research activities are grounded in the analysis of the construction or industrial context – learning to ask unconventional questions here yields research opportunities that otherwise remain opaque. The analysis yields a general research direction that guides the following, open-ended experimentation phase. Here physical and digital experiments produce many ideas in rapid sequence. Rough prototypes, even those produced manually, provide early feedback on opportunities, but also help failures to emerge quickly. The evaluation criteria derived through the analysis are used to filter out ideas for further development, and prototyping is used iteratively to answer questions that gain specificity as the research proceeds. The work, while focused on bringing design to bear as a value on ceramic material systems, is embedded in the industrial context, but not dominated by it.

Acknowledgements

Research was supported by the Harvard Graduate School of Design, Design Robotics Group through funding provided by ASCER Tile of Spain. Additional guidance provided by the Harvard Ceramics Program. Other research collaborators included Prof. Panagiotis Michalatos, Anthony Kane, Amanda Lee, and Justin Lavallee. Student work included coursework from *Construction Automation* (2010) taught by Professor Martin Bechthold and *Material Processes and Systems: Ceramics LAB* (2011) taught by Prof. Martin Bechthold and Nathan King at the Harvard Graduate School of Design.

References

Andres J et al. 1994, *First results of the development of the masonry robot system ROCCO: A fault tolerant assembly tool, in Proceedings of the 11th ISARC*, Elsevier, Amsterdam, pp. 87-93.

Andreani S "Flowing Matter: Robotic Fabrication of Tectonic Ceramic Systems", *Proceedings of the XXth ISARC* Eindhoven.

Bechthold M et al. 2011, "Integrated Environmental Design and Robotic Fabrication Workflow for Ceramic Shading Systems", *Proceedings of 28th International Symposium on Automation and Robotics in Construction ISARC2011*. Seoul.

Bonwetsch T et al. 2007, "Digitally Fabricating Non-Standardized Brick Walls" *Manubuild Proceedings*.

Fiori M 2007, "Italian Tile Face Rising Costs" *Ceramic World Review; Technology, News, Markets*. vol. 71 pp. 46-5.

Gramazio F and Kohler M 2008, *Digital Materiality in Architecture*, Lars-Müller, Baden.

Ikeda Y and Harada T 2006, "Application of the Automated Building Construction System using the conventional construction method together." *Proceedings of the 2006 Conference for Construction Automation ISARC*, pp. 722-727.

King N et al. 2012, "Robotic Tile Placement: Tools, Techniques and Feasibility", *Proceedings of the XXth ISARC* Eindhoven.

Laurel B (ed.) 2003, *Design Research*, MIT Press, Cambridge.

Tanijiri H et al. 1997, "Development of automated weather-unaffected building construction system.", *Automation in Construction* 6, pp. 215-227.

Pritschow G et al. 1996 "Technological aspects in the development of a mobile bricklaying robot", *Automation in Construction*, vol. 5, no. 1, pp. 3-14.

Clemens Neugebauer, Martin Kölldorfer

Fabricating the Steel Bull of Spielberg

Introduction

In 2010, Red Bull commissioned Clemens Neugebauer and Martin Kölldorfer to make a large-scale sculpture for their racetrack in Spielberg, Austria. While there had been a previous proposal in 2004 to construct a 23 meter high bull jumping over the very last curve of the race track to provide an appropriate setting for TV coverage of racing events etc., that particular project was cancelled along with the entire racetrack due to environmental concerns. However, the idea of a Red Bull racetrack at Spielberg was revived in 2008, though at a significantly smaller scale. Consequently, in 2010 the idea of a Red Bull landmark emerged again, aiming to create a sculpture that would be big enough to be seen from the nearby motorway – a considerable distance of nearly two kilometres away – by occupants of passing cars. The final design was developed in extensive discussions with the client, settling on a bull that would be different from the classic Red Bull logo, and an arch through which the bull would jump. These considerations also resulted in the relocation of the project, from the last curve

Figure 1 Steel bull at Spielberg, initial concept rendering

to one of the especially highly visible loops of the racetrack.

As neither we nor the client could quite comprehend the scale of the sculpture, Red Bull required an elaborate mockup before any more funds could be allocated to the project. In order to best represent the sculpture's design, we created a 17 meter wide and 16 meter high two-dimensional prototype, consisting of 250 square meters of red carpet fabric and weighing 0.6 tons. This prototype was then raised by a crane, shown to the client for a few minutes and then quickly disassembled and destroyed in order to keep the project confidential.

Technical Project Stages

As is the case with many projects, the very first stages of the project consisted of hand-drawn sketches. However, we quickly

Figure 2 Robot milling foam mold

moved on to cooperate with 3D expert Richard Maierhofer to create parallel a three-dimensional model that would ensure the feasibility of the current design. For the central bull sculpture the concept was to directly use the planar triangles of the 3D model as construction elements. These triangles would be made out of Corten steel, a special type of steel whose surface quickly oxidises. Then, in order to break up the "digital" look of the structure, amorphous holes were to be cut into each of the 1800 triangular elements.

The next step required a full structural calculation of the bull's support structure. As any support structure would be visible from the outside, Martin Kölldorfer had the idea of relying on biomimetics and shaping the support structure similar to the skeleton of a bull.

The client's wish to have the bull jump through an arch proved to be a significant aesthetic challenge. Here history offered a wide array of different arches, from the Arc de Triomphe to the Michelin Arch. To counterweight the "masculinity" of the steel bull, we decided on producing a "female" counterpart that would be non-symmetric and freeformed.

The Fabrication of the Red Bull Arch

Due to the projected free-form shape, we quickly realized that only a casting process could exactly reproduce the designed form. This gave us a wide – though expensive – range of materials to choose from. Finally, we settled on aluminium, as a clear contrast to the rusty brown bull. Gold was also added to the list of materials, to coat the horns of the bull itself. This mix of materials was

Figure 3 Foam molds, ready for aluminium casting

Figure 4 One of 83 cast aluminium parts.

ideally suited to the artistic concept, with the plated horns using the oldest manufactured metal, the bull consisting of iron – a material used by humanity for thousands of years – and the graceful arch consisting of the relatively young material aluminum.

Aluminum also gave us the idea of actively involving the customers of Red Bull in the sculpture, by using aluminum from used Red Bull cans from around the world, turning the fingerprints and the DNA that would be left on a can – along with the can itself – into a piece of art. Unfortunately, this concept proved impossible to realize: while it would have been easy for Red Bull to acquire 5 million used cans, the structural engineer gave us the bad news that molten, recycled cans would have to be alloyed with magnesium if they were to be used for the arch. Therefore, the amount of Red Bull cans in the aluminium arch was reduced from 5 million to a symbolic number.

The final 3D model of the arch consisted of 83 geometrically unique parts, requiring a unique cast for each piece. After considerable research we decided to use a modern form of lost wax casting, the so called lost foam casting. This is a type of evaporative pattern casting, which takes advantage of the low boiling point of foam: Instead of having to melt the wax out of a mold, foam simply evaporates. Further advantages are that very complex casting – that would otherwise require cores – is possible, it is dimensionally accurate, and maintains an excellent surface finish.

When the initial 3D model was created, the goal was to create elements that are as large as possible, to avoid having to weld an excessive number of parts. However, this proved to be problematic in the long run, as it proved difficult to find a company that is capable of milling 3x3x1m positive molds out of expanded polysty-

Figure 5 Mounting of aluminium-cast arch segments

rene. While some companies were capable of processing large parts with five-axis CNC milling machines, the costs were prohibitively high, as the companies were not used to dealing with such geometrically complex parts and subsequently submitted only very high tenders.

We finally started to look for alternatives that would allow us to fabricate these parts ourselves. Even though we did not have high demands regarding accuracy – a tolerance of 1mm would have been enough, as cast parts shrink by up to 2% anyway – finding a large scale machine at an affordable price seemed impossible, until we came across the idea of using an industrial robot.

The purchase of a second-hand robot – costing a fraction of a five-axis CNC machine – was quickly finalized, however we only realized too late that an industrial robot would not automatically include software to generate the robotic toolpaths. The search for a powerful and affordable robot solution finally brought us into contact with the Association for Robots in Architecture, which provided us with the necessary software and training to deal with the robotic fabrication ourselves.

Despite the odds that are stacked against someone who buys a robot without any knowledge of CAM, we succeeded in producing the required molds within about 6 months – significantly longer than the initially projected 5 weeks. This was mostly due to the many things that had to be set up around the robot to provide an efficient workflow – from an industry-sized vacuum cleaner to a custom built hot wire saw for cutting 5 cubic metre polystyrene blocks. Another challenge was finding the proper balance between automatic and

Figure 6 Traces of robotic toolpaths on the finished arch

manual labor, e.g. evaluating if grinding by hand is faster than having the robot mill extremely smooth surfaces.

At the end, it took seven articulated lorries to bring the massive polystyrene molds to the foundry in Germany – several molds were damaged during the 850 km transport, some were even destroyed and had to be produced again. The most critical phase of the casting process was putting the large molds into the sand, without causing any damage to the soft EPS or – even worse – distorting the geometry of the parts.

The most significant problems during the production of the molds resulted from the size of the individual parts, as the largest, industrially produced blocks of EPS have a volume of 5 cubic metres – not enough for our purposes. Gluing blocks together was the only viable solution, though not ideal, as the polyurethane foam used for gluing both impeded the milling process and was even noticeable on the final, cast aluminium pieces.

Conclusion and Outlook

We consider the robot to be a fantastic tool for artistic purposes, a beloved slave that is never in a bad mood – we provide the robot with an initial design and, without complaint, he mills the molds for casting. Working with the robot is similar to how successful sculptor colleagues in former times and the present run their workshops, from the Greek Phidias to Erwin Wurm and Anish Kapoor: The students do all the rough work, while the masters just design and finish the artistic product. And this seems to be the most important aspect for us: in the case of industrial serial production, products have to be machined to tolerances of tenths of millimetres, but as artists we have the luxury of being able to decide – or having to decide – whether we want to have the absolute accuracy of a machine, or just want the rough shape and then finish it ourselves. Similarly, CAD/CAM processes allow artists to work at multiple scales, as a machine can be used both for small scale mock-ups and full scale objects – a workflow that is especially important in monumental projects such as the arch to allow the artist to accurately judge the proportions of his design.

In conclusion, while a robot may not be an artistic tool, it is for us an ideal tool *for* artists that is capable of producing or preparing elements of all shapes and sizes.

maltech

Joshua Bard, Steven Mankouche, Matthew Schulte

Morphfaux

Recovering architectural plaster by developing custom robotic tools

Materially Informed, Custom Robotic Tooling

Plaster encourages multiple types of tooling because it undergoes a number of material transformations during its curing process. Many traditional approaches to shaping plaster deploy a constellation of tools and techniques (both additive and subtractive) as plaster transitions from liquid, to paste, to a fully-cured solid.

Extending craft-based practices of applied architectural plaster encouraged the use of a six axis robotic arm equipped with workstations positioned along a 15m linear external axis. Custom end of arm tooling was designed and fabricated to shape plaster in its multiple states. Four approaches were developed to investigate plaster's multiple physical states– liquid, paste, semi-cured, and fully-cured. Two of these approaches resulted in new robotic end of arm tooling.

Robotic Profiling

Running molds have historically been the primary vehicle for producing decorative molding using architectural plaster. Profiles can be run in situ with rails temporarily affixed to existing walls and ceilings or lengths of molding can be fabricated offsite then adhered on site in a bed of wet plaster. Because repeatability in the process is essential, running molds tend to be straight or comprised of simple geometric elements (e.g. lines, circles, ellipses). A robot has no need for fixed rails and can shape profiles (with reliable repeatability) freely in three dimensions according to any geometric trajectory that can be digitally described (Fig. 1).

In addition to constraints of global form, typical running molds maintain constant cross-sectional profiles along the run. Where local variation in the profile is desired (e.g. dentil molding), elements

Figure 1 Left: custom robotic profiling tool; right: plaster robotically applied in lab

need to be cast or else tooled subtractively. By adding a further axis to the profiling tool, variability can be introduced into the cross-sectional profile along its path. The tool has two interchangeable plates actuated by a stepper motor. As plates slide past one another the resulting profile morphs along the length of the molding (Fig. 2). These innovations open up new territories of curvature, inflection, and variation in the language of decorative plaster molding and contribute to other similar additive CNC fabrication techniques such as foam deposition (Gramazio 2008; McGee 2011) and ceramic extrusion (Peter Webb 2011).

Robotic Wire Saw

A robotically-mountable wire saw enables ruled surface cutting of semi-cured and cured, wet plaster. The saw provides a rough cutting throat of 24"x18" and uses a round wire blade for cutting in any direction. The tool is designed to section large plaster blocks with very little waste. Because plaster can fully adhere to itself across cold joints the block may be regenerated with a fresh plaster pour and continue to be shaped through subsequent cuts (Fig. 3). The design of Morphfaux's wire saw makes parameters related to wire lag tunable through an integrated blade tensioning system, custom guide wheels adjacent to the cutting mouth, and a variable frequency drive to modulate cut speed. The robotic wire saw developed during *Morphfaux* is similar to many CNC and robotic hotwire cutters (Verde 2011; Pigram 2011). But unlike hot wire cutting, wire saws open up a material palette beyond light weight foams that includes high density foams, autoclaved aerated concrete, and aluminum. The integration of a diamond abrasive wire opens new possibilities for ruled-surface cutting of stone and concrete which are typically machined via traditional milling.

Tool Development

Although the tools developed during the research have led to promising results, they have also exposed areas of need for future development. Presently, the variable profiling tool relies on a separate Arduino pro-

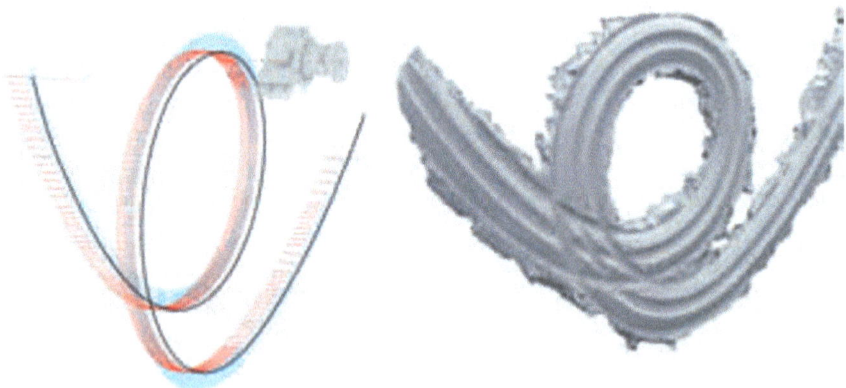

Figure 2 Decorative intersection of tool trajectory

cessor to control its external axis. Using an integrated controller activated within the robot's protocol would be more ideal. The research team has recently contacted USG (United States Gypsum) to partner in developing an automated delivery system to integrate with the profiling tool. The wire saw would benefit from adjustable guides wheels to allow greater cutting speeds.

Robotic Mobility and Onsite Construction

Robotic fabrication offers real potential for a reinvigorated sense of craft in architecture and promises to further disrupt the widespread acceptance of homogenized industrially produced building components. Recent trends in robotics suggest that custom fabrication will increasingly take place in situ at the site of construction. In the case of Morphfaux a full-scale lath wall was built around the work envelope of an industrial robot (Fig. 5). Plaster was applied directly in the three dimensional orientation of the wall without flattening the geometry to a series of cut sheets whose component parts would then be reassembled in the space of the final artifact. Work at this scale required use of the robot's external axis, a 15m track bisecting the work space of the lab. This configuration proved viable during the research but Morphfaux anticipates increased mobility in robotic fabrication where plaster is applied directly in a variety of construction sites.

References

Gramazio F and Kohler M 2008, *Digital Materiality in Architecture*, *Lars Müller*, Baden.

Millar W 1899, *Plaster Plain and Decorative. A Practical Treatise on the Art and Craft of Plastering and Moulding*, B.T. Bastford, London.

Pigram D and McGee J 2011 "Formation Embedded Design", in J. Taron (eds) *ACADIA 2011: Integration Through Computation*, Zurich, Switzerland, pp. 1-10.

Verde M and Hosale M 2011, "Investigations in Design & Fabrication at Hyperbody" in R. Glynn and B. Sheil (eds.) *Fabricate: Making Digital Architecture*, Riverside Architectural Press, Cambridge, Ontario.

Webb P and Pinner M 2011, "Terra Therma" in R. Glynn and B. Sheil (eds.) *Fabricate: Making Digital Architecture*, Riverside Architectural Press, Cambridge, Ontario.

Credits

Funding: Research through Making Grant, A. Alfred Taubman College of Architecture and Urban Planning, University of Michigan; Office of the Vice Provost of Research, University of Michigan.

Consulting: Hofmann Plastering, Saline MI; Gary Schultz, Spider Technologies; Super Matter Tools

Figure 3 Left: custom robotic wire saw; right: successive cuts from plaster block

David Pigram, Iain Maxwell, Wes Mcgee,
Ben Hagenhofer-Daniell, Lauren Vasey

Protocols, Pathways, and Production

Introduction

Difference and repetition are a recurring thematic within architectural discourse. Through the championing of digital techniques (algorithmic, associative or other) and numerically controlled fabrication methods, contemporary practice seeks an expansion of the linear and highly standardized protocols of industrial production. Algorithmic design methodologies, when coupled to robotic fabrication, enable an explicit and bidirectional traversal of the modern division between design and making (Fig. 1). This paper describes one such method which modifies the familiar marching cube algorithm to take advantage of its latent possibilities for fabrication efficiency. A recently completed project, "The Clouds of Venice", serves as a case study for a new integrated mode of production, one that increases the quality and number of feedback relations between design, matter and making.

Figure 1 Robotic fabrication of the "Clouds of Venice" installation

Crossing the Divide

The contemporary exploration of the architectural potentials of robotic fabrication and algorithmic design techniques perhaps most significantly permits a (re)affirmation or expansion of the role of the architect as master builder by challenging Leon Battista Alberti's 15th century division between design concept and building. This division has significant consequences which delimit the phase-space of possibility, i.e. the kinds of buildings that can be designed and produced. This is the architectural equivalent of Ludwig Wittgenstein's famous dictum (1972) that "the limits of my language mean the limits of my world". For architecture this means that representational constraints are at least as significant as the more recognised limitations of structure, time, and budget (Benjamin, 1991). Traditionally architects use the drawing as their mediating device and so, as noted by William Mitchell (2001), "architects tend to draw what they can build, and build what they can draw." The concurrent shifts from shop drawings to the direct generation of instruction code and from highly linearised production chains to flattened systems of intense feedback represent significant factors in one of the contemporary transformations of our discipline.

The "Clouds of Venice"

Supermanoeuvre in Collaboration with Matter Design

Commissioned for the Australian Pavilion's exhibition for the 2012 Venice Biennale, the installation explores the realisation of a diffuse spatial condition via an ultra-high population assembly of mass-customised, robotically fabricated, steel rod elements. Fundamental to the realisation of the work was the development of a bespoke algorithmic design approach – premised upon the Marching Cube (MC) algorithm – coupled to customized file-to-factory software, robotic handling and fabrication processes. The tectonic – cold-forming (bending) steel rod – forms part of a larger and ongoing research trajectory into algorithmic design strategies coupled to robotic fabrication undertaken at Taubman College, University of Michigan.

Marching Cube Method as an Open Ended Method of Standardisation

The MC method consists of an ordered set (3D grid) of cubes located within a spatial lattice of vertices - where each cube comprises 8 such vertices. Each vertex within the lattice attains a scalar value via sampling the system's input data set. The data set is arbitrary but is commonly a set of weighted 3D points. Thus the lattice denotes a scalar field of values either above or below a desired threshold (isolevel) from which the status of any cube can be extracted: inside, outside or partly intersected. Those cubes that cross the threshold will contain part of the isosurface which is defined in a procedural way. Since each of the cube's eight vertices can be either "marked" or "unmarked", there are 256 (2^8) possible conditions. In levering aspects of reflective and rotational symmetry, the MC algorithm elegantly reduces that count to 15 (Fig. 2). The first scenario is considered trivial, as all of the cube's vertices lie either above or below the desired isolevel and as such produce no geometry. For each of the remaining 14 cases, between 1 and 5 triangular facets (faces) will be added to the resultant mesh object (isosurface). The final process of the algorithm establishes the actual moments of intersection that define each triangular facet by either linearly-interpolating (stepwise or infinitely scalar) or simply referencing the mid-point of the 'cut' edge itself. Within the context of computer graphics and surface reconstruction, aspects of resolution and edge-interpolation enable trade-offs to be made between the generation of smooth, detailed and seamlessly shading volume representations and computation time. When considered from the perspective of the constraints of architectural fabrication – economic, material, geometric or other – it is this precise ability to restrict possible outcomes that provides the possibility of geometric standardisation.

The tectonic system employed by the Venice case study project does not limit itself to the facet-based geometries of surface representation. Indeed such a strategy was quickly dismissed on the grounds that character would remain hostage to the most obvious formal characteristics of the MC algorithm. Instead, the project operates at a level of abstraction from the isosurface through opportunistically and sinuously tracing the 14 unique grids it provides. The tracery flirts with the intersecting surface patterns and, subject to encoded situation, either the inside (those being below the iso-level) or outside (those being above the iso-level) of the cube edges themselves. Thus the 14 scenarios yield 28 possible "types" with their selection and placement being directed by an extended implementation of the MC algorithm. In collapsing possibility to definable "families" we were able to design and tune the traceries for visual quality as well as test fabrication viability (Fig. 3). Once the generalised patterns were settled upon, opportunity for mass-variation was reopened through applying highly orchestrated deformations to the global MC lattice. The deformations were used to account for the unusual geometry of the exhibition space and improve pre-assembly, packaging and shipping efficiencies as well as to achieve specific architectural desires such as reinforcing circulation routes and establishing more legible grains and densi-

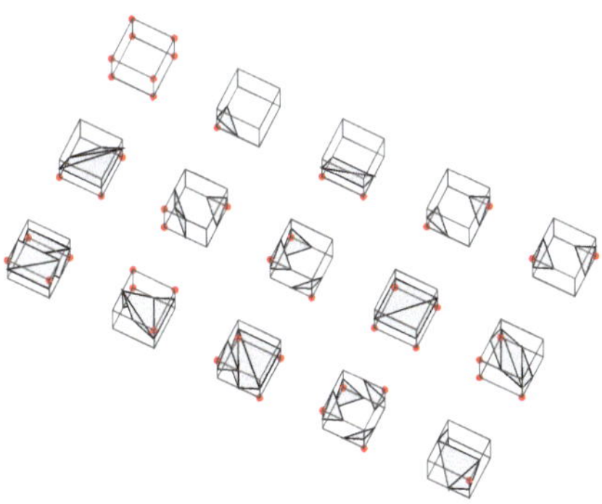

Figure 2 15 marching cube scenarios

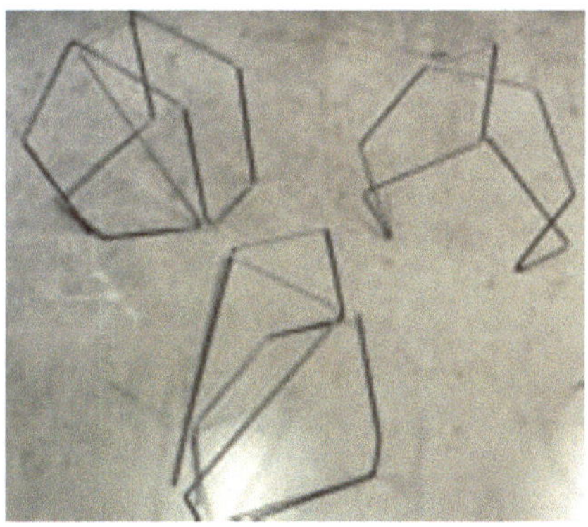

Figure 3 Marching cube traced steel instantiations

ties within the structure. The result is that the installation consists of 1,000 unique parts extensively pre-assembled and organised through the spatial grid of the MC setup itself.

Robotic Bending

The case study project celebrates the constraints of creating structure through bundling, imparting a one dimensional material with surface- and volume-like qualities through principles of aggregation and variegated densities. Research and development pertaining to tool design, multi-station fabrication processes and sequences continues previous trajectories that saw the realisation of a custom-made, free-standing, robot-tended, CNC bender, controlled as an additional axis of the robotic system.

Using this, the robot controls the angle and plane of each bend and the distance between bends, thus creating a series of complex and unique parts. The sequence of operations requires a precise choreography of external clamps, the robotic gripper, and the motion of the arm. The algorithmically generated instruction code makes typical construction drawings redundant and interestingly, also bears little resemblance to the typical point to point motion programs of CNC machines.

The process itself is an adaptation of well established wire/tube bending strategies that are used in the mass production of formed parts. What is significant about this application is the development of the tool and process coupled with a generic platform, i.e. robotic manipulator. This allows continued feedback of process parameters into the generative algorithms, which breaks with the traditional workflow of CNC bending and forming. The tight, often simultaneous, development of fabrication hardware (bending dies, hydraulic grippers, integrated servo) (Fig. 4) and software (KRL code generation from modeled or scripted geometry), means that one can intercede with either aspect equally. For example, the limiting acute angle that can be bent can be made smaller by rebuilding the bender itself in a more compact fashion (Fig. 5), or by replacing any acute angle generated beyond that limit with a double bend (Fig. 6). We implement both approaches simultaneously, and as a result, an expanded fabrication dexterity and wider palette of formal possibility develops far faster and with more immediately testable output, than pursuing either avenue alone ever would.

Figure 4 Bespoke robot actuated bender

Figure 5 Typical bend detail

Conclusion

As demonstrated by the "Clouds of Venice" installation, the potential offered by the complete and bi-directional integration of robotic fabrication and algorithmically generated form and instruction code allows for a radical transformation of industrial production processes. Algorithmic techniques enable non-hierarchical, non-linear and explicit negotiations between an enlarged set of architectural intentions and the material substrates and fabrication concerns through which they operate. Flattening, reducing or eliminating the steps of translation necessary to the transition from design to fabrication provides the possibility of creating feedback loops between design intent and fabrication logic while removing the limitations associated with prevailing representational techniques. Critically, this drive towards streamlining the translation steps is not a reductive exercise, merely acting in the service of efficiency; instead it establishes a critical reciprocity between designing and making. A novel approach that shifts disciplinary concerns away from object-centric notions of artefacts, including how such things are made, towards a deeper concern for the disciplinary implications of the structures underlying the production processes themselves.

References

Benjamin Andrew, "Notes on the Surfacing of Walls: Nox, Kiesler, Semper," in *Machining Architecture*, ed. Lars Spuybroek (1991; repr., London: Thames & Hudson, 2004), pp. 342–51.

Mitchell William J, "Roll Over Euclid: How Frank Gehry Designs and Builds," in *Frank Gehry, Architect*, ed. J. Fiona Ragheb (New York: The Solomon R. Guggenheim Foundation, 2001), p. 354.

Wittgenstein L, *Tractatus Logico-Philosophicus*, trans. David Pears and Brian McGuinness, (London: Routledge, 1972), p. 56.

Figure 6 Assembly detail, showing instances of single and double bends

Jens Cortsen, Silvan Oesterle, Dorthe Sølvason, Hanno Stehling

From Digital Design to Automated Production

Complex-shaped concrete sub-constructions with steel reinforcement

Introduction

This project paper presents an overview of ongoing research from within a larger European project into the development of CAD tools for the design and realization of "non-standard" concrete structures. The European research project combines the knowledge and resources of architects, designers, concrete technologists, civil and structural engineers and robot experts with the practical experiences of key players in the construction sector in a 4-year collaborative venture. Fourteen academic and industrial partners will develop a set of new technologies including digital design and fabrication tools, new formwork and reinforcement systems to radically change the way concrete is currently produced and used. With a construction process spanning a broad range of expertise, collaboration through an effective digital workflow is vital to the successful execution of free-form concrete structures. The state of research widely applied in academic institutions is to directly link design and fabrication in an iterative feedback loop (Bechthold, 2011). The software tools that enable this type of interaction are often custom-made and project specific (Oesterle, 2009).

Software Framework: Design Tool

In any construction process there exists a network of organizations connected both upstream and downstream in the process. The collaboration in non-standard building practices called-for requires the integration

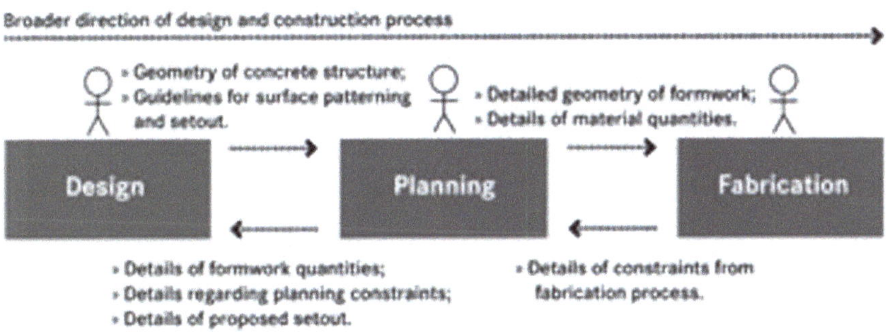

Figure 1 Workflow diagram for non-standard concrete construction process, showing key use cases design, planning and fabrication

of this range of expertise. The research presented here develops a use case model that addresses the key steps in decision making for concrete construction projects from design through to fabrication (Williams, 2011). The following parties have been identified through experience gained in the development of a number of key non-standard projects: architect - planner - fabricator (Fig. 1). A flexible software framework derived from the use case model enables a seamless flow of information between these parties. It provides a specific software tool for each use case party but all software tools load the same technology data. In this sense the tools provide a use case specific view of the construction process (Fig. 2).

The focus of this project paper is on the fabrication use case and the fabrication tool implementing it. The tool allows construction ready geometry to be assigned to different fabrication cells using a new data format called Open Fabrication Language (OFL). It employs open standards and file formats and is intended to be open and extensible.

Fabrication cells are provided to the tool as plugins. Each cell can accept or decline individual operation types and can also have additional constraints like a maximum size of the working area. The tool provides feedback about the applicability of geometry assigned to a cell and the location of constraints and limits. Output from the fabrication plugin can be any kind of fabrication data. It is not covered by the tool anymore and therefore is not restricted in any way. It can describe anything from a direct machine connection to a manual process and to network communication. In this project paper we will focus on the robotic fabrication plugins for double curved milled formwork and steel reinforcement structures.

Robotic Formwork Fabrication Plugin

The fabrication of non-standard formwork

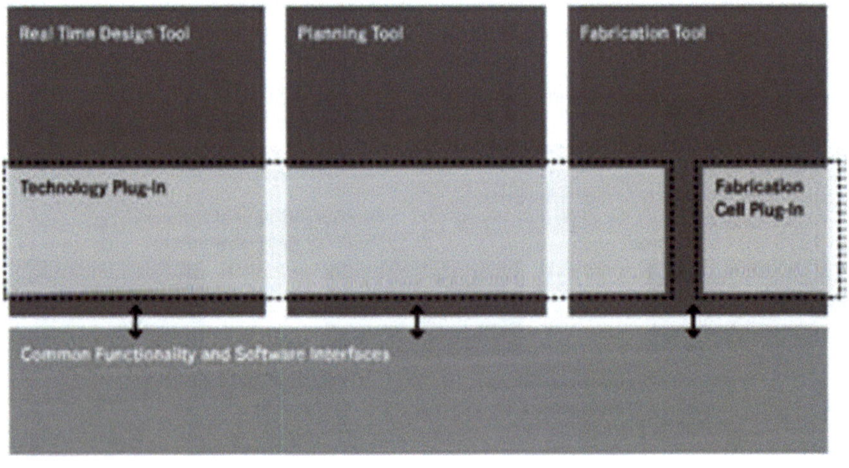

Figure 2 The arrangement of the software framework, showing the relation of tool and construction technology plug-in

can be automated with a 3-5 axis milling machine or by robot milling. Robot milling is chosen because of the lower acquisition cost and higher flexibility of the robot. The material is expanded polystyrene (EPS), which is rigid enough to resist the forces arising when casting concrete. Robot milling is still not widely used due to the complexity and the time-consuming process of preparing the robot movements for milling the formwork. Therefore one of the aims of this project is not only to use the robot for automation but also to encapsulate the complete fabrication cell into a plugin, that communicates with the fabrication tool and automatically calculates the robot milling path for each formwork element based on the information about the formwork geometry provided by the fabrication tool.

The milling fabrication cell plugin software is designed as a tree structure where the root node is the communication and control program. The root node communicates with a third party computer-aided manufacturing (CAM) program for planning the milling paths through the application programming interface (API). The result is then transferred to an off-line milling simulation framework (Fig. 3) through an API. The framework has a post processor for calculating a valid and collision-free robot configuration during the complete milling process from the CAM result. The last leaf node is the communication to the robot cell for production of the actual formwork.

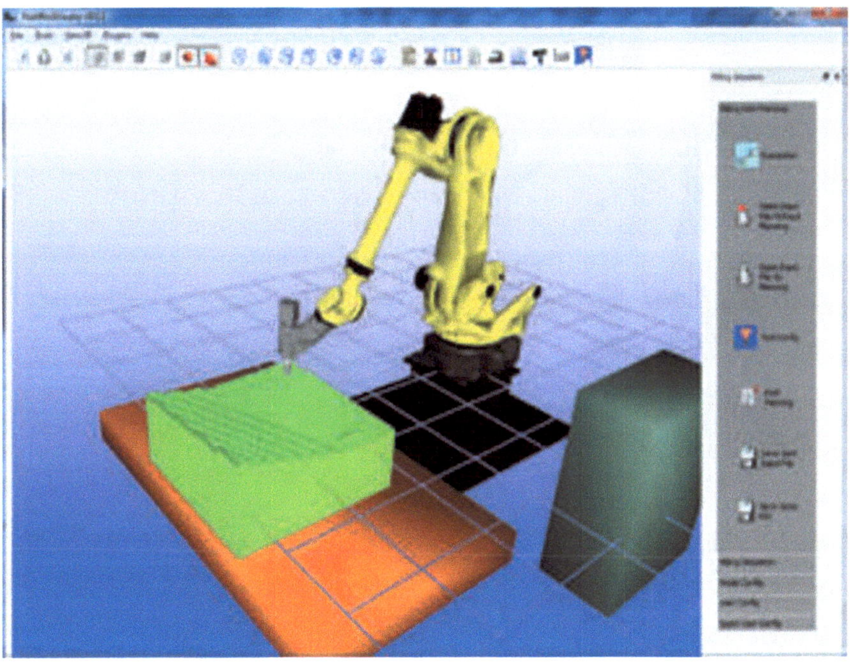

Figure 3 The milling post processor with simulation and generation of collision-free robot trajectories

Robotic Steel Reinforcement Fabrication Plugin

The fabrication of the associated reinforcement structure for a given formwork is a major problem and one that is very difficult problem to solve. A reinforcement machine that can bend a double curved reinforcement bar and then join the bars to a complete reinforcement structure is not available on the market. In this project paper we present our approach to solving this problem (Fig. 4).

The major challenge is to map the 3D CAD model from the OFL format to manufactureable data. In order to make the rebars approximate the NURBS curves it must be determined where the rebars must be bent and by how much. The rebars are to be binded with steel wires at the intersection points. The rebars must be bent between the intersection points to ensure stable bindings. Furthermore, the distance between two bends must be sufficiently large to leave space for the gripper to bend up to 100 degrees in a single bend (Fig. 5).

Another challenge is mathematical modeling of the rebar during bending and movement. The rebar will deflect due to material properties, gravity and physical impact on the rebar from robot movement when attached to the robot gripper. We use an off-line simulation toolbox (Cortsen, 2012) based on a dynamical simulator. This

Figure 4 Two robot reinforcement fabrication unit. The robot to the right is #1 and the floor mounted gripper is the bending tool. The robot to the left is #2 and handles the binding of the rebar grid. In the upper right corner a manually produced 3D reinforcement grid is shown.

toolbox enables path planning and simulation of the robot movements while taking rebar deflection into account and at the same time avoiding collision and maximizing the robot speed in any time step. The simulator handles motion planning for both robots and makes it possible to have two robots working concurrently to ensure optimal performance.

References

Bechthold M, King J, Kane A, Niemasz J and Reinhart C, 2011, *Integrated Environmental Design and Robotic Fabrication Workflow for Ceramic Shading Systems, Automation and Robotics in Building Construction*, Seoul, Korea

Williams N, Stehling H, Scheurer F, Oesterle S, Kohler M and Gramazio F 2011, "A Case Study of a Collaborative Digital Workflow in the Design and Production of Formwork for 'Non-Standard' Concrete Structures." *International Journal of Architectural Computing 9* (3): pp. 223–240.

Oesterle S 2009, "Performance as a Design Driver in Robotic Timber Construction." In *Proceedings of The Fourteenth Conference on Computer-Aided Architectural Design Research in Asia* (CAADRIA), pp. 663–671. Douliou, Yunlin, Taiwan: National Yunlin University of Science & Technology.

Cortsen J, Jørgensen JA, Sølvason D and Petersen H G 2012, "Simulating Robot Handling of Large Scale Deformable Objects: Manufacturing of Unique Concrete Reinforcement Structures", *IEEE International Conference on Robotics and Automation*, St. Paul, USA

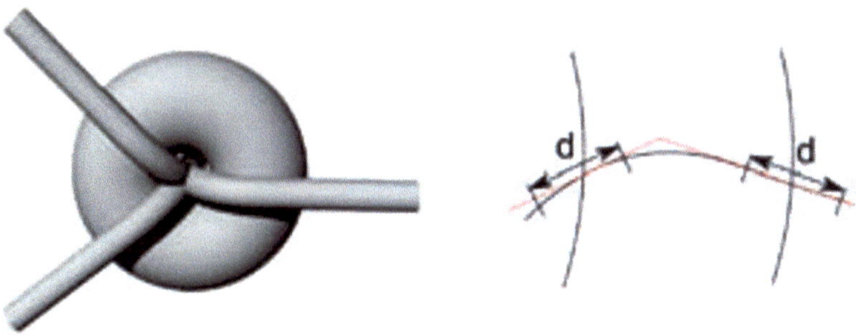

Figure 5 Left sub-figure shows the bending tool as torus for control of correct bending radius according to standards with three rebars showing the 360 degrees bending possibility. The right sub-figure shows the sub-grid for the bending algorithm, where d is the minimum distance between two neighbor bends to ensure sufficient space for the gripper. The lines are the rebars as nurbs curves and the dotted line is the calculated manufactureable reinforcement bar.

KUKA

Jan Brüninghaus, Carsten Krewet, Bernd Kuhlenkötter

Robot Assisted Asymmetric Incremental Sheet Forming

Surface quality and path planning

Introduction

For manufacturing small batch sizes of sheet metal products a new technology called asymmetric incremental sheet forming (AISF) has been developed since the early 1990s. The main advantage of this process is its high flexibility, since no or little product specific tooling is needed. Therefore free formed unique parts or small batches can be produced cheaply and speedily. Economic studies show a potential usage for batches up to approximately 500 parts (Tuomi and Lamminen, 2004). Since the forming forces needed are not dependent on the dimension of the part AISF can be used to produce even very large parts with inexpensive machines.

In the realm of architecture Trautz and Herkrath (2009) examined AISF for manufacturing different elements of a double-layered, facet-like folding structure. Katajarinne mentioned the production of metal façade elements as a use case for AISF [2]. There was also a project on responsive skin where AISF was used to manufacture a mold, which is then used for injection molding UPM Profi – a recycled paper composite – at UCLA Architecture & Urban Design [3].

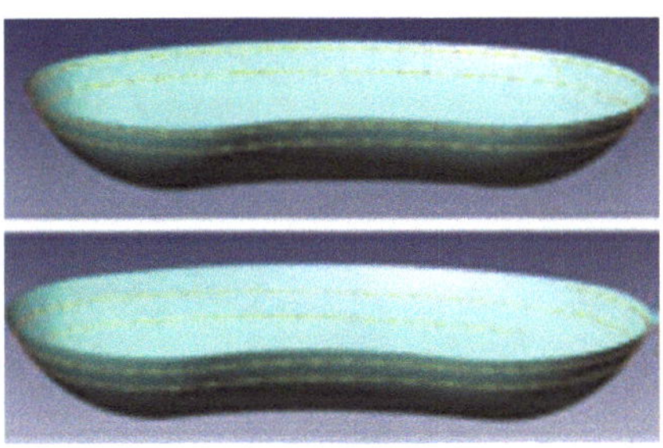

Figure 1 Different tool movement strategies

The AISF Process

A fixed sheet is deformed step-by-step by a small, mostly spherical, generic tool. It travels along the surface of the final part geometry. This can be done with a layer or with a helical strategy (Fig. 1).

Fig. 2 shows the different process variants used nowadays. SPIF and TPIF can be used with a machine with at least 3 axes. The left picture in Fig. 3 shows a SPIF setup with a KUKA Quantec KR 210 R2700 prime at IRPA. For DPIF-P and DPIF-L two synchronized machines with at least 3 axes are needed. Robots are capable of such a kind of synchronized movement (i.e. using ABB MultiMove or KUKA RoboTeam technology). The center picture in Fig. 3 shows a DPIF-L setup with an ABB 6620 and an ABB 4400/60 using ABB MultiMove at IRPA.

All kinds of metal such as steel, aluminum, copper, titan, and even some plastics, can be formed. Composite products such as sandwich panels or polyurethane based color-coated metal sheets

(Katajarinne and Vihtonen and Kivivuori, 2008) are also possible.

The formable sheet thickness depends on the forces the machine can apply. Forming forces are highly dependent on material, wall angle, infeed and tool diameter. Since the sheet is fixed, AISF causes material thinning in formed areas. For simple forming strategies, as discussed here, the sheet thickness t at every point is dependent on the corresponding wall angle. The relationship can be approximately defined by the so called cosine's law, where t_0 is the initial thickness:

$$t = t_0 \cos \alpha$$

All materials have a certain maximal wall angle that can be formed. For many steel or aluminum alloys angles up to 60-70° can be formed in one step. For steeper angles multi-stage strategies can be applied.

Surface Quality

Tool diameter and infeed have the greatest influence on surface quality. Larger diameter and smaller infeed leads to finer surfaces. A lubricant should be used to minimize friction and enhance results.

To prevent tool marks on the part, two sheets can be used in combination. In this setup the additional one lies on top of the other. Both are fixed in the blank holder. The tool only interacts with the upper one and the lower one is formed indirectly by the deformation of the upper one. Since the additionally sheet is used only for forming the real product sheet, it is called a dummy sheet (Skjoedt and Silva and Bay and Martins and Lenau, 2007).

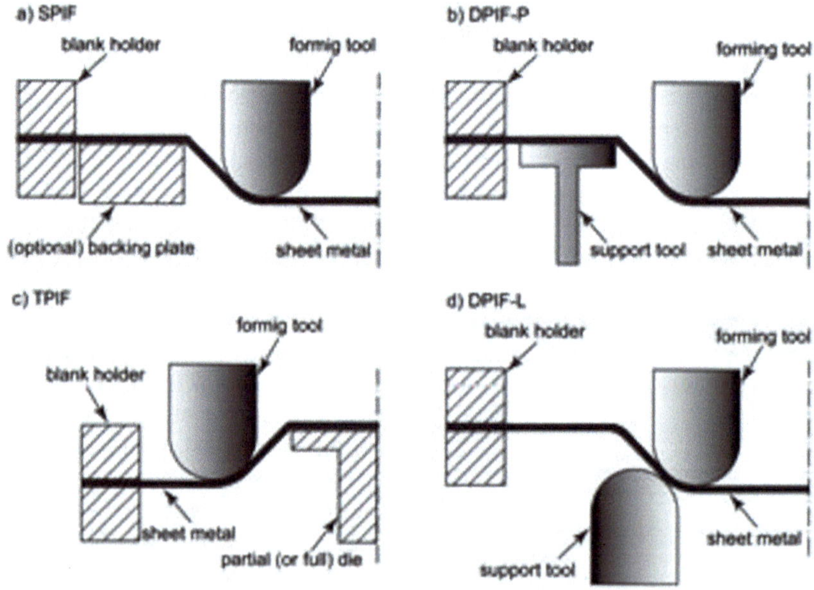

Figure 2 Process variants used in AISF

To visualize typical AISF results some parts with a simple geometry, a cone frustum with 60° wall angle, are formed with SPIF. As material DC01 (thickness: 0.5 mm) was used. The tool (case hardened X155CrMoV12-1) had a spherical tip (diameter: 12mm) and was moved non-rotating on a helical tool path (Fig. 3 right).

In the initial sheet the direction of rolling is clearly visible, but disappears in the formed part. At the inner surface a wavelike structure is visible as a result of the tool movement. On the outer surface a small amount of orange peel effect appears, so the surface is somewhat bumpy (Fig. 4). The inner surface has a glossy appearance, while the outer surface has a matt finish.

Fig. 5 shows the same cone frustum formed with a dummy sheet and an infeed of 0.5 mm. While both sides of the dummy have a glossy appearance, the other part has a matt finish on both surfaces.

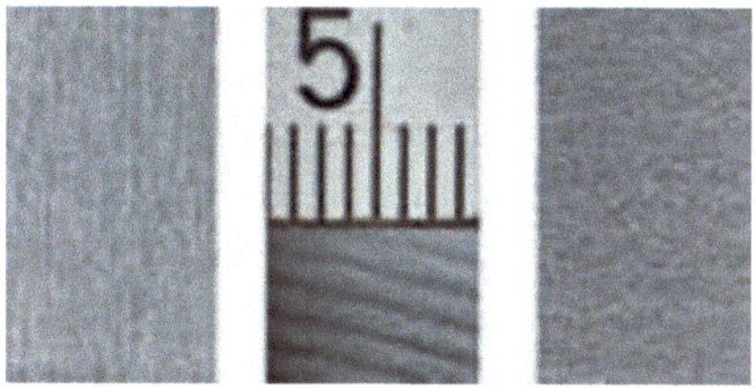

Figure 4 Cone frustum, infeed 1 mm: initial sheet (left), inner surface (middle), outer surface (right)

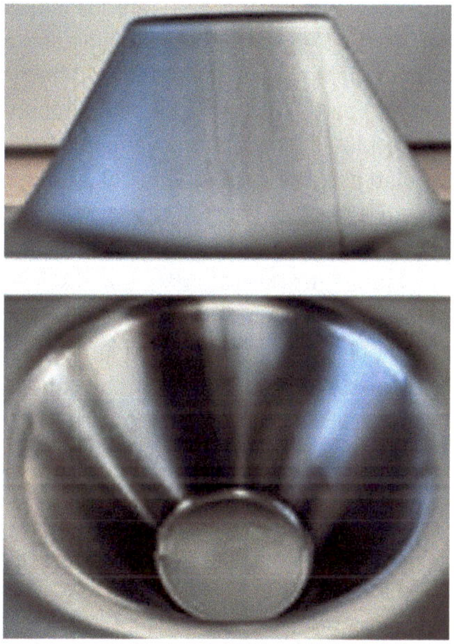

Figure 5 Cone frustum, infeed 0.5 mm, using dummy sheet: produced part (top), formed dummy sheet (bottom)

Path Planning

Since many researchers use milling machines, CAM systems are used mostly in literature for path planning. In SPIF and TPIF surface milling strategies are suitable for AISF. Although KUKA has introduced KUKA.CNC, for most robots the generated

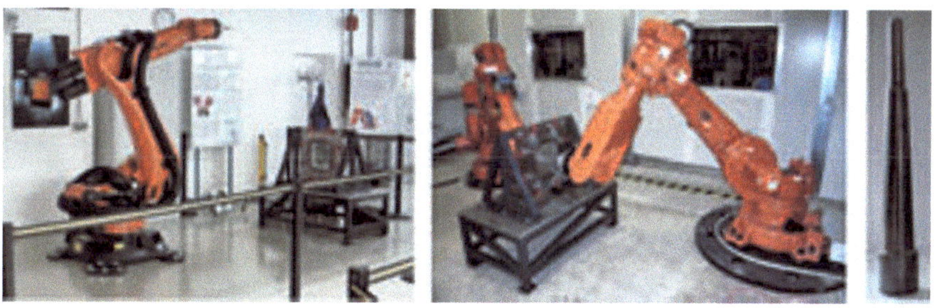

Figure 3 Forming set up used at IRPA: SPIF (left), DPIF-P (center), forming tool (right)

paths have to be translated in the native robot language. As for DPIF-P and DPIF-L no direct analogous process in milling exists, thus CAM systems cannot be used.

A dedicated path planning system for AISF was developed at IRPA. It uses the plug-in functionality of the offline-programming and simulation system FAMOS robotic. The virtual AISF cell is built up in the system and parts are loaded as parametric CAD data. After defining some forming parameters, as i.e. the tool diameter, the paths for both cooperating robots are calculated. The robot movement can be simulated in the system, and collision detection is possibly. The FAMOS robotic post processors take care of generating the real robot programs (ABB MultiMove and KUKA. RoboTeam is supported) and programs can directly be transferred to a manufacturing cell without further manual modifications (Meier and Brüninghaus and Buff and Hypki and Schyja and Smukala, 2009).

Further work at IRPA will detach the path planning algorithms from a specific frontend. Access to geometric libraries as well as GUI and post processing will be encapsulated by interfaces (Fig. 6). So every system that implements the defined interfaces can interact with the path planning library and everyone can work in their preferred environment.

References

Katajarinne T, Vihtonen L and Kivivuori S 2008, "Incremental forming of colour-coated sheets", *International Journal of Material Forming*, vol. 1 no. 1, pp. 1175-1178.

Meier H, Brüninghaus J, Buff B, Hypki A, Schyja A and Smukala V 2009, "Tool Path Generation for Free Form Surfaces in Robot-Based Incremental Sheet Metal Forming", *Key Engineering Materials*, vol. 410-411, pp. 143-150.

Skjoedt M, Silva MB, Bay N, Martins PAF and Lenau T 2007, "Single point incremental forming using a dummy sheet", *Proceedings of the 2nd International Conference on new Forming Technology*, Bremen, Germany, pp. 267-276

Trautz M and Herkrath R 2009, "The application of folded plate principles on spatial structures with regular, irregular and free-form geometries", *Proceedings of the International Association for Shell and Spatial Structures (IASS) Symposium 2009*, Valenda, Spain.

Tuomi J and Lamminen L 2004, "Incremental Sheet Forming as a Method for Sheet Metal Component Prototyping and Manufacturing", *Proceedings of the 10 èmes Assises Européennes de Prototypage Rapide*.

[2] http://chem.aalto.fi/en/current/news/view/2012-04-25

[3] http://www.aud.ucla.edu/student_projects/technology_seminars/289_technology_seminar_responsive_skin_38.html

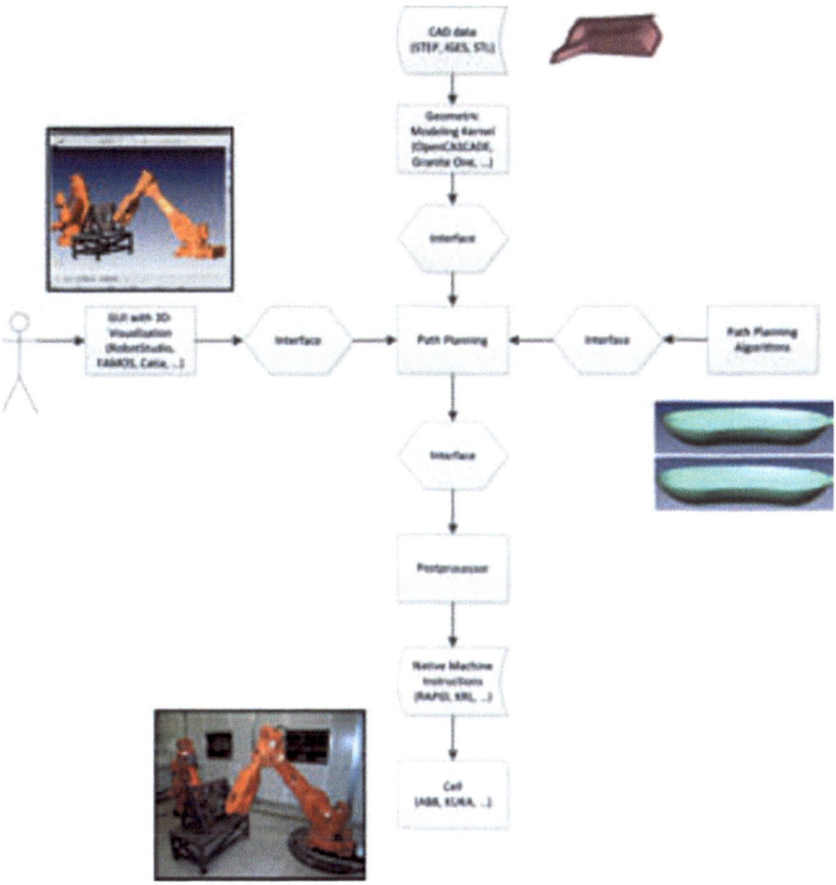

Figure 6 Path planning workflow

Elizabeth Tsai, Michal Firstenberg, Jared Laucks, Yoav Sterman, Benjamin Lehnert, Neri Oxman

CNSILK

Spider-silk inspired robotic fabrication of woven habitats

Introduction

Robotic fabrication and manufacturing technologies are traditionally known for their benefits in automation processes of full-scale design and construction (Seyam 2003). However, their value as design content and material generators has only recently been explored. CNSILK aims to explore the potential of integrating advanced robotic manufacturing techniques through custom developed end-arm tooling with biologically inspired design principles to create a biologically inspired digital design fabrication platform. With a special focus on robotic deposition of tensile fibers inspired by Aranaeidspider web construction, this design approach seeks to establish a pilot case study in biomimetic fabrication. By integrating on-the-fly material generation, multi-axis fiber control, and a seamless assembly method, the platform described herein aims to mimic natural construction methods in achieving woven architectural structures that are continuous in morphology and physical property through the implementation of a multi-axis tensile digital fabrication platform.

Biomimetic Inspiration

Of all known silk producing insects and animals, spiders make the most extensive use of silk in a variety of structural and functional roles (Zhao et al., 2005). When examined under magnification, the structure of orb-spider silk is surprisingly complex (Fig. 1). Web architecture is determined by the constraints of the spider's environment as well as its energy balance. The structural integrity of an orb web, for example, is dependent on the surrounding objects to

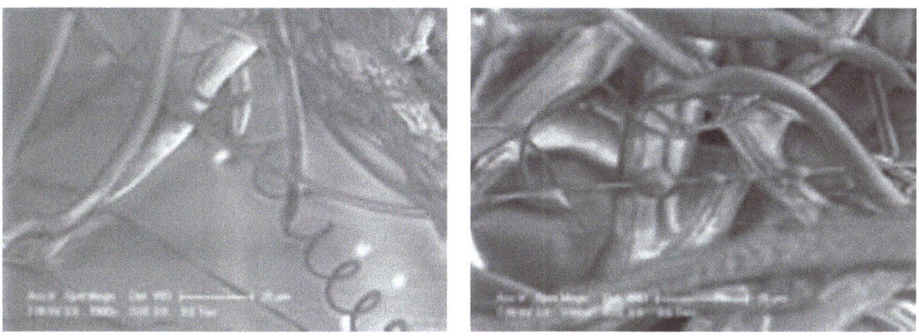

Figure 1 Environmental scanning electron micrographs of spider silk fibers illustrating the different structures present including spring-like motifs (left) and droplets of glue (right)

which it is attached as well as on the silk material, which must be structurally optimized given the high metabolic cost of generation (Gosline et al., 2004).In spinning a web, spiders vary not only the physical structure of the web and its morphological features, but also its material composition.

From the perspective of digital design fabrication, web spinning represents a form of natural additive manufacturing in which the end product is informed by multiple environmental factors and material optimization (Oxman, 2010). In contrast to industrial additive manufacturing methods, which generally print with compressive elements, spider webs are composed of largely tensile elements with seamless structures and functionally graded material compositions.

Methods & Progress

Experiments were conducted using two-types of materials. Initial tests were done with the robotic arm (KUKA KR5 sixx R850 industrial 6-axis robotic arm) using yarn wrapped onto a steel frame structure. Further experiments were conducted by weaving with nylon 6,6 synthesized and drawn from the interface of a two-phase system. This process implies the integration of material generation as part of the fabrication process, seeking to weave with fibers synthesized immediately prior to deposition.

Yarn Weaving

Four-ply cotton yarn was used as the weaving material and a galvanized steel frame with hooks spaced 0.0508 m by 0.706 m was used as a weaving frame. An attachment to the arm was built to contain a spool of yarn; to allow the holder to maneuver around the pegs, the yarn was passed through a 0.05 m hollow tube attached to the end of the holder

Nylon 6,6 Synthesis

Nylon 6,6 was synthesized from adipic acid and hexamethylenediamine solutions. A 10% w/v solution of hexamethylenediamenein water with 1% w/v NaOH and a 10% w/v solution of adipoyl chloride) in hexane was used. The synthesis reaction occurs at the interface between the two solutions

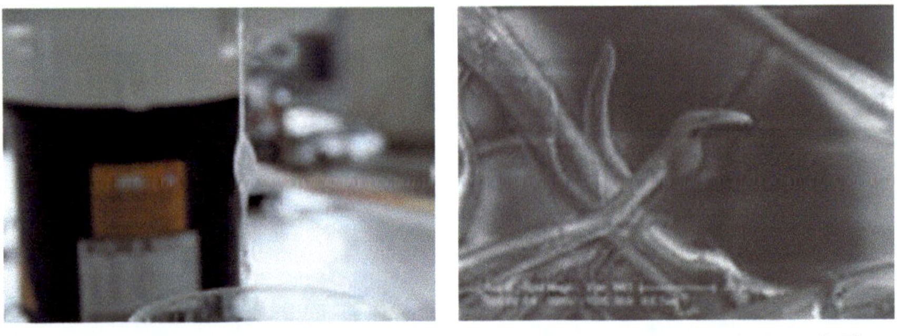

Figure 2 Nylon 6,6 being synthesized and drawn out of a 2-phase liquid system by the robotic arm (left) and an electron micrograph of spider silk containing micro-droplets of glue (right). Different thicknesses and bubbles may be created by varying the draw rate and technique.

and, if the product (nylon) is continuously removed from the interface, the reaction is driven forward, resulting in a single nylon strand.

Three methods of weaving with nylon were explored. In the first method demonstrated in Fig. 2 strands of nylon were drawn from the two-phase nylon prepolymer solution by the robotic arm. Variations in thickness were easily controlled by modifying the speed at which the strand was drawn, with faster speeds resulting in thinner strands. Thicker and stronger fibers can be created by using wider containers for the nylon synthesis and by adding certain agents such as detergents and glycerol to modify surface tension; Hollow cavities in the strand can be created by changing the drawing sequence and occasionally reversing the draw direction. These cavities may potentially be filled with additional materials or glue in analogy to the drop of glue found on some spider silks.

In the second method (Fig. 3) the nylon prepolymer solution was placed on the table in front of the robotic arm and a weaving frame with pegs was attached to the arm. Nylon strands were wrapped onto the weaving frame by twisting and manipulating the robotic arm in six axes. In the third method (Fig. 4), the frame was stationary while a shallow container of the nylon prepolymer solution was manipulated such that different hooks were submerged into the solution successively, resulting in multiple strands.

Figure 3 Robotic Arm manipulating the weaving frame that the nylon is being wrapped around. The nylon prepolymer solution is resting on the table below the arm.

Figure 4 Nylon threads attached to hooks on a steel frame formed by dipping the hooks into a nylon prepolymer solution

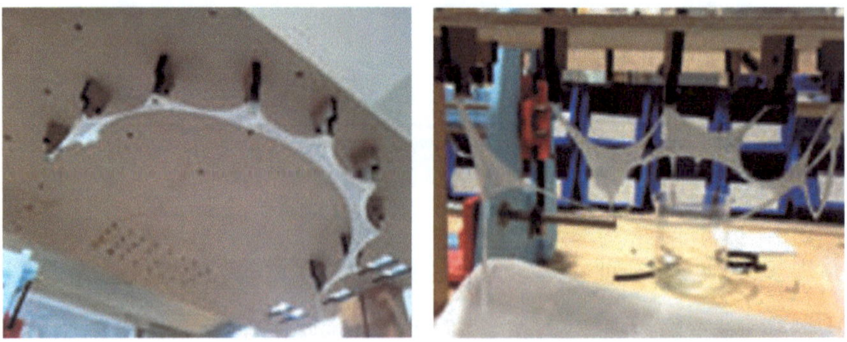

Figure 5 Images of nylon 6,6 membranes created by turning and dipping a shallow container of nylon prepolymer onto adjacent pegs (left), and by formulating the nylon prepolymer solution using 50% detergent solution (right)

Creation of Nylon 6,6 Skins

By using a method of turning and drawing thin, skin-like membranes can be created. A fixed frame was used and a shallow container of nylon prepolymer solution was successively dipped and rotated onto pairs of adjacent pegs (Fig. 5, left). Formulating the nylon prepolymer solution with 50% /v of detergent resulted in stronger and larger nylon skins when drawn using multiple or cylindrical hooks (Fig. 5, right). Various agents may be further added to the aqueous solution to alter the optical and mechanical properties of the resulting nylon skin.

Conclusion & Future Explorations

Drawing inspiration from the spider silk system, CNSilk seeks to develop a material synthesis and fabrication approach for structures whose architectures, responses, and process of fabrication are informed not only by a static design but also by the properties of the fiber itself and its surrounding environment. At its core, CNSilk explores the concept of material synthesis as an integral part of the fabrication process, where tensile members are dynamically generated to adapt to current environmental factors. The intent of this approach is to allow the material system to function in a less intensive way than electronically driven methods, embedding the technology into the system rather than it functioning as a supplement. Continued development of this research could potentially be adapted to integrate material and density gradients so as to incorporate both structural skin and apertures through either a single or multiple materials.

References

Oxman N 2010, "Material-based design computation", PhD thesis, Massachusetts Institute of Technology.

Gosline JM, DeMonet ME, Denny MW 2004, "The structure and properties of spider silk", Endeavour, Vol 10(1), pp. 37-43.

Seyam AM 2003, "Formation of textile structures for giant-area applications", Mat. Res. Soc. Symp. Proc., Vol 736, pp. D1.3.1-D1.3.12.

Zhao HP, Feng XQ, Yu SW, Cui WZ, Zou FZ 2005, "Mechanical properties of silkworm cocoons", Polymer, Vol 46, pp. 9192-9201.

Tom Pawlofsky

Rhino2krl

A simple CAD to robot interface for fast process prototyping

Individual Robotic Production Processes

For a designer the aura of a robot is fascinating: the combination of an articulated robot with an (electronic) hand tool, with automation-components or with self-made hardware seems to allow the fast implementation of individual industrial production processes.

Rhino2krl - Overview

To make this potential available, "rhino2krl" tries to meet the lack of an offline robot-programming tool that easily integrates into everyday design-workflow [1,2].
Rhino2krl is a simple Cad-plug-in to generate a readable and adaptable robot control code. It follows one basic idea: to select curves and to send them to the robot as tool-paths. The main features are:

1. Integration as plug-in into a powerful and affordable CAD application (rhino3d)
2. Input-curves can be drawn, imported or generated
3. Geometrical information and machining parameters are interpreted by the robot control
4. Integration of different hardware set ups
5. Interaction with other plug-ins, such as parametric design, cad-cam interfaces or nesting.

Figure 1 Robotic log-cutting using a chainsaw, programmed in Rhino2krl

Rhino2krl - Interface

For all commands, rhino2krl uses curves to define the movement of the Tool-Center-Point (TCP). A changing spatial direction of the tool can be defined by additional input like a surface, a guiding curve, or block-instances that represent the tool.
It is possible to combine different kinds of movement or tool-changes into one job.

Figure 2 Longtime exposure "technocrafts"

Figure 3 The "spinning sphere" lampshade fabrication

Control Code - KRL

Rhino2krl outputs a special "KRL-code" (Kuka Robot Language) that approximates the tool paths by lines and arcs [3][4]. Contrary to a long list of linear movements, it exports the geometric information together with several parameters. This information is interpreted by a custom function of the robot control to calculate the resulting movement.

This setup allows different influences on the job at the robot's interface (HMI) - without switching back and forth from the CAD environment. The parameters influence the work-piece and tool coordinates, the speed-settings, the strategy to approach or depart from a tool-path and the robot's curve-approximation at kinks. In addition, there is only a single, identifiable line of code for every CAD curve in the main KRL program. This concept of interpreting the geometry at the robot control greatly accelerates the process of fitting the digital world to the real world. Furthermore, it enables a directed approach of optimizing a single robot job or an entire production process.

Examples of Application

The longtime exposure *"technocrafts"* shows simple planar tool-paths as example of a "3D movement". The robot moves a bulb and switches it via the digital output (Fig. 2). *"Spinning sphere"* is a robotic reproduction of a lampshade known from the 1960s. It uses a tool-path that is generated by a custom Java-based application and that is sent to the robot via rhino2krl as *"perpendicular to surface - movement"*. The origin of a cotton thread that is soaked with resin is defined as an "external" tool-center-point, which means that the robot moves the work-pieces – in this case a blown up ball that can be removed after the resin has dried (Fig. 3). *"ZweiRaumwand"* and *"brickolage"* are examples of the *"flank-movement"*. The process of these two projects approximates the shape of complex and curved walls by custom-cut bricks of lightweight concrete (Fig. 4). A robot moving a jigsaw performs the cutting. For each contour, the bevel is locally adjusted. This process minimises waste, and requires a machine time dependent only on the surface area and not on its complexity

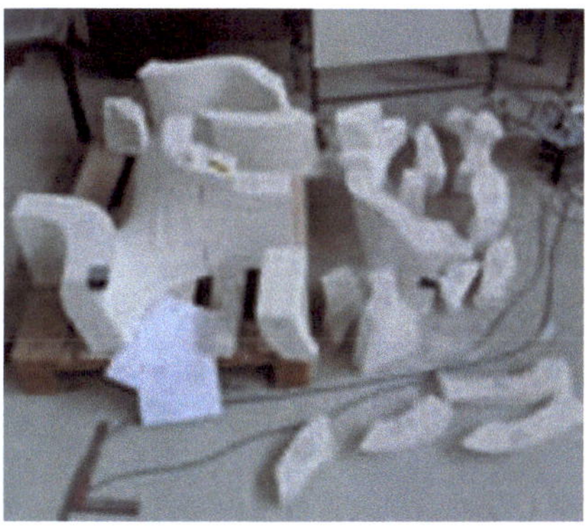

Figure 4 Design and joint pattern of "ZweiRaumwand" (left) and "Brickolage" (right)

or the amount of material that has to be removed. The resulting surface offers a precision that satisfies architectural needs and shows an attractive joint pattern as texture. The author has designed "ZweiRaum-Wand" by using custom, Java-programmed modeling software, based on voxels. The sculpture consists out of 80 bricks and has served as a preliminary study for brickolage (Figs. 5–7). *"Brickolage"* is the result of a workshop about digital fabrication at the Chair for CAAD, ETH Zurich [5][a]. A reaction-diffusion algorithm implemented in Java generated the surface of the sculpture. The production data are the result of Python programming and a parametric definition in "grasshopper". 1400 different bricks were cut by means of the rhino2krl interface (Fig. 5, 6). After the above experiences, a chainsaw moved by the robot seamed to be reachable challenge for the *"robot's chainsaw stool"*. The usage of logs directly from the forest offered an exciting contrast to the high-tech robot (Fig. 7). As a preparation for a digital fabrication seminar the author implemented the process. Then the stools and the tool paths were designed together with the student, using manual CAD-techniques [b]. Rhino2krl's "direction-by-block-instance - movement" was used to generate the control-code.

Conclusion

Rhino2krl enables easy access to the great potential of robotic fabrication. But the vision of custom production processes also involves many technical challenges of the physical world. Therefore the robot remains a complex tool that can be mastered only by persistent efforts. The profit of these efforts are the use of a strong tool that connects the digital and the physical world.

References and Credits

[1] http://en.wikipedia.org/wiki/Industrial_robot#Robot_programming_and_interface

[2] http://de.wikipedia.org/wiki/Industrieroboter#Programmierung

[3] Kuka Roboter GmbH 26.09.2003, KR C2 / KR C3 Programmierung Experte, KUKA System Software (KSS) Release 5.2,Version: 00

[4] Kuka Roboter GmbH 05.07.2007, KUKA System Software 5.2, 5.3, 5.4 Bedien- und Programmieranleitung für Systemintegratoren, Version: 1.1

[5] http://www.mas.caad.arch.ethz.ch/mas1112/category/m2/

[a] Supervision by Mathias Bernhard (design and scripting), Manuel Kretzer (design and material), Tom Pawlofsky (process development, robot interface and programming)

[b] Thanks to Markus Gläser, Grit Werner, Tibor Weissmahr, www.hfg-karlsruhe.de

Figure 7 The log after machining: waste, first stool, second stool, toolpath-planing 2 of 14 cuts

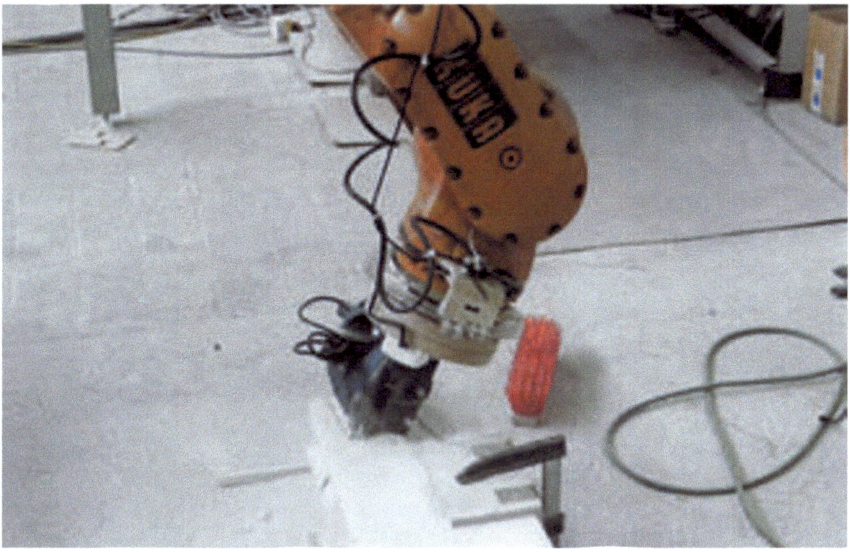

Figure 5 Robot cutting a lightweight concrete block with a jigsaw, production process for "ZweiRaum-Wand" and "Brickolage"

Figure 6 "Brickolage", final sculpture

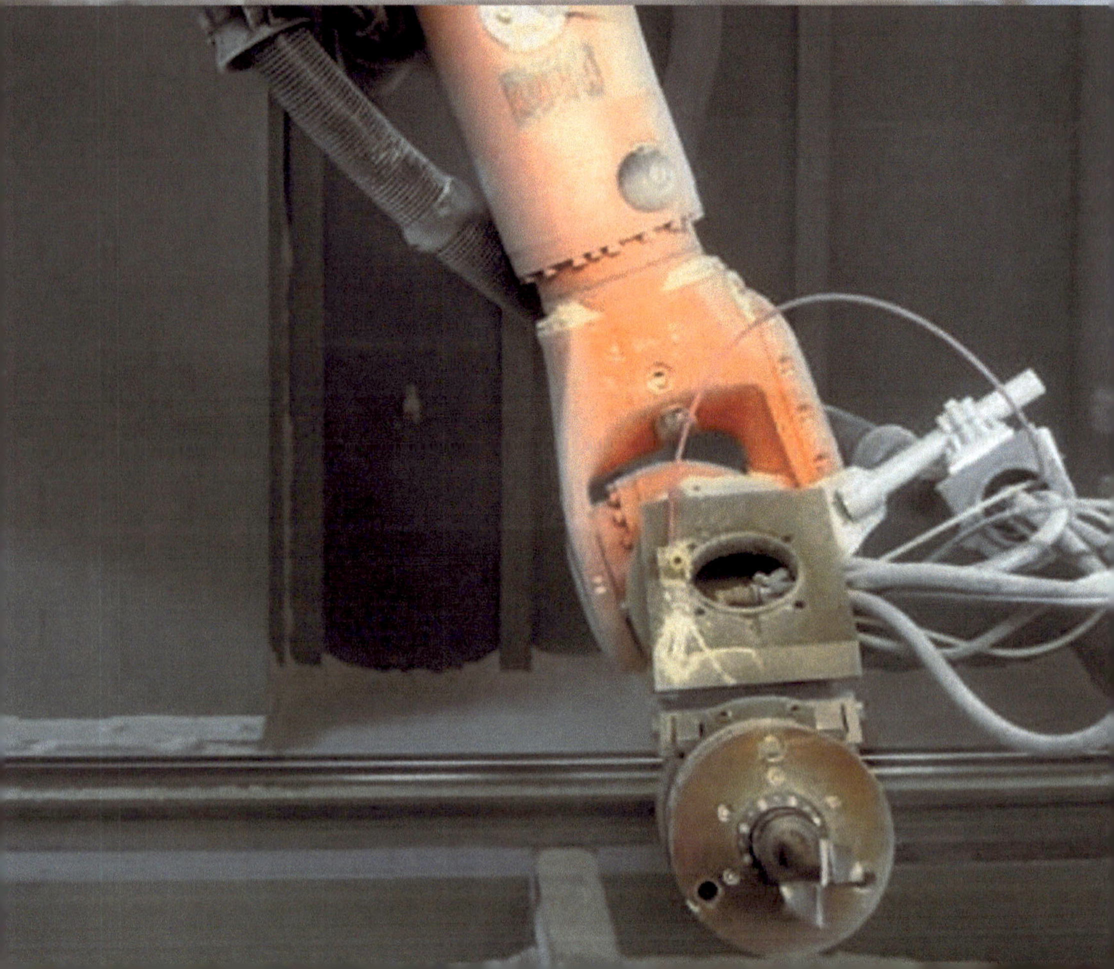

Baris Çokcan

Robotic Fabrication for Düzce Teknopark

Streamlining fabrication through versatile machines

Düzce Teknopark, a technology center for Düzce University, has been developed by Il Architects int, a small Austrian/ Turkish architecture firm based in Istanbul.

Infinity of science and the convergence of industrial and academic research provided the inspiration to develop a multifunctional twisting, three-dimensional loop emerging from a toroidal form. The environmentally self-sufficient and self-supporting building envelope houses office spaces and research labs for industry and university spin-offs. Incubating new technologies and innovations requires highest security standards where each floor has to be encapsulated. In the foyer the double façade twists like a conch shell into a voluminous entrance space, creating a pocket within the double-façade. In analogy to the fertilization process, the penetration of the building by researchers, clients, and visitors is articulated in the atmospheric climax of the entrance hall as a meeting area "in-between."

The cost driver for a freeform façade such as Düzce Teknopark's is not the surface cladding itself but rather the joining of the single elements. Therefore, in the case of Düzce Teknopark with its overall surface area of roughly 8000 m^2 and more than 7000 m of joints, the edges of the single elements largely define the costs, which in turn are influenced by the efficiency of the fabrication process of the structural members. Düzce Teknopark requires the definition of the fabrication process for more than 21km of cutting, joining and mounting, making streamlining the design to fabrication process a highly desirable goal. A flexible CNC fabrication with a robot lends itself particularly well to such diverse tasks. Due to its many degrees of freedom and the possibility of using multiple end-effectors, re-clamping of the single quadrilateral elements is minimized and the processes of sawing, milling, drilling etc. can be executed in one task.

Today the flexibility of digital tools enables designers to quickly implement design and fabrication constraints by fluently adapting parametric schematas during the design to fabrication process. Inhouse customized tools incorporating, for instance, wood manufacturing and machinic know-how about sawing, milling, drilling holes etc. have for a long time been the sole property of the building industry. The knowledge of how to incorporate these strategies within the design and robotic fabrication process allows small architecture firms to convince clients and contractors of their technological capabilities and removes creative limits from the design to fabrication process.

Yaron Elyasi

Free Molding Technology

The Concept

A computer guided robotic arm, combined with a plastic extruder generates a unique method whereby molten plastic is allowed to fall freely, producing one-of-a-kind, artistic creations. Despite being a technological robotic-guided process, Free Molding Technology veers from standard template production by allowing random elements to affect the resulting object.

The Material: Plastic as a Material for Art and Design Products

Using plastic for art and design artifacts has not always been considered appealing and artistic. Traditional materials - metals, glass, and ceramics – are considered more "worthy" of an artist's time and talent.

One reason for this is the way that plastics are generally produced. Plastic products are mainly designed for functionality and mass-production requirements leave little room for individual expression. Furthermore, until the middle of the 20th century, craftsmen did not have the option of using plastic as a sculpting material and were forced to use traditional materials.

Another reason that consumers often show disrespect and contempt for this material, is its mass-produced, cheap, polluting image. Yet, the tremendous ecological damage caused by plastic is not due to the plastic itself, but to the social behavior that wrongly treats plastic as disposable. In fact, plastic offers myriad properties that are ecologically superior to traditional materials: Plastic can be processed between 180-240°C, whereas traditional materials require a minimum of 1000°C, making them high energy consumers and more harmful

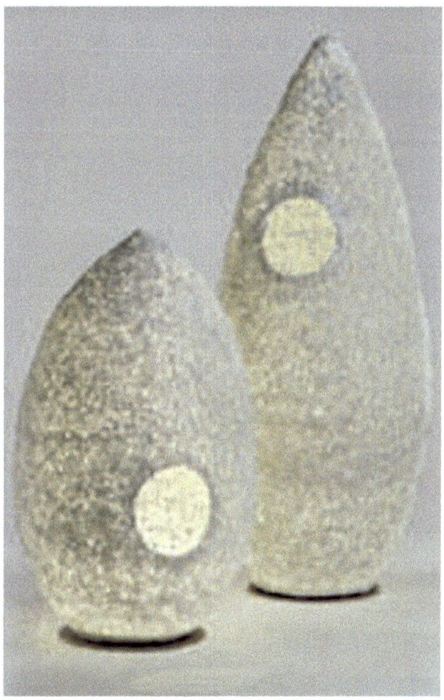

Figure 1 Cocoon lighting, 2012

to the environment. Traditional materials are taken directly from Mother Nature's resources; plastic is a byproduct of petrol. Once the prejudice is overcome, plastic can be considered a strong, intriguing substance for use in artistic projects and design objects.

The Innovation: Free Molding Technology

Early in his design education, Yaron Elyasi, embarked on a project using recycled plastic as his medium. His intended final project was, unfortunately, aborted when all 2.5 million bottle caps (collected over 4 years) were accidentally discarded, forcing him to rethink his project.

An unsuccessful quest to carry on with his project by hand-sculpting plastic emerging from industry machines led him to explore other ways to mold plastic. This led to his first plastic extruder: a mechanical meat grinder with external heating, with which he started making plastic "pottery" and exploring other technical possibilities.

Natural phenomena on a small scale formed a basis for possible molding techniques. One used a whirlpool of water as a mold: molten plastic was poured into the turbulence, producing a random, chaotic form frozen in time. His "whirlpool mold" experiments led him to explore the aesthetics of chaos, and to remove the strict rationality governing the plastic industry. By removing meticulous planning, the artist had freedom to create products different from industrial norms, as well as from each other. The next step was experimenting how the stream of molten plastic could be turned into textures derived from natural linear designs: fingerprints, wind, water ripples, sound and light waves, etc. A texture

Figure 2 Robotic fabrication process in RobotWorks

library was formed with variations in density, line thickness and type of plastic.

Progressing further, he combined plastic extrusion technology with a methodology where plastic was poured onto, rather then injected into, a mold: the "mold" thus lost its intrinsic meaning and became a "three-dimensional canvas." The placement of the plastic stream was random, the structure evolving depending on the circumstances, like wax dripping from a melting candle. Random, uncontrollable factors – wind, gravity, the composition and temperature of the raw material, the temperature of the surrounding air or the "canvas" – all affected the final result.

By manipulating the "mold" while the plastic flowed onto it, different patterns and textures were formed; upon attaching itself to the "mold," the melted material achieved a form and volume that maintained its given structure upon solidification. The result was magical: suddenly, plastic – typically characterized by cold smooth surfaces that mirrored the mold – received a new, organic, spontaneous appearance. The manufacturing method's unique aesthetic value has been exploited in Tom Dixon's famous work, "Fresh Fat Chair" (2004) [1], made by plastic extrusion without the robotic integration.

Introduction of Robotic Control

Up to this point items were produced individually. However, marketplace requirements demanded a method whereby items could be effectively mass produced, while still maintaining individuality, and where the pattern could be better planned and controlled. This was achieved by introducing robotic control.

A prototype was built combining the studio's existing plastic extrusion machinery, a six-axis robotic arm and a Solidworks-based application called RobotWorks [2]. The RobotWorks application is well-known in industry, but not so familiar in the design and art world. Using RobotWorks, a three-dimensional guiding path is applied by the designer to drive the robotic arm.

Integrating robotics,specially designed molds and plastic extrusion resulted in Free Molding Technology (in final patent registry phase). The designer feeds a pattern into a computerized program that controls the extruder›s velocity, and a robotic arm replaces the human one, maneuvering the mold beneath the stream of melted plastic, thus allowing different patterns and textures to form. Using Free MoldingTechnology, small mass-produced lines of different designs can be easily implemented,

Figure 3 Eco Reflect bowl, 2011; Lava bowl from recycled mobile phones, 2007

producing similar products yet maintaining each and every one's individuality.

On one hand, introducing random factors to the process, even though robot-controlled, generates unique artifacts. As in nature, no two are completely identical. On the other, the computer interface presents the ability to introduce precise, specific designs, patterns and logos into the production process.

The aesthetic achievement is profound. Each object is one of a kind. Evidence of the success of this "breaking the mold" method was provided by the audience's admiration during the designer's first exhibition at The Israeli Trade Fairs Center in 2001. The Milano Mobile show and others throughout Europe followed.

Figure 4 Darbuka stool, 2006

The Future

Consumer Participation in the Design Process

Free Molding Technology stands at the forefront of catering to an emerging worldwide trend of mass customization and "Do-It-Yourself" home decor and interior design [3], [4] by refining the ability to customize products industrially so that the consumer will be able to participate in the design. This will be an evolutionary leap affecting the entire industry.

Extending the Method to Other Materials

Etto studio [5] focuses on design using recycled and raw plastic, but this technology has potential for use with other materials (glass, metal or wax), and, eventually, with clay and with polymer mixed with wooden fibers.

References

[1] Tom Dixon http://www.tomdixon.net/products/fresh-fat-chair

[2] 2 Nathan Naveh, http://www.robotworks-eu.com/support/whitepaper.htm

[3] "The Future is Fabulous: Digital Fabrication, Datascaping and Design Futurism"

http://papers.ssrn.com/sol3/papers.cfm?abstract_id=1594323

[4] http://cubify.com/apps.aspx

[5] http://www.ettostudio.com

Federico Diaz

Outside Itself

Interactive installation assembled by robotic machines untouched by human hand

Art is not art if it is created by the human hand – Manifesto Nero, Federico Diaz Venezia 2010

Outside Itself is an interactive, mathematically programmed, robotically-produced, light-responsive installation that grows and morphs as a life force unto itself. It can also be described as a self replicating, site-specific sculptural installation shaped by viewer-inspired data which gives new meaning to the term "going viral".

Imagine thousands of black spheres being created and morphing according to changes in ambient light generated by the fluid interactivity of viewers (refer also to the previous project Geometric Death Frequency 141 created in 2010 for the Massachusetts Museum of Contemporary Art, Fig. 1). The balls are assembled by two precisely calibrated robots into an exponentially-shifting composition. Each ball represents an individual "photon". Optical sensors monitor the available light at the site,

Figure 1 Datasculpture Geometric Death Frequency 141, 2010, MASS MoCA Massachusetts Museum of Contemporary Art

creating a data stream that controls the robots. The surrounding light is affected not only by the passage of time from day to night, but by the number of viewers surrounding the installation, their movements, and even the color of their clothing.

Although the installation is produced without being touched human hand, it is completely interactive. The mathematical program enables the two robots to build and, together, to arrange about 2,000 of the 5-centimeter-diameter balls every 12 hours, completing a large, continuously shape-shifting construction over a period of several months (Fig. 2).

The shape and composition of the installation respond directly to its immediate surroundings. Like an infinitely adaptable organism, or society itself, it constantly reflects its environment. For the artist the robot is like an "outstretched hand of our senses", that extends human ability beyond the limitations of the body, in the same way that society now uses technology to simulate or stimulate experience, or to create "social networks". Technology is relied upon to communicate and achieve what the body cannot–to go beyond, to go "outside" of oneself.

Acknowledgements

Jindrich Bartosek, robots developer
Barbara Holomkova, coordinator
Lukas Kurilla, software and architecture
Martin Licko, 3D graphics
Stepan Malovec, graphic designer
Lukas Matocha, coordinator
Tomas Mosansky, robots supporter
Krystof Pesek, interactive software
Outside Itself was curated by Alanna Heiss

Figure 2 Datasculpture Outside Itself, 2011, 54. Biennale di Venezia / Italy

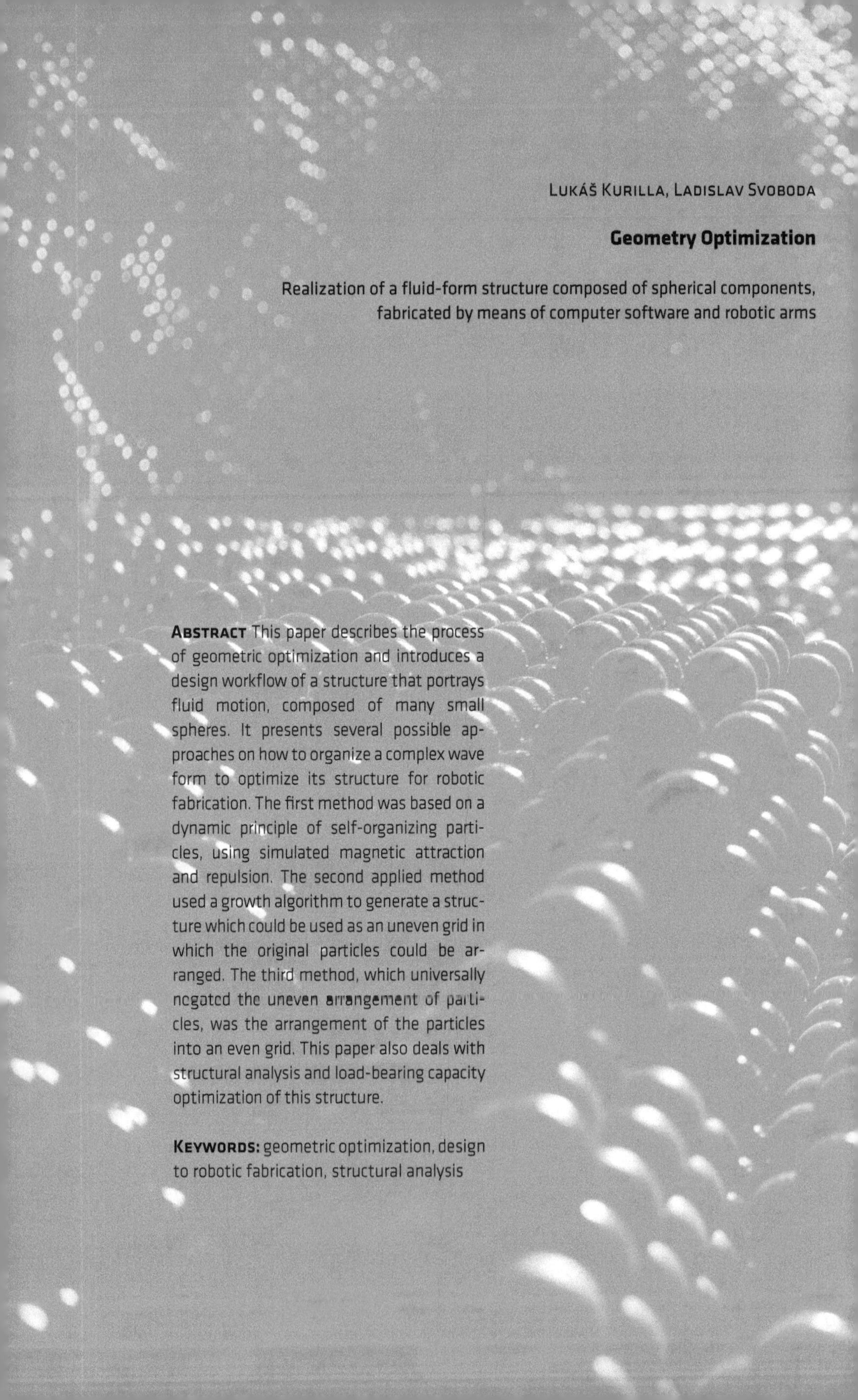

Lukáš Kurilla, Ladislav Svoboda

Geometry Optimization

Realization of a fluid-form structure composed of spherical components, fabricated by means of computer software and robotic arms

Abstract This paper describes the process of geometric optimization and introduces a design workflow of a structure that portrays fluid motion, composed of many small spheres. It presents several possible approaches on how to organize a complex wave form to optimize its structure for robotic fabrication. The first method was based on a dynamic principle of self-organizing particles, using simulated magnetic attraction and repulsion. The second applied method used a growth algorithm to generate a structure which could be used as an uneven grid in which the original particles could be arranged. The third method, which universally negated the uneven arrangement of particles, was the arrangement of the particles into an even grid. This paper also deals with structural analysis and load-bearing capacity optimization of this structure.

Keywords: geometric optimization, design to robotic fabrication, structural analysis

1 Introduction

The fluid-form sculpture, described in this paper, titled Geometric Death Frequency - 141, was designed by Federico Díaz as a 2-year exhibition project for the MASS MoCA museum in Massachusetts. The sculpture consists of approximately 420 thousand spherical elements made from ABS plastic material, each being 4,7cm in diameter and weighing 9g. The entire structure is 10 meters long, 5,4 meters wide and 4 meters high. This project took a year to produce, including 3 months of robotic manufacture, and was then transported by boat to the U.S.A. where the sculpture was completed (Fig. 1).

Our aim was to fabricate a complex wave form, simulated using the RealFlow programme [1]. During the project we considered and tested several ways of producing wave forms. First, we focused on the use of generated mesh geometry to formulate the surface of the wave. Then, we considered the particles that we would need to generate this wave. The author decided to try using spheres as "particles," and, after an impressive visual simulation, we decided to pursue further development in that direction. The original arrangement of the particles from the simulation could not be used directly as the basis for production. Distances between the particles were varied, so many spheres would intersect, and most would not connect to anything, levitating in space. Therefore it was necessary to develop geometry optimization to rearrange the original structure of the particles, enabling us to replace them with spheres so that each sphere would touch its neighbours at one point (Fig. 2).

2 Robotic Fabrication

As we began planning fabrication with spherical components, a new question emerged - how to maintain accuracy in a structure consisting of this number of spheres. A similar case was solved in the construction of a brick wall, Bonwetsch (2007). A robotic arm was used in the construction to control the production of

Figure 1 Realization of a fluid-form structure composed of spherical components, Geometric Death Frequency - 141

Figure 2 Fluid simulation: a) surface area based on mesh geometry; b) volume of the wave, represented by particles; c) particle position dilemma

a wide variety of components, and to reduce potential faults, which multiply with every added component. In sphere-based structures the threat of inaccuracy is even higher than in line-based geometry. Robotic fabrication reduces it. We also considered fabricating polyurethane foam spheres, extruding them in a process similar to the project "The Foam" [2], Bonwetsch (2008). However, this technique was very problematic, so we finally decided to use prefabricated spheres, which we joined together by spot-gluing.

For the fabrication we used two industrial robots, KUKA KRC 16 with controller KRC2 (Fig. 3). We controlled the robot through a plain text file, which was generated by simple Java application Robo.d [3] specially developed for this project. Each line of generated file carried information about a single sphere, its position, and vector points of bonding and gluing. This information was further processed and converted into KUKA Robot Language (KRL) using an application written in C# by Jindrich Bartošek, which mainly calculates

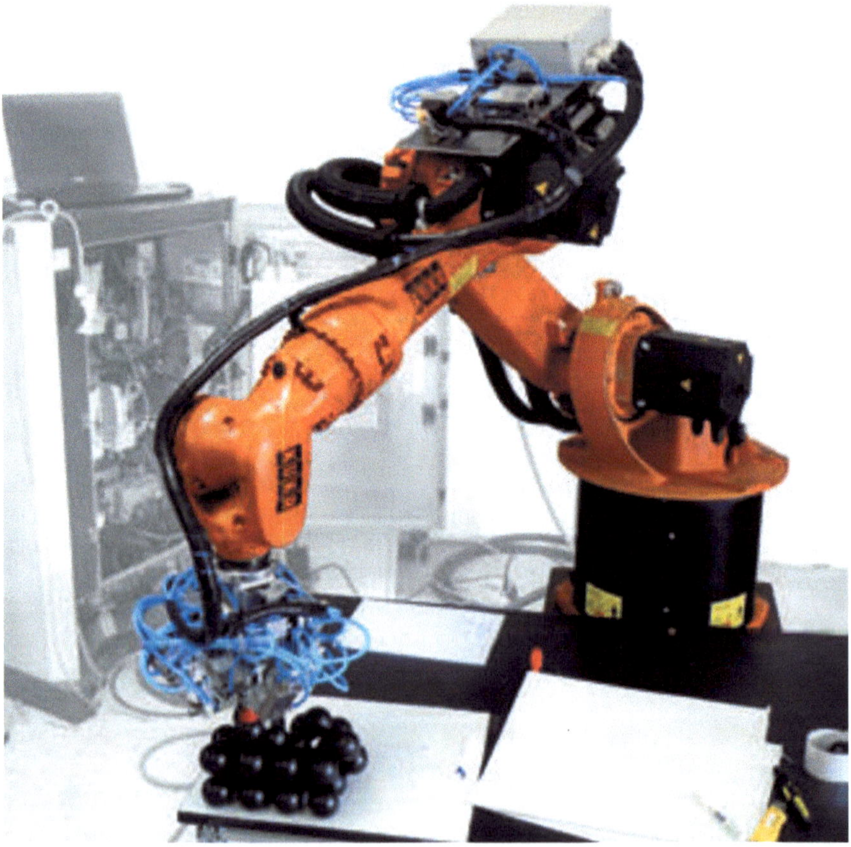

Figure 3 Testing phase of fabrication with KUKA KR 16

vectors for rotation of the robot. He worked for the firm Blumenbecker [4], with whom we consulted throughout production, and which also ran and serviced the robots.

It was necessary to develop a special cupped head to grasp the spheres. We applied a liquid accelerant on each sphere to speed the bonding process. At the beginning it was necessary to hold the spheres above the jet applying accelerant and rotate them throughout the application. Later four jets were added to the head, which allowed direct application of a liquid accelerant to the grasped spheres. This improvement shortened the robotic arm's trajectory and thus fabrication time. In the optimized solution the robotic arm grasped each sphere from a feeder and placed it under the jet, where glue was applied to the sphere at all the necessary points of contact with the other spheres in the structure (Fig. 4). Finally, the robotic arm placed the sphere into position, according to the coordinates and held it until it stuck firmly to its surrounding neighbours in the structure. The robotic arm released the sphere by gently blowing it away. An adhesive and an activator were developed by the firm Henkel in order to shorten the bonding process, to provide the required mechanical properties, and to ensure resistance to external conditions. To avoid collision between the robot and structure, we determined that a vertical stacking process was ideal. The time needed for gluing a sphere depends on the angle of connection. The more an angle diverges from the vertical axis, the more time consuming it is (Fig. 5a). But the most critical problem is the range limit beyond which the robot is no longer able to glue two spheres together - any angle horizontal or lower to the position of the neighbouring sphere. Any attempt to do so would weaken the structure, providing no support for further subsections of the structure.

Figure 4 Detail of the special cupped head in glue application process

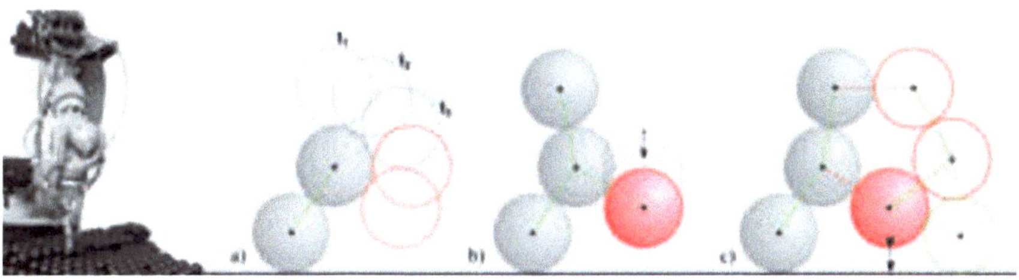

Figure 5 Robotic fabrication: a) angle fabrication possibility and time consumed (t1<t2<t3); b) vertical stacking problem; c) first layer problem (red - vanished, green - added sphere reducing vanished spheres)

In the first step of fabrication, the robotic arm glued the first layer of spheres to a planar surface foundation (Fig. 5c). To prevent the problem outlined above, and maintain structural integrity, it was important that spheres in this first layer lie exactly on the same plane.

Another factor complicating production was the fixed position of the robot, where the dimensions of the robotic arm limit the maximum size of what it can build. Therefore, the entire structure had to be divided into smaller production sections. In breaking up the structure, ensuring that each separate piece of this first layer had the exact same height, so the next level would lay on the same plane, provided a further complication.

3 Geometry Optimization

Geometry optimization can be divided into two phases. The first is the optimization of the distances between the spheres, through a computer algorithm, in order to connect with neighbouring spheres. The second phase is a quality control measure, to ensure that the resulting structure is feasible and structurally sound. This second phase is integrated as a part of the optimization algorithm and continues into post-process optimization. The aim of this process was to arrange the spheres into the desired geometry, with as little change to the original simulated wave form as possible. During this process, several methods for optimizing the structure geometry were tested in order to ensure the feasibility of the structure.

3.1 Self-organization through Simulated Magnetic Fields

This first method was based on a dynamic principle of self-organizing particles, using simulated magnetic attraction and repulsion. By giving each particle in the algorithm a positive or negative "charge", it naturally led to each particle having as many neighbours as possible while reducing particle overlap and discontinuities in the structure. It was a method that should ensure a dynamic configuration of the spheres and verify manufacturability of the resulting irregular structure.

Being a non-linear task, this first method was repeated numerous times, resulting in an approximate design that still required further optimization. If any problematic particles still remained in the resulting structure they were omitted in further processing.

This optimization algorithm is similar to the relaxation algorithm. A vec-

Figure 6 Self-organization method: a) first layers of the structure; b) vanished, omitted particles in the structure; c) magnetic principles in the algorithm

tor shift is calculated for each particle in the structure , the length and direction of which depends on the locations of surrounding particles. If the distance is smaller than the diameter of the sphere, then they must be overlapping, so the particle is repelled. If the distance to the neighbouring particles is greater, the particle is attracted, until they touch. After several iterations the structure starts to coalesce, problematic intersections disappear, and particles gain a number of neighbours.

Differences between various iterations could be observed in a log containing the number of overlapping particles, and the number of particles with 0, 1, 2, 3, 4 or more neighbours (Fig. 7, right). Particles located in the first layers were aligned on the same plane and were prevented from moving vertically. This ensured that planes created in the resulting structure would meet production requirements. (Fig. 6a).

When checking for feasibility (Fig. 6b), the structure resulting from this process had many elements missing from the original concept, which was not ideal. Due to the dynamic form of this structure it was quite a complicated task to bring back these elements without further limiting what we could add in the final stages (Fig. 7, left) .

Additionally, a wide range of different gluing angles made fabrication times unpredictable, with frequent defects, dropped spheres, etc. Fabricated sections were fragile and incohesive, making transport and handling very complicated.

In the following optimization methods we focused more on durability of the sections and specification of angle limits. These methods were based on the arrangement of the spheres into a defined grid. The grid made it easier to add the missing spheres at specific places in the structure and thus reduce the number of vanished spheres in the post-process optimization.

3.2 Growth Algorithm

On this optimization algorithm we cooperated with the Department of Cybernetics, Faculty of Electrical Engineering CTU in Prague. The basic principle of this solution was based on mathematical definitions of fabrication rules. The optimization algorithm was then used as a method to define fabrication limits as an uneven grid "grows"

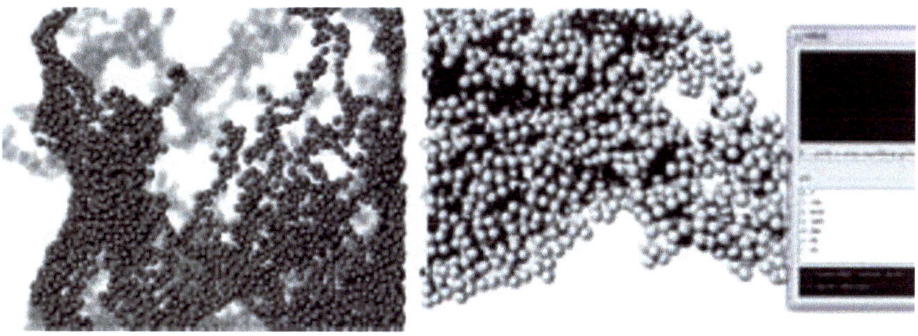

Figure 7 Self-organization: (left) vanished spheres in the structure; (right) approximation result of optimization

higher. Using growing rules in the algorithm we generated a structure that could be used as an uneven grid in which the original particles could be arranged.

The rules of the algorithm are similar to an L-system. Unlike an L-system, the generated structure does not fork, but if the distance between a pair of "parent" particles is within the specified limit, they automatically create a "child", another particle that sits on top of them, resulting in a new layer. The "child" particle can be linked to its parents at various angles. The angle is chosen at random but cannot interfere with the existing structure. So, if one "child" particle overlaps with another, it automatically readjusts its position to fit neatly with the other "children". Randomly selected angles further contributed to the unevenness of the resulting structure.

Each parent can have only one child with the same partner, but can have an unlimited number of partners. These rules ensure that each of the spheres in the structure is glued to two parent spheres and after creating its own child, each sphere has 0 DOF (zero degree of freedom) as it is linked to at least three spheres. This enhanced the durability of the fabricated sections. Specifying the distance limits between the parents (Figure 8) reduced the number of problem angles and thus reduced the error rate of production. At the beginning of the process the first production layer was randomly generated using the growth algorithm to create further layers. We assumed that in this way it would be possible to create a grid of any size. However, in the upper layers it was no longer possible to create additional offspring. As can be seen in the picture, as the structure grows, there are less and less particles, resulting in a pyramid-like structure, instead of a solid, regular tower. Therefore, it is not possible to use this algorithm to define the grid.

3.3 Simple Tiling

Despite the fear of losing the dynamic impression of the original concept and shifting to a clearly perceptible uniformity, we finally decided to incorporate a uniform grid to obtain better control over optimization and

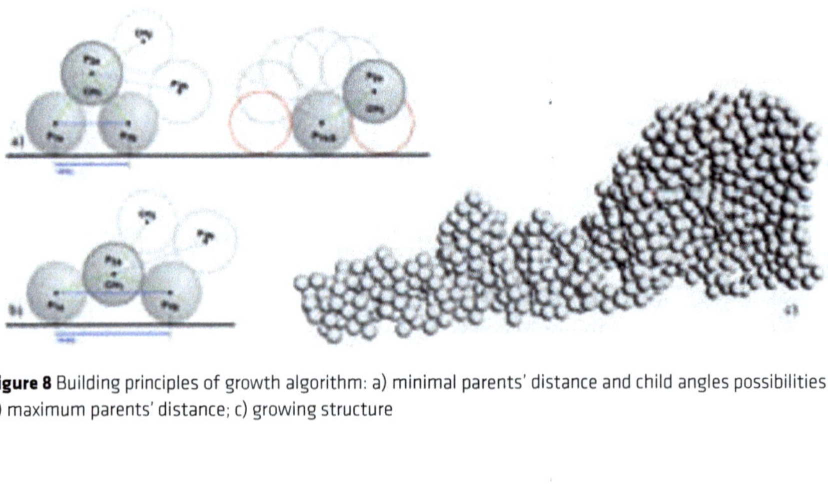

Figure 8 Building principles of growth algorithm: a) minimal parents' distance and child angles possibilities; b) maximum parents' distance; c) growing structure

production. The grid was created by tiling a pattern consisting of two spheres, which formed a rectangular grid with a sphere placed in the centre of the resulting cube. Unlike other patterns this model had a lower number of spheres per meter cubed, and was perceived as airy and less uniform. In the resulting structure there was one constant angle, allowing for faster and more stable construction.

The original particles from the animation were arranged into the grid creating the wave model with a equal distance between neighbouring particles. The problem with the first production layers was eliminated, and uniformity of the grid allowed us to fabricate sections which were strong, interchangeable, and fit seamlessly. There still remained the problem of completing the final details of the design, the irregularities that did not fit into this algorithm due to a lack of support and so vanished (Fig. 9). Therefore it was necessary to add these spheres and limit further disappearances (the post-optimization process). Unlike a self-organized irregular structure, the grid made it easier to put the spheres in the required positions. The post-optimization algorithm rules tried to provide a pair of parents for each sphere, which reduced errors and increased the density and rigidity of the structure. Another advantage of this uniform solution was that it gave us the ability to turn a section upside down, which also reduced sphere disappearances.

4 Structural Analysis

The resulting wave structure was designed for outdoor conditions, where it would have to handle a large amount of snow. We were afraid about load-capacity of some parts of the object, so we decided to conduct a structural analysis. For this we used a software tool based on FEM structural analysis, which we have been developing, MIDAS [5, 6] Svoboda (2012).

4.1 Analytical Model

We had to find the best way to create an analytical model, incorporating additional properties of the structure such as materials and cross-sections. In the analytical model, all the spheres were transformed into beam finite element mesh with nodes placed in their centers (Fig. 10c). The bearing capacity of the beams, normal and bending stiffness were obtained experimentally by the load test of several cantilever girders consisting of ten axially aligned spheres. The measured quantities were verified by a detailed FE analysis of a three-dimensional model with the spheres and glue joints precisely resolved. To analyse the structure it was necessary to make several changes in the analytical model including removing disconnected and levitating elements. For this reason we applied a "virus" algorithm in which one "infected" sphere spread to all the other spheres with which it could connect (directly or through a chain) (Fig. 10a). Unconnected spheres were removed, otherwise it would not be possible to analyze the structure using FE method.

The structural analysis was calculated based on the weight of the structure, plus additional snow. The snow load was applied to the top nodes of the structure (Fig. 10b). For the sake of numerical analysis speed-up, only the arch-shaped part of the structure comprising about 250

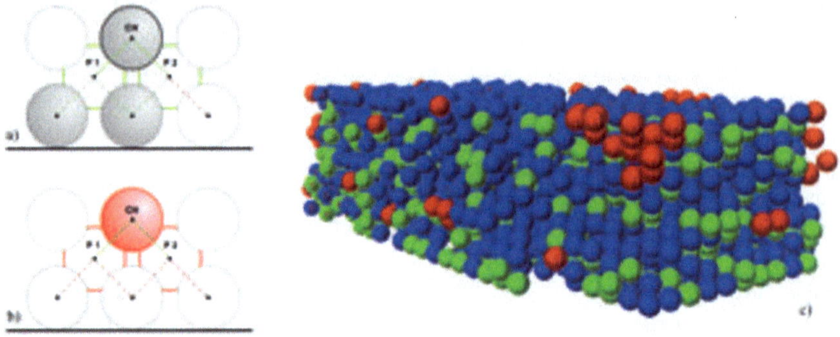

Figure 9 Post-optimization process principles: a) add two parents (green) if possible; b) if not possible remove child (red); c) fabricated section, post-optimization process result.

thousands spheres was considered as critical. The analytical model was exported to a plain text file and sent to MIDAS structure analysis software, which processed it and computed the results in VTK file-format. The resulting mechanical quantities were converted into a concise bit of information on regions with exceeding load-bearing capacity (red beams) (Fig. 11a) and was displayed using the Paraview software [7].

4.2 Structural Optimization

Based on the results of this analysis, we optimized the form to increase the load bearing capacity of the structure. In order to strengthen the structure, we first merged different animation frames of original fluid simulation, Díaz (2010). On one hand, the load capacity increased thanks to a thicker structure. On the other hand, there was still the problem of adjusting the shape of the arch to a more ideal form.

We decided to reinforce the arch form. We considered steel rod supports, but were discouraged by complications to do with calculations and the poor synergy of combining these materials. Therefore we created support for the structure with spheres in a parabolic shape, which we merged with the fluid-form of the arch. This ensured sufficient load bearing capacity for the most exposed part of the structure.

Figure 10 Analytical model: a) result of "virus" algorithm; b) top nodes with applied snow load; c) spheres transformed into beam finite elements

5 Conclusion

This paper presents several possible approaches to geometry optimization. Some of the techniques were not successful due to the complexity of the task. However, they showed potential for future use in the optimization of general complex forms.

We began with the most ambitious technology of self-organization, which was able to preserve the original concept of the object to the greatest extent. Through the growth algorithm, which can be used to create uneven grids, we developed a more manageable technique using uniform grids. Using the uniform grid we achieved the desired results. Based on the knowledge and experience obtained we can expand and improve these optimization techniques even more.

Structural analysis was performed as a study. It would be useful to employ homogenization techniques for similar structures, as this might save time. Creating a less time consuming analysis would allow one to use an optimization algorithm to automate the arch shaping process and merge the geometric and structural optimization processes.

Structures created from spheres, when compared with those formed of rectangular components have a more interesting visual impact and also higher geometric variability in composing their structure. In

Figure 12 Snow-covered structure, load-bearing capacity verification in real world conditions.

addition, the sphere shape has a specific physical property, specifically the hollow body, which was used in our case too. A structure made of hollow spheres has a surprisingly adequate load-bearing capacity and it is light, both of which are good properties for a self-supporting structure. The structure, after some surface treatment, could also have good insulating properties.

Concerning robotic fabrication, changes made in robotic fabrication might allow the entire structure to be fabricated in one place (without needing to divide the structure into prefabricated sections). Simi-

Figure 11 Arch: a) analysis results; b) arch support with parabolic shape; c) resulting structure

lar fabrication processes are used in Flight Assembled Architecture [8]. However, in the vertical stacking method a complex structure will certainly contain some problematic areas that would be impossible to fabricate, and the whole process would have to be carefully planned to avoid structural problems and possible collisions between the robot and structure.

In the future, for more ambitious projects, the industrial robots currently available would not be sufficient. It would be necessary to design custom robots that would be able to run throughout the structure, building it gradually in layers. They might also be able to recycle the waste directly, reform it, and use it as a building material [9].

6 Discussion

The implementation of robotics in architectural fabrication effects the planning of a project at every stage. It defines new production data and architectural documentation. It also allows one to apply basic principles of micro-scale, where objects are made up of small components with their own logic configuration, to our human macro-scale. We can talk about particles to spheres, as if they were bits to atoms. If we can manage a particle cloud, then we can fabricate new kinds of materials with different properties, all using robots. New ways of projecting and designing also change requirements for architectural software tools. Newly developing software tools for architects should be more like the environment we know from games (like Mine Craft or World of Goo).

They should use dynamic game engines, where "live structure" would be created and would respond to the designer´s impulses. Live structure could be changed based on the optimization processed in the background. Other analyses, depending on the designer´s decision, could be made on the structure to mark problematic areas and to make the designer aware of them and allow him/her to react. The whole structure might function as a large interactive information model in which the designer realizes further relations and consequences of his/her actions directly during the design process. Such digital tools would ensure manufacturability of a designed structure in any form.

Acknowledgements

The authors thank Federico Díaz for his creative impulse and for allowing them the privilege of working on the GDF project. We would also like to thank Henri Achten, Matej Lepš, Jan Novák, Jan Zeman and the CTU in Prague for their collaboration and manuscript consultation. We also would like to thank Jindrich Bartošek and the firm Blumenbecker for their collaboration and use of their robot. Our thanks also to Martin Šetina and Stathis Zervas for their financial support. We would like to extend our thanks to Justína Kurillová and Thomas Arthur Smith for manuscript and language correction. Finally we gratefully acknowledge the endowment of The Ministry of Industry and Trade of the Czech Republic under project FR-TI1/568 and GA CR under project 103/09/H095.

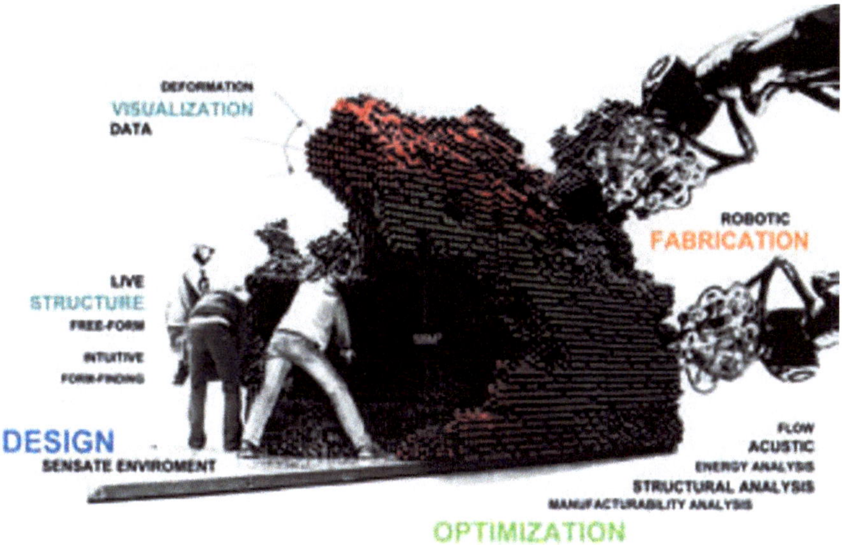

Figure 13 Design to fabrication, illustration of a digital design tool with a specific design workflow

References

Bonwetsch T, Gramazio F, Kohler M 2007, *"Digitally Fabricating Non- Standardized Brick Walls"*, *Manu-Build*, conference proceedings, Rotterdam. D. M. Sharp, 2007, pp.191-196.

Bonwetsch T, Baertschi R, Oesterle S. 2008, "Adding Performance Criteria to Digital Fabrication Room - Acoustical Information of Diffuse Respondent Panels", *Silicon+Skin, Biological Processes and Computation, Proceedings of the 28th Annual Conference of ACADIA*, Minneapolis. 2008, pp. 364-369.

Díaz F 2010, *Geometric Death Frequency -141*, Edizioni Charta, Milano, ISBN-978-88-8158793-3, pp. 40-41.

Svoboda L, Novák J 2012, "A simple framework for integrated design of complex architectural forms", CoRR, abs/1203.2499, 2012, <http://arxiv.org/abs/1203.2499>

[1] http://www.realflow.com/rf_casestudies.php?cs=casestudies/cs_34massmoca.php.

[2] http://www.dfab.arch.ethz.ch/web/e/lehre/137.html.

[3] http://robo-d.kurilluk.com.

[4] http://www.blumenbecker.cz/en/.

[5] http://mech.fsv.cvut.cz/~da/MIDAS/en/index.html.

[6] http://www.igend.cz/en.

[7] http://www.paraview.org/.

[8] http://www.dfab.arch.ethz.ch/web/e/forschung/209.html.

[9] http://singularityhub.com/2012/04/23/3d-printing-robot-produces-chairs-and-tables-from-recycled-waste/.

Karola Dierichs, Tobias Schwinn, Achim Menges

Robotic Pouring of Aggregate Structures

Responsive motion planning strategies for online robot control of granular pouring processes with synthetic macro-scale particles

Abstract Loose, designed macro-scale granulates can be used as architectural material systems. Combined with a digitally-controlled emitter-head the pouring process can serve as an alternative to known additive manufacturing techniques. The potential of macro-scale granulates lies in their ability to re-configure as well as in being a functionally graded material. Given that these loose granulates merely display probable rather than certain behavior, the use of responsive motion-planning becomes a critical aspect. The research presented here introduces the field of synthetically produced architectural granulates. An overview of the current state of the art of robotically poured granulates is given. Within this context, the proposed robotic pouring process for designed granulates is outlined. The established feedback loop consisting of optical sensing, parametric motion-planning, and robotic actuation is described in detail. In conclusion, an outlook for further research is given.

Keywords: aggregate architectures, synthetic macro-scale particles, gunctionally graded materials, robotic pouring, responsive motion planning

Introduction

Architectural systems commonly seek to form stable and permanent assemblies, where both the form of the individual element and that of the overall structure can be defined in a very precise geometrical manner. In recent years, however, architects have come to investigate so-called granular structures or aggregates. These consist of large masses of elements that are only in loose frictional contact. If the individual grains are synthetically produced, the resulting granular structures can be calibrated to suit specific architectural requirements, such as structural and environmental performances. Designing with these aggregate structures requires the architect to observe the evolving formation rather than to precisely define it (Dierichs and Menges, 2012).

Using a six-axis robot as a pouring device for designed aggregate structures both renders the pouring process precise and offers the opportunity of pouring patterns which are otherwise hard to achieve. These robotically poured aggregate structures can be seen as alternative forms of digital additive manufacturing, which allow for reconfigurable, functionally graded material systems.

In pick-and-place robotics elements are individually positioned in the overall structure in a very controlled manner (Bonwetsch, Gramazio and Kohler, 2007). In contrast to this, robotic pouring implies that the designed granules are flowing out of an emitter, such as a linear magazine or a flow-controlled pump. The resulting aggregate formations are consequently predictable only in terms of probability rather than certainty. The use of interactive robot-control thus becomes a necessity through which observation and interaction with the pouring process can be achieved. So far, responsive motion-planning in the architectural context has been used mainly in combination with assembly systems such as in the ECHORD project currently conducted at the ETH Zurich [1]. The use of interactive robot-control in combination with a loose aggregate consisting of synthetic macro-scale particles, however, is novel.

In its first part, the paper will give a brief overview on the notion of aggregate architecture. Secondly, robotic aggregate pouring and interactive robot-control will be defined, and current tools and technologies described. In the third part, the newly developed technologies will be presented focusing on sensing strategies for highly coarse synthetic macro-granulates. In conclusion areas of further research into robotic pouring with interactive motion-control will be discussed.

Aggregate Architectures

Aggregates are defined as large amounts of elements in loose contact (Cambou 1998; Duran 2000). In nature sand or snow are considered granular – or aggregate – systems (Bagnold 1954; Ball 2004; Nicot 2004; Rognon, Chevoir and Coussot, 2008). In architecture there are only very rare examples consciously deploying loose granular matter as architectural systems in their own right (Hensel and Menges 2008a, 2008b; Dierichs and Menges, 2010). However, especially if the grains are designed and produced synthetically, aggregates are a very relevant area of research into architectural material systems, due to their capacity to continuously reconfigure as well as their innate potential for functional grading throughout a structure (Fig. 1).

In an architectural context, loose granulates have rarely been deployed. The few examples range from building physics, where the aggregate serves as an insulator, to vernacular architecture, where it is mainly used as a filler in walls (Houben and Guillaud, 1994; Hausladen, de Saldanha and Liedl, 2006). It is also known as a large-scale sand-mold in building construction (Treib 1996; Dierichs 2010; Kohler, Gramazio and Willmann, 2012). Several geo-engineering applications consider not only the geo-technical, but also the architectural and urban landscape aspects of their interventions (Hensel and Menges, 2006d; Trummer 2008; Hensel, Menges and Weinstock, 2010). The most directed research into loose granulates as a material system in an architectural context was conducted initially under Frei Otto and was later developed in Diploma Unit 4 at the Architectural Association and the GDA Studio at Rice University (Gaß and Otto, 1990; Hensel and Menges, 2006a, 2006b, 2006c; Hensel and Menges, 2008a, 2008b). The field of research into loose granulates as an architectural material system in their own right is thus relatively unexplored.

One of the most relevant branches within this area is the use of so-called designed granulates. This implies that the individual particle is customized to meet specific architectural performance criteria, such as fast frictional interlocking or heat insulation (Hensel and Menges, 2006b, 2006c; Tsubaki 2011; Dierichs and Menges, 2012). The research presented here is situated within this specific area of research into designed architectural granulates.

State-of-the-Art: Robotic Pouring of Granulates and Responsive Motion-Planning

Robotic Pouring of Granulates

Over the past seven years robotic pouring of granulates is being explored mainly using the aggregate as a formwork or alternatively operating in a geo-technical context. The following will give a more in-depth overview of the different projects as well as a comparative evaluation with regards to

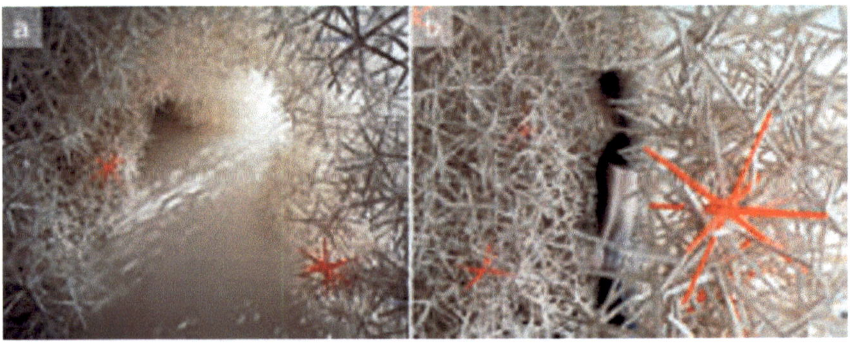

Figure 1 Aggregate structures consisting of synthetic macro-scale particles: a) aggregate vaults formed by up to ten thousand synthetically produced particles, b) building-scale aggregate structure

(i) the type of the granulate, (ii) the role of the granulate in the design, (iii) the use of self-organizational capacities of the material and (iv) the motion-sensing techniques involved.

The first project to be mentioned in this context was conducted at the IFF (Institut für Industrielle Fertigung und Fabrikbetrieb) at the University of Stuttgart. Its aim was to develop automated systems for the direct modeling of large-scale sand-molds using additive and formative manufacturing methods. The developed processes use materials that can be poured, like sand, as a mold for industrial parts. Both the pouring process itself and tools for compacting and modeling the aggregate are developed (Schaaf 2005). The context is not architectural but industrial production. Self-organization in the aggregate is not searched for as a design tool, but considered as a necessary production parameter. The granulate serves as a formwork, not as the design result itself.

The second project was conducted at the ETH Zurich in 2011 under the supervision of Fabio Gramazio and Matthias Kohler as well as Christophe and Yael Girot. The course investigated the potential of digital fabrication processes for landscape design using materials like sand. Deploying machines fitted with sensors, the students were able to achieve processes of landscape formation that enabled a feedback between the process of pouring and the evolving formations (Kohler, Gramazio and Willmann, 2012; [2]). This project thus situates itself within the context of landscape design. The self-organization of the naturally occurring material sand is explicitly looked for in the process and is supported by the use of sensor-equipped pouring devices.

As a successor to this project the same team at the ETH developed an additive manufacturing technique for reusable concrete formwork made of sand in 2011. The project bears great similarity to the research work conducted in 2005 at the IFF. In this case, however, the self-organizational capacities of the formwork are explicitly looked for and the context is clearly architectural, resulting in three prototypes of 1 meter by 2 meters for a retaining wall (Kohler, Gramazio and Willmann, 2012; [3]). Again the material sand is used as a formwork, sensing is not explicitly used according to the project description, however self-organization of the material is part of the design process.

In 2011 very similar studies into robotically-poured sand formwork were initiated at the Institute for Advanced Architecture of Catalonia (IAAC) under the supervision of Marta Malé-Alemany. The most significant difference to the experiments conducted at the ETH lies in the fact, that the sand itself is solidified in its top-layer or through liquid injection with the use of glue, wax, plaster and the like ([4]; [5]).

To conclude, the robotic pouring projects conducted so far use exclusively the natural granulate sand (i), the aggregate serves either as a recyclable formwork, whose top-layer can also be solidified, or it is part of a geo-technical landscape formation (ii), self-organizational behavior is explicitly being looked for in three of the projects, which are conducted in a design-oriented context (iii), motion-sensing is used only in one landscape design application (iv).

Responsive Motion Planning in Architectural Robotics

Optical sensing in industrial robotics is currently used for object recognition which might trigger subsequent predefined routines that are executed by the robots such as in sorting or handling of objects on an assembly line. Alternatively, the orientation of a specific object in space can be determined in real-time to interactively adjust manipulator alignment with the object (Xie et al., 2008).

In the architectural context, optical sensing is used to incorporate tolerances that might accrue over the course of an assembly process [1]. However, in both contexts sensing and robotic execution are clearly separated and consecutive data processing steps. In the case of aggregates a more fluid approach is desirable and required in order to provide instantaneous feedback and the possibility of minute adjustments, especially with regard to synthetic macro-scale particles.

Responsive Motion Planning Strategy for Online Robot control of Granular Pouring Processes with Synthetic Macro-scale Particles

The use of a digitally-controlled emitter-head such as a six-axis industrial robot as a pouring device for designed granulates can be regarded as a novel, alternative technique to known additive manufacturing - such as fused deposition modeling (FDM) (Oxman 2010; Soar and Andreen, 2012). In FDM a continuous stream of heated polymer is extruded from an emitter head and deposited on a height-adjustable printer-bed to form individual horizontal layers. As the polymer cools down, the layer solidifies into a permanent configuration and forms the basis for the next layer. Existing FDM techniques on the architectural scale usually replace the polymer by a mineral material such as concrete which solidifies after deposition. In contrast to FDM and other 3D-printing techniques, synthetic granular aggregates are free of additional binders and do not rely on polymerization. Robotic pouring of non-convex designed granulates

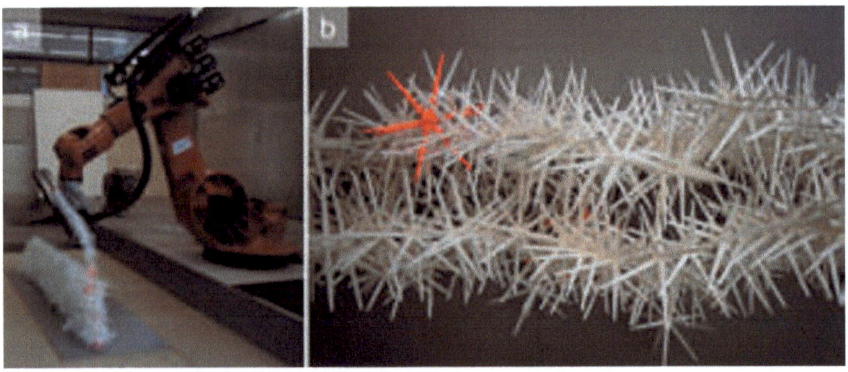

Figure 2 Robotic pouring of designed granulates: a) robotic pouring using a magazine emitter-head, b) poured structure using a linear KRL-controlled pouring path

thus offers the possibility of fully recyclable structures, as no binder is used and the particles interlock merely through friction. Moreover - depending on the material used - these poured aggregate structures can be light in weight as opposed to the relatively heavy additive manufacturing products (Fig. 2).

Compared to the state of the art overview presented in the previous two sections, the robotic pouring process presented here is novel in two respects. Firstly, it uses a designed, synthetic granulate instead of a natural one like sand. Secondly, the poured granulate does not just serve as a mould, it acts as the actual architectural structure itself, mainly due to the fact that the non-convex particles interlock and allow for forming self-supporting vault-structures or walls. Given that the material really is poured, a high degree of self-formation is involved in the process, which is explicitly being looked for in the design process. The resultant formations are thus predictable only on the level of probability, not certainty. Consequently responsive motion-planning forms an integral part of the process, as it allows the robotic pouring path to interact directly with the emerging formation rather than following a pre-defined pattern only.

Interactive, online robot control denotes the responsive steering of the robot through data collected at runtime through a continuous sensing process. This approach is suggested as an alternative to traditional robot programming techniques that consist of either step-by-step online teaching by a human operator or offline programming, e.g. as part of a CAD-CAM workflow (Brell-Çokcan and Braumann, 2011). In the first approach, human visual control allows for deliberate and interactive robot programming, in what is termed the lead-through method (Pan 2012), without the need for further simulation as the robot is programmed by actively executing the program. A downside of this approach is that it quickly becomes very labor-intensive and therefore is only applicable for simple automation tasks; additional disadvantages are the limited flexibility and room for customization once the program has been taught (Pan 2012). The second approach of offline programming allows for very intricate and parametrically adjustable motion paths. The disadvantage being that this approach requires extensive simulation of the complex robot kinematics for collision avoidance, singularities, outreach-positions and so forth. Interactive online programming, in turn, offers the opportunity to combine the inherent advantages of both aforementioned approaches by implementing an optical sensor-based feedback mechanism akin to the human eye, commonly termed visual serving (Xie et al., 2008), in combination with custom generative motion path algorithms that allow for intricate, yet highly adaptive robot control at runtime.

Utilizing a commercially available 3D scanner, different strategies have been evaluated within the course of this research project with respect to their potential for establishing the following feedback loop: optical sensing, collecting and evaluating sensor data, adapting the motion plan based on a task-specific strategy in a parametric design environment, execution on the robot and renewed sensing.

In the implemented approach, the sensor data is streamed through an open-source plug-in to a dedicated computer for data-processing. There the raw data set consisting of an unordered point cloud is filtered and converted into input data for robot-control. The task-specific motion strategy is updated with the current data and the new target points can be calculated. Consequently the Cartesian coordinates of the motion path are translated from the task space to the joint space of a specific robot with respect to its particular kinematics. A data feed to the robot operating system is established and the motion path is executed by the robot. The strategy is similar to the "hit and withdrawal" approach proposed by Solvang (2008) in that the XY-position of the tool center point is given and the current Z-coordinate is evaluated independently - in the case presented here this is determined through the optical 3D-sensor. In both cases, the motion paths are subsequently adjusted to maintain a constant offset from the work-piece. For the purpose of simulation this loop has initially been implemented by replacing the actual robot with a second computer running a custom inverse kinematics plug-in. In this case, the aggregate pouring process is simulated by a rigid body simulation from the gaming environment. Rigid body dynamics as a mathematical model is especially suited to calculate the behavior of hard, non-convex particles (Pöschel and Schwager, 2005; Featherstone 2008). The simulated model then serves as an input for the calculation of the motion paths.

The use of a highly coarse non-convex granular material renders the motion-planning especially challenging, as no smooth surface per se is established that can serve as a datum of height, but instead a median highest point needs to be established while the tool is moving across the aggregate as it is being poured. For that purpose the motion-planning tool developed uses a three-fold strategy to scan the

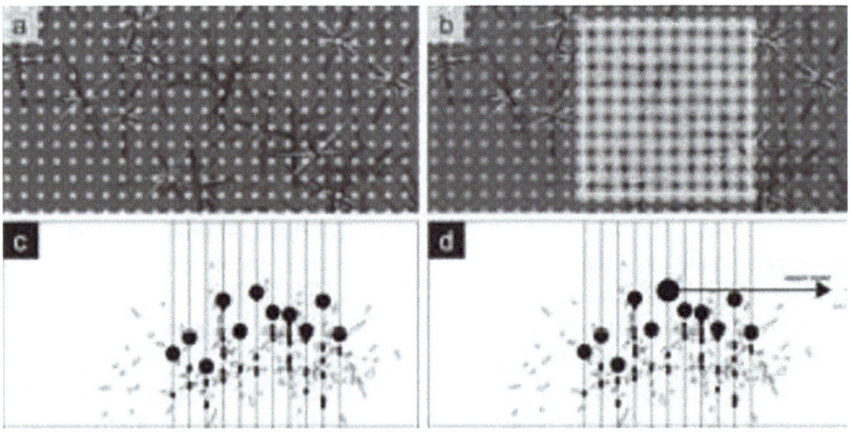

Figure 3 Responsive motion-planning strategy for online robot-control using macro-scale non-convex granulates: a) a point-field is established, b) a selected area is scanned for intersections, c) within each intersection line the highest point of all intersections is established, d) within the entire scan-field the highest point is selected and chosen as the datum for the tool-point. This process is repeated over and over to establish the tool path.

synthetic granulate. Initially a scanning-point grid representing the visual field of the 3D scanner is established (Fig. 3a).

From this point-field, rays are emitted in the directions of the positive Z-axis and the resulting mesh-ray intersections are read consecutively in an area measuring 15 cm x 15 cm (Fig. 3b). The following step is a filtering operation, establishing the highest intersection point within the currently scanned field (Figs. 3c and 3d). This point becomes the datum, i.e. the Z-coordinate for the current target point of the tool-path. This three-fold scanning operation is repeated over and over as the TCP is moved along in the XY-plane and the tool path is established (Fig. 4).

The motion-planning tool developed thus enables a fluid scanning process of the macro-scale particles, where minute adjustments resulting from the poured configuration are quickly incorporated into the tool path.

Contributions and Further Research

Interactive motion control has been established as a crucial aspect for robotic pouring of loose granular structures, both on a design methodological and practical level. Consequently, a loop between a 3D scanner and a six-axis industrial robot has been established using a parametric modeling software. An interactive motion-planning strategy has been developed to digitally simulate a tool-point in relation to an aggregate consisting of synthetic non-convex particles.

Further research at a technical level is being conducted into the development of a flow-controlled emitter-head for the non-convex particles that allows for a constant stream of granules. This effector will be linked to the online motion-control tools presented here. At a design level only distance sensing has been established so far. However, the aim is to establish more intricate interaction patterns of the pouring process with the configuration being poured. For that purpose the RBD simulation will be streamed to the parametric online motion tool of the robot in order to investigate viable pouring and interaction patterns before implementing these real-time.

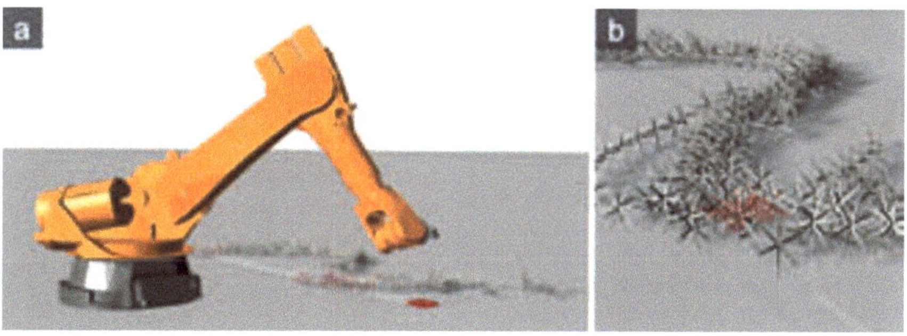

Figure 4 Simulation of the tool-path using the online motion-planning strategy established. As a basis of this simulation a rigid-body dynamics generated model of the designed granulates is being used: a) real-time simulation of the robotic kinematics, b) detail of the robotic motion paths.

Acknowledgements

The authors would like to thank Sean Ahlquist, German W. Aparicio and Steffen Reichert under whose tutelage initial research on the Kinect was carried out at the ICD, University of Stuttgart, as well as Sebastian Bullinger and Richard Gomez, students at the Faculty of Architecture and Urban Planning, University of Stuttgart, who took this research further into possible robotic applications. They would also like to thank Michael Preisack from the RoboLAB, Faculty of Architecture and Urban Planning, University of Stuttgart, for his support with the first robotic pouring tests.

References

Bagnold RA 1954, *The Physics of Blown Sand and Desert Dunes*, 2nd ed. 2005, Dover Publications, Mineola.

Ball P 2004, *The Self-Made Tapestry: Pattern Formation in Nature*, 2nd ed., Oxford University Press, Oxford.

Bonwetsch T, Gramazio F and Kohler M 2007, "Digitally Fabricating Non-Standardised Brick Walls", *ManuBuild Conference Proceedings*, Rotterdam, The Netherlands, pp. 191–196.

Borenstein G 2012, *Making Things See: 3D Vision with Kinect, Processing, Arduino, and MakerBot*, O'Reilly, Sebastopol.

Brell-Çokcan S and Braumann J 2011, "Parametric Robot Control", *Proceedings of the ACADIA Conference 2011*, Banff/Calgary, Canada, pp. 242–25.

Cambou, B (ed.) 1998, *Behaviour of Granular Materials*, CISM Courses and Lectures 385, Springer-Verlag, Wien/New York.

Dierichs K 2010, *Verfahren zum Herstellen eines Formkörpers*, German Patent No. 102008049408, Filed September 29, 2008, Issued April 8, 2010.

Dierichs K and Menges A 2010, "Natural Aggregation Processes as Models for Architectural Material Systems", *Proceedings of the Design and Nature Conference 2010*, Pisa, Italy, pp. 17–27.

Dierichs K and Menges A 2012, "Aggregate Structures: Material and Machine Computation of Designed Granular Substances", *Architectural Design, Special Issue: Material Computation - Higher Integration in Morphogenetic Design (guest-edited by A. Menges)*, vol. 82, issue 2, pp. 74–81.

Duran J 2000, *Sands, Powders, and Grains: An Introduction to the Physics of Granular Materials*, Springer-Verlag, New York/Berlin/Heidelberg.

Featherstone R 2008, *Rigid Body Dynamics Algorithms*, Springer Science + Business Media, New York.

Gaß S and Otto F (eds.) 1990, *Experimente / Experiments, Form – Kraft – Masse 5 / Form – Force – Mass 5*, Mitteilungen des Instituts für leichte Flächentragwerke (IL) Universität Stuttgart 25 / Information of the Institute for Lightweight Structures (IL) University of Stuttgart 25, Karl Krämer Verlag, Stuttgart.

Hausladen G, de Saldanha M and Liedl, P 2006, *ClimaSkin: Konzepte für Gebäudehüllen, die mit weniger Energie mehr leisten*, Callway, München.

Hensel M and Menges A (eds.) 2006a, "Hani Fallaha - Adaptive Pneumatic Shelters 2003–2004" [Project Description] in Hensel M and Menges A (eds.), *Morpho-Ecologies*, AA Publications, London, pp. 232–241.

Hensel M and Menges A (eds.) 2006b, "Eiichi Matsuda - Aggregates 01 2003-2004" [Project Descritpion] in Hensel M and Menges A(eds.), *Morpho-Ecologies*, AA Publications, London, pp. 262–271.

Hensel M and Menges A (eds.) 2006c, "Anne Hawkins and Catie Newell - Aggregates 02 2004" [project description] in Hensel M and Menges A (eds.), *Morpho-Ecologies*, AA Publications, London, pp. 274–283.

Hensel M and Menges A (eds.) 2006d, "Gen Takahashi - Aggregates 03 2005-2006" [project description] in Hensel M and Menges A (eds.), *Morpho-Ecologies*, AA Publications, London, pp. 286–295.

Hensel M and Menges A 2008a, "Aggregates", *Architectural Design, Special issue: Versatility and Vicissitude, Performance in Morpho-Ecological Design* (guest-edited by M. Hensel and A. Menges), vol.78, issue 2, pp. 80–87.

Hensel M and Menges A 2008b, "Materialsysteme 05: Aggregate", *Arch+, Form Follows Performance: Zur Wechselwirkung von Material, Struktur, Umwelt*, no. 188, pp. 76–85.

Hensel M, Menges A and Weinstock M 2010, *Emergent Technologies and Design: Towards a Biological Paradigm for Architecture*, Routledge, Abingdon.

Houben H and Guillaud H 1994, *Earth Construction: A Comprehensive Guide*, Intermediate Technology Publications, London.

Kohler M, Gramazio F and Willmann J 2012, "Die Operationalität von Daten und Material im digitalen Zeitalter", *Positionen zur Zukunft des Bauens - Methoden, Ziele, Ausblicke*, detail - Institut für internationale Architektur-Dokumentation GmbH & Co, München, pp. 6–19.

Nicot F 2004, "Constitutive Modelling of Snow as a Cohesive-granular Material", *Granular Matter*, vol. 6, issue 1, pp. 47–60.

Oxman N 2010, *Material-Based Design Computation*, Dissertation for the degree of Doctor of Philosophy in Architecture: Design and Computation, Massachusetts Institute of Technology, Cambridge Massachusetts.

Pan, Z, Polden, J, Larkin, N, Van Duin, S, and Norrish, J 2012, "Recent Progress on Programming Methods for Industrial Robots", *Robotics and Computer-Integrated Manufacturing*, 28(2), Elsevier, pp. 87–94.

Pöschel T and Schwager T 2005, *Computational Granular Dynamics, Models and Algorithms*, Springer-Verlag, Berlin/Heidelberg.

Rognon, PG, Chevoir, F and Coussot, P 2008, "Rheology of Dense Snow Flows: Inferences from Steady State Chute-flow Experiments", *Journal of Rheology*, vol. 52, issue 3, pp. 729–748.

Schaaf W 2005, *Automatisiertes Modellieren großflächiger Sandgußformen*, Jost Jetter Verlag, Heimsheim.

Schöpfer M, Schmidt F, Pardowitz M, and Ritter H 2010, "Open Source Real-Time Control Software for the KUKA Light Weight Robot", *Proceedings of the 8th World Congress on Intelligent Control and Automation*, Jinan, China, pp. 444–449.

Soar R and Andreen D 2012, "The Role of Additive Manufacturing and Physiomimetic Computational Design for Digital Construction", *Architectural Design, Special Issue: Material Computation - Higher Integration in Morphogenetic Design (guest-edited by A. Menges)*, vol. 82, issue 2, pp. 126–135.

Solvang B, Sziebig G and Korondi P 2008, "Robot Programming in Machining Operations" in Ceccarelli M (ed.), *Robot Manipulators*, InTech, Rijeka/Shanghai, pp. 479–496.

Treib M 1996, *Space Calculated in Seconds: The Philips Pavilion, Le Corbusier, Edgar Varèse*, Princeton University Press, Princeton/New Jersey, Chichester/West Sussex.

Trummer P 2008, "Engineering Ecologies", *Architectural Design, Special issue: Versatility and Vicissitude, Performance in Morpho-Ecological Design (guest-edited by M. Hensel and A. Menges)*, vol.78, issue 2, pp. 96–101.

Tsubaki K 2011, "Tumbling Units - Tectonics of Indeterminate Extension" in Gail, P and Meredith, M (eds.), *Matter: Material Processes in Architectural Production*, Routledge Press, Taylor & Francis Books, London, pp. 187–203.

Xie SQ, Haemmerle E, Cheng Y and Gamage P 2008, "Vision-Guided Robot Control for 3D Object Recognition and Manipulation" in Ceccarelli, M (ed.), *Robot Manipulators*, InTech, Rijeka/Shanghai, pp. 521–546.

[1] European Clearing House for Open Robotics Development,http://www.echord.info/wikis/website/home, accessed July 15, 2012.

[2] Gramazio & Kohler, Architektur und Digitale Fabrikation, ETH Zürich 2011a, Prozedurale Landschaften 1, ETH Zürich, 2011, http://www.dfab.arch.ethz.ch/web/d/lehre/208.html, accessed April 2, 2012.

[3] Gramazio & Kohler, Architektur und Digitale Fabrikation, ETH Zürich 2011b, Prozedurale Landschaften 2, ETH Zurich, 2011, http://www.dfab.arch.ethz.ch/web/d/lehre/211.html, accessed April 2, 2012.

[4] IAAC, Institute for Advanced Architecture of Catalonia 2011, RS 3/DS3: Digital Tectonics, Team 1, 4 and 6, http://www.iaacblog.com/digitaltectonics/category/teams/, accessed July 8, 2012.

[5] IAAC, Institute for Advanced Architecture of Catalonia 2012, RS 3: Digital Tectonics, Sand Bot, http://www.iaac.net/students-research/digital-tectonics-8/rs3-sand-bot-116, accessed July 8, 2012.

ALEXANDRE DUBOR, GABRIEL-BELLO DIAZ

Magnetic Architecture

Generative design through sensoric robots

ABSTRACT Digital fabrication is emerging quickly in the architecture industry and is beginning to reinforce the building process. This emergence brings with it a focus on natural, recyclable and local materials overlapping with global access to data and digital tools. We are now able to explore the possibilities of merging these technologies and developing a new system of design. What can examining the collaboration of these techniques tell us about the future of architecture? As this field is gaining momentum in exploring the process of production, we were led to focus on a system in additive manufacturing. The research of magnetic architecture explores the use of different sensors and various digital tools to understand the possible trajectories that digital fabrication can provide for a future design process.

KEYWORDS: digital fabrication, sensors, generative design, additive manufacturing, robotic coding

Context

There now exists the option not only to build architecture but also to manufacture it. The techniques of digital production assist the architect in the construction process, because the information generated in the design process is used to manufacture the various parts of a building. We are even seeing machines that define this genre in practice and in academia. Digital technology has thus gone beyond the representation stage to take its place precisely in the production phase of architecture.

Neri Oxman (2007) describes the "factory to file" protocols: "Machine execution should not merely be regarded as a service tool for materializing design but rather as an opportunity to inform the design process as one which integrates machine-logic across all scales of production. Material choice and fabrication methods are not innocent decisions, but are rather pre-determined factors which guide the design both with respect to artifact and process from start to end."

Challenging the traditional norms of linear file-to-factory production processes, we studied the potential of linking the collection of material data with mechanic control. Implementing this information into a generative design gave way to new opportunities in approaching digital fabrication within architecture.

Comprehension of Magnetic Properties : Focus on Control, Limits and Simulation

The research on magnetic architecture aims at developing a new building process that focuses on an iron-based material controlled in a magnetic field. One goal of this research is to develop a freeform additive process able to face the challenge of 1:1 rapid prototyping. In contrast to existing 3D printing technology with great precision (µm) but slow feed rate (mm^3/h), the building scale should achieve a much greater speed (m^3/h) regardless of the precision (cm). With the control over iron in a magnetic field we saw potential to increase the flexibility of the additive process. Fig. 1 was one of the first experiments in collecting data from this formation to design a nozzle for a machine with multiple axis movement in the system we envisioned.

The experiment focused on setting up this process for the building scale using recycled and granular material. The iron filings used throughout our experiments were collected from industrial waste. Adding iron filings permitted us to increase the structural capacity of a given material and allowed us to manipulate its behavior through magnetic forces. Digitally controlling both material deposition and magnetic forces leads to new design opportunities. This technology will allow us to build freeform architecture without using scaffolding or support material.

The main complexity arising from the combination of the chosen material and the digital fabrication technology was the difficulty in virtually simulating

Figure 1 Iron-based material controlled by magnets

a process like iron inside a magnetic field. Even if we succeed in barely simulating it (Fig. 2), the final shape is never precisely duplicated.

Material Investigation : Exploration of Mixtures and their Behavior

Moving forward, different types of mixtures were tested to help solidify the iron in its position while in the magnetic field. The idea was to create one form and have the magnets move to create the next one. This movement would be controlled by one, if not both, of the axes. At every scale each material would pass through *the round* or fail due to several performance factors (Fig. 3). The material's behavior, as the scale increased, provided information on how to address the path for the additive manufacturing. Could we start construction at any given point, in any given position?

Increasing the size led to the use of electromagnets and eventually it became clear how versatile the mixture could potentially be. From clay to concrete to sand, different materials were able to go to the maximum distance between the magnets. However, the time for the material to solidify was an important factor. Though each material could meet the performance requirements, each had its own time for solidification. A mixture of liquid plastic that solidified in approximately two minutes and thirty seconds became the final solution for this stage in the research.

Design of Custom End-effector : Digital Control over the System

To develop the envisioned control system, we needed an end effector capable of manipulating the distance and strength of the magnetic field. Detachable shields to create different connection configurations were also needed for design variations. The evolution of the end-effector became possible only through an iterative loop of tool design and production needs. This involved responding to the behavior of the material in the magnetic field and understanding the material shape after solidification so as to then incorporate this information in the tool design. The initial reason for using a magnetic field to control material was to explore the flexibility in the additive process. The final design (Fig. 4) was the logical solution at this point to begin the exploration of the additive manufacturing opportunities. The manipulatable magnetic field could now be moved into the desired orientations with the assistance of multiple

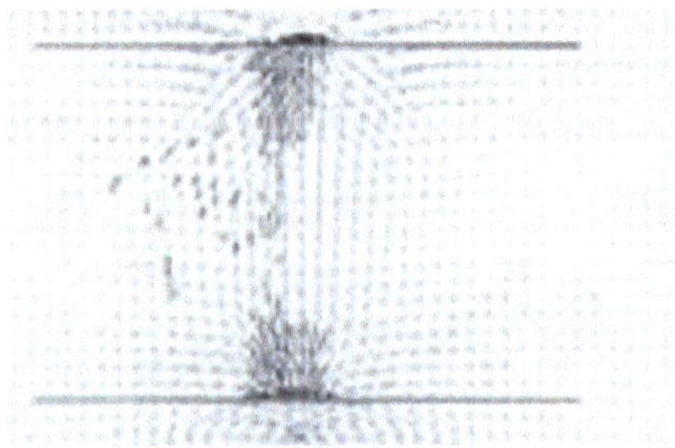

Figure 2 First attempt to simulate the material behavior in the magnetic field

Figure 3 First material samples made with magnets

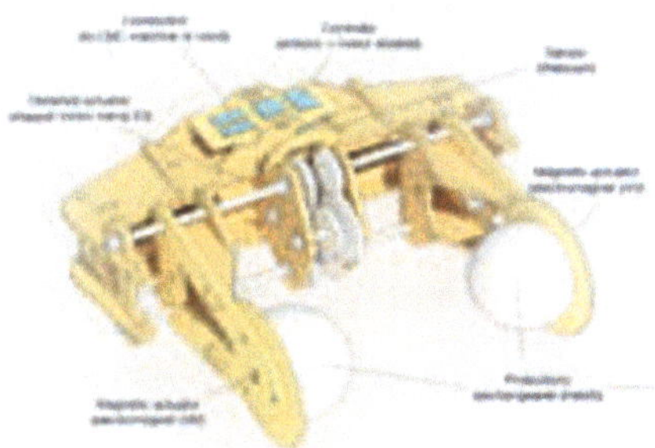

Figure 4 Final nozzle design with various detachable shields

axis movement. A 3-axis computer numerically controlled (CNC) was first used to test the 2D connections but produced limitations in the development of 3D networks. Using a 6-axis industrial robot provided a wider range of positions, leading to previously unexplored territories (Fig. 5).

Top-down Process : Generative Design Incorporated in Robotic Production

To explore the robot capacities the idea of building a free-form shape arose. Looking at the opportunities given by the digital world, we chose to generate a 3D network using a parametric model that considered the limitations of the tool. The software allowed us to optimize the result through genetic algorithms. To evaluate the suitability of each result, two types of analysis were processed: environmental (sun exposition) and structural (load repartition). To simplify the script each column element was represented by a line. The output of such generative design created an optimized 3D network (Fig. 6).

At this point, Jerome Frumar's comment seemed relevant: "*Iterative reduc-*

Figure 5 Final end-effector mounted on the industrial robot

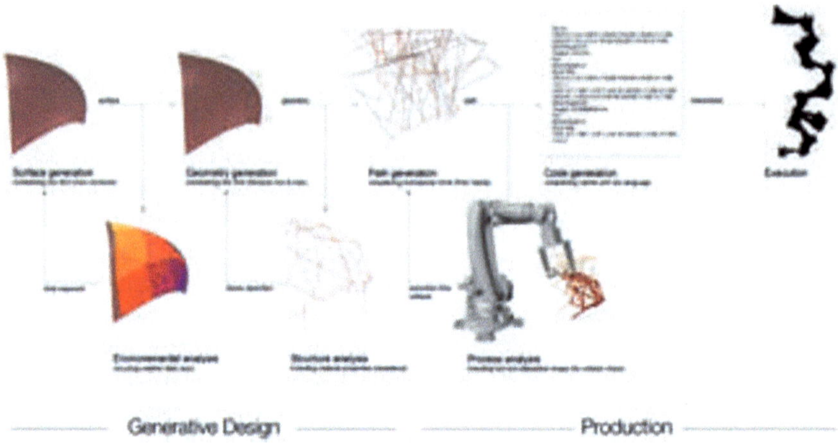

Figure 6 Generative process of a desired 3D network (software used : Grasshopper, Galapagos, Geco, Karamba and KUKA|prc)

tion of irrelevant material using evolutionary-based computational processes returns architecture to its modernist musings. By distilling objects according to functionalist principles are we not effectively practicing modernist theory in an evolutionary fashion?"

The schematic representation of the 3D network (lines) was then *read* to create the proper instructions for the robot and the custom tool. Firstly, each line was sorted to start the construction from the two foundations (the vertical lines on the bottom of Fig. 6 – structure analysis) and continued incrementally, assuring connection between the previous and next column. Then, for each particular element, a routine was applied:

1. Open the claw
2. Move fast to the secure position (tool in alignment with the column to print, but with a safety distance).
3. Move slowly to the first position in a linear movement (less likelihood of collision).
4. Close the claw to the desired element size
5. Initiate the printing process (turn on magnetic field and deposit material)
6. Wait for drying time
7. Open the claw
8. Move slowly back to secure position
9. Move fast to next column secure position.

Finally, a similar generative algorithm is used to optimize the path generation for the industrial robot. The fitness evaluated here is the path length, for time optimization, and the presence of collision, for feasibility. The input variables are the security distance and the angle of approach of the robot.

An interesting process was to merge these two logics into one, including the production logic from the initial script, leading to a generative design totally aware of the technology involved in the final construction. Apart from some mechanical problems and many iteration designs, we were not faced with huge problems in automatizing the process. Yet, when we started building the desired shape with material, we realized how difficult it was to predict its behavior.

To connect the new formation with the previous one, the precision of the digital model and the robot execution were not matching the imprecision of the material behavior. We ended up having to manually adapt the robot's position to absorb this differences.

Bottom-up Process: Use of Sensors to Enhance Robotic Intelligence

In observing the top-down process, it became clear that there was more data to be extracted from the entire system. Sensors became a way to approach the collection of this information. Through artificial vision we were able to start dissecting our formations and to add logic to the next positioning (Fig. 7). There were various approaches behind developing the logic for this sensor. This data driven design could have an outcome of several logics, depending on the performance needs. In this phase of research, structure and connections became a focus in analyzing the outcome.

Figure 7 Live screen shot of incremental coding using artificial vision.

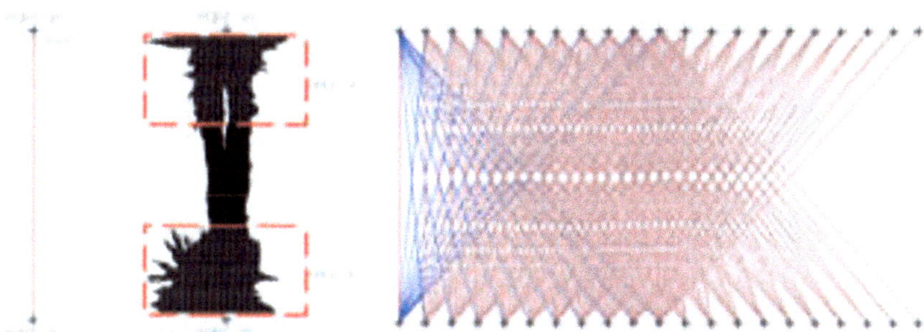

Figure 8 Generative design rules

It was understood that the previous top-down process did not take into account that the material was self-computing, unpredictable and self-aligning within its magnetic setup. Therefore we designed a law in which the camera would dissect the silhouette of the formation and through a calculation of points decide the maximum distance and angle for the next formation to be attached. Fig. 8 shows the division of the silhouette and an overlap of the possible combinations of two steps after a formation.

Figure 9 One of the final models.

Applying the position of Manuel DeLanda (2002): "We may now be in a position to think about the origin of form and structure, not as something imposed from the outside on an inert matter, not as a hierarchical command from above as in an assembly line, but as something that may come from within the materials, a form that we tease out of those materials as we allow them to have their say in the structures we create."

To create a system with live feedback of material behavior, we began to explore the possibilities of what sensors could give to an architectural design through a generative process (Fig. 9). The use of artificial vision is only a small step in incorporating sensorial logic into robotic control. Is it possible with sensor technology to bring architectural consideration down to such micro scale where nature begins to control design again?

Conclusion

Using sensors, this experiment has shown how an architectural scale structure can be built by incremental design. However, from a structural point of view reaching an optimized building through a bottom-up process is questionable. We believe in the possibility of mixing both a global and a local approach in order to arrive at a fully optimized process.

Nevertheless, we also believe that in-situ generative design can reach an unexpected optimum using other sensors such as light-, humidity- or temperature sensors. This field of "in-situ generative design" has not yet been widely researched, therefore further and deeper research should lead to new design opportunities. According to Jasper Morrison: "Design is not done with rules, but with intuition. Intuition never lies." So, can we provide rules for our own intuition? Can we code intuition into an architectural design approach?

Magnetic architecture is a research platform employed to explore the use of generative designs at different scales of analysis, before and during production. This platform endeavors to incorporate new technologies to enhance the capabilities of additive manufacturing in a process in which the two scales, local and global, can coexist throughout the design. Through the investigation of a digitally controlled additive process, merging these technologies offered many possibilities for future research that are explained more thoroughly on our website, magneticarchitecture.org.

With architecture manufacturing through digital fabrication, the industry needs more research in the direction of implementing technologies outside of its classic borders. This in turn brings the design of the architect back into the factory. The definition of the architect is currently being reformatted and the emergence of digital fabrication is playing a role in that reformatting process.

Acknowledgements

Envisioned by: Alexandre Dubor, Gabriel-Bello Diaz, Akhil Kapadia and Angel Lara, at the Institute for Advanced Architecture of Cataluña.
Guided by: Marta Malé-Alemany as main tutor.
Supported by: Santiago Martìn Laguna, Jordi Portell Torres, Miquel Lloveras Corvalan, Guillem Camprodon Pujol as faculty team.
Inspired by: Tokujin Yoshioka , Jolan Van der Wiel, Physalia & Gerardo del Hierro, and the awesome power of magnets.

References

De Landa M 2002, *Material Complexity*, Presentation at Digital Tectonics, University of Bath, UK.

De Landa M 2007, "Opportunities and Risks", *Domus*, no. 901, pp. 192-193.

Frumar J 2007, "Code To Craft: Beyond the Voxel", *Proceedings of the 4th International Conference of the. Association of Architecture Schools of Australasia (AASA)*, University of Technology, Sydney, Australia.

Morrison J 2006, *Everything but the Wall,* Lars Müller, Baden.

Oxman N 2007, "Digital Craft: Fabrication Based Design in the Age of Digital Production" , in *Workshop Proceedings for Ubicomp 2007*: International Conference on Ubiquitous Computing, Innsbruck, Austria.

Nathan King, Jonathan Grinham

Automating Eclipsis

Automated robotic fabrication of custom optimized metal façade systems

Abstract Stemming from ongoing research into modular automation strategies and in conjunction with the development of Virginia Tech's Lumenhaus this research explores the potential for the robotic fabrication of a novel, complex, high-performance metal shading system. The Eclipsis facade system utilizes innovative circular geometry in the form of laser-cut holes to produce a series of folded tabs at calculated degrees to create a specialized yet customizable high-performance facade system. The system's algorithmic logic is designed for customization taking into account varying degrees of privacy, spatial condition and local environmental performance criteria including daylighting and ventilation. The logic of each semi-perforation is simple and controlled by parameters defined only by dimension, location, rotation, and the angle of the folded tab. Complexity emerges in the part-to-whole relationship where the aggregation of instantiated parametric geometry provides regulated infinite-variation of tab size and rotation thus eliminating the possibility of any standardized fabrication method or industrial process. This paper presents a design-driven approach that merges Computer Aided Design and Robotic Manufacturing through the implementation of an "Integrated Environmental Design-to-Robotic Fabrication Workflow" and explores a prototypical comparison between manual and automated-robotic fabrication strategies for commercialization (Bechthold 2010).

Keywords robotic folding, automated workflow, metal, façade, optimization

Introduction

This paper presents a pragmatic design research approach that was undertaken during the design and fabrication of the Eclipsis facade system. Eclipsis was developed as a responsive architectural component for Virginia Tech's Lumenhaus, the winner of the 2010 Solar Decathlon Competition in Madrid, Spain and is an assembly of individually operable layers including enclosure, insulation, and shading screens (Fig. 1). The system facilitates the realization of an architectural pavilion that also serves as a dwelling, while acting as the primary instrument in the address of many pragmatic considerations and performance criteria (Virginia Tech, 2009). The facade described in this paper is the only differentiated component within the system and must resolve primary performance concerns with the demanding architectural intentions of the project as a whole. The Eclipsys Facade System is envisioned as a facilitator of responsive architecture through its inclusion in both highly customized-dynamic and static facade systems.

To address the manufacturing of the proposed complex system, several prototypes were created using a series of fabrication process ranging from industrial metal-working techniques such as punching, braking, and die-cutting to integrated robotic manufacturing. After industry consultation and prototypical testing surrounding the folding of the shading tabs two feasible fabrication strategies emerged: manual folding and automated industrial robotic folding. Manual folding is feasible on small production runs but presents significant limitations to the flexibility and economic viability of the system in the context of larger scale construction. Successful integration of industrial robotic tab rotation facilitates maximum variability within the system but presents the added complication of the programming of thousands of unique robotic movements that would be difficult be overcome in traditional programming workflows. The ongoing research discussed in this paper explores the integration of a previously published automated Rhinoceros-based robotic fabrication workflow with the development of high performance tab-based metal shading systems (Bechthold, 2010).

Eclipsis System: Shutter Screen

The primary function of the outermost layer

Figure 1 Building integrated façade system; Virginia Tech Lumenhaus

of the façade system is the control of sunlight while maintaining view and privacy. RhinoScript–a Visual Basic (Vb) coding environment designed for, and embedded in, the Rhino 3D modeling software–allowed for the development of variable criteria to produce a pattern that could be cut by industrial numerically controlled processes. A simple, repetitive, circular pattern was chosen that could be parameterized to meet the multiple performance criteria. An innovative aspect of this approach is the circular geometry of laser-cut holes with tabs folded at calculated degrees (Fig. 2). The folded tabs have four variables: the diameter of the circular cut, the planar orientation of the tab to the surface, the degree of tab rotation away from the surface, and the thickness of the tab material. These variables are articulated to block and reflect sunlight, and to create controlled views and spatial conditions (Grinham 2011).

As a research platform, the system's algorithmic logic was designed for large scale customization. Shading, refracted light, privacy and ventilation were woven through a complex definition designed within Grasshopper. Additional design criteria introduced include security, privacy, view and the development of a particular quality of light. Grasshopper, along with Visual Basic programming language, allowed for the development of subtle complexity with extensive design versioning. The logic of each tab is simple; the complexity emerges from the swarm logic of the part-to-whole relationship (Grinham 2011).

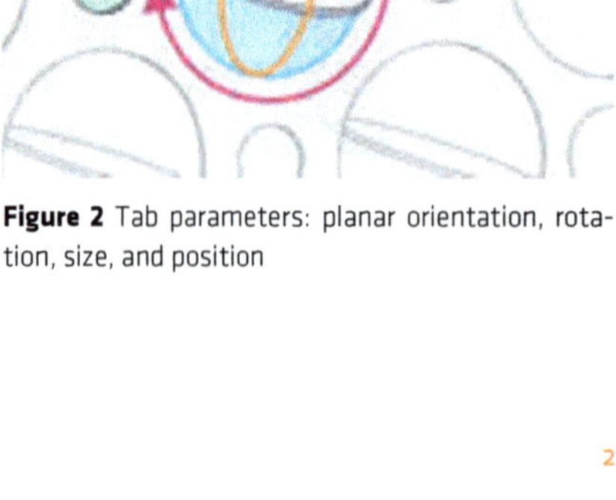

Figure 2 Tab parameters: planar orientation, rotation, size, and position

Parameterizing Programmatic Requirements

In the open pavilion typology of the building, program is used to define the relationship of open and closed. These parameters were diagramed through the narrative of a typical morning routine. First, the occupant begins their day in the privacy of the bedroom; upon rising the tabs are opened to allow views and light. The user then proceeds through the core of the house and finally sits at the table for breakfast, where once again there is a clear view outside (Fig. 3). The effectiveness of the screen to shade while providing dynamic light is an integral aspect of the project. The computational design tool allows for refracted light to shift with the sun while also producing a smooth transition between lighting conditions, and allows for the customization of the proportional relationship of the scale of opening and perpendicular rotation (Grinham 2011).

A diagonal grid (diagrid) was adopted that allowed for a tight 'packing' of circles that increased views and light transmittance and for the weaving of a secondary condition: a pure, non-tabbed, cut hole. Through multiple material prototypes, it was determined that a variable tab thickness was required to sustain tab rotation beyond sixty degrees. This adjustment was allowed within the Grasshopper definition by reducing the arc length for circles with a rotation angle greater than sixty degrees (Grinham 2011).

Structural Integration

In addition to integrated program-based tab positioning discussed above the definition also incorporates the dimensioning of structural framing in relation to panel size. The macro façade surface used to define the location and position of individual perforations and tabs must be divided into modules relative to stock material, machine size, and locations of building supports. A return is allowed on the perimeter of each perforated panel that enables enhanced structural rigidity and hidden attachment points for an integrated structural frame. Each module is laser etched based on location within the façade to enable expedited site installation and organization.

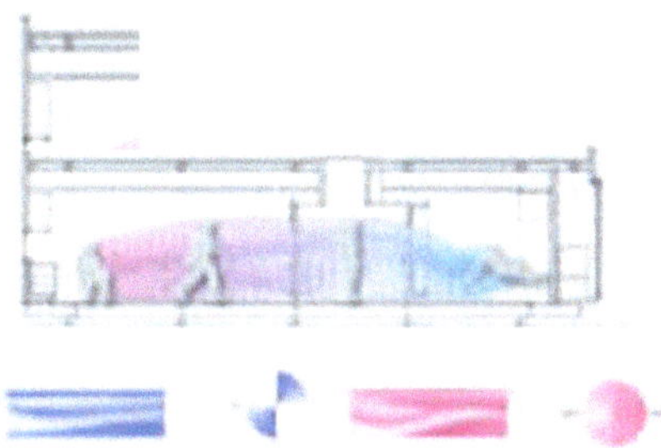

Figure 3 Program-based surface parameterization: Top: morphed privacy and view diagrams; Bottom right: privacy driven topographic diagram used to drive system porosity; Bottom left: topographic diagram used to map solar orientation and drive planar tab rotation.

Prototyped Manufacturing Strategies

Three computer numerically controlled (CNC) methods of steel fabrication were explored. The first method, die cutting, was evaluated based on the fabrication consultants, A. Zahner's, in-house operations and extensive application of die cut panel systems. Die cutting was ultimately dismissed due to the limited variations of circle patterning (six diameters) and the inability to punch concentric arches while leaving a material tab, which is needed for the folded system. The second method, water-jet cutting, provided infinite geometric variations; however the water jet produced a large kerf, reducing the stability of the tab and resulting in a heavy "halo" effect in direct sunlight caused by stray abrasive materials. CNC laser cutting was chosen because of the process' ability to make tightly controlled cuts, unlimited geometric patterning, and laser etching capacity, a technique envisioned to inscribe data to simplify onsite assembly and facilitate manual folding. Prototyping, along with industry collaborators, provided a key understanding of feasibility (Grinham 2011). Later full-scale mock-ups, described below, tested both performance and multiple fabrication strategies (Fig. 4).

Manual Folding Strategies

Given the complex nature of the rotation of the circular tabs and the limitations imposed by traditional industrial CNC programming strategies, a manual system of folding was developed that introduced several design parameters representing tradeoffs between performance and feasibility. First, a controlled set of nine rotational increments was established, ranging from ten degrees to ninety degrees. This allowed for simple calibration "wedges" to be produced to calibrate the angle of tab rotation. In order to understand the polar rotation of each disk a "paint by numbers" approach was adopted where each disk would have its corresponding angle laser-etched on the exterior surface.

Etchings provided information relating to the degree of rotation, and the rotation's polarity—the direction the disk rotated perpendicular to the surface (Fig. 5) (Grinham 2011).

Robotic Folding

Due to industry limitations in industrial robotic programming workflows, tab rotation presents the single most challenging aspect of the industrial production of the shutter screen. The automated programming strategy proposed here would be integral to the commercialization of the system. In an attempt to resolve the part complexity with the need for industrialized production a series of tests resulting in several full-scale robotically folded prototypes were created using a 6-axis industrial robotic work cell and an integrated, automated programming workflow (Fig. 6).

Automated Robotic Production Strategies

Computational Design-to-Robotic Manufacturing

Automated robotic workflows have become relatively common and previously published work addresses aspects of automated programming in relation to sheet metal manipulation. For example, "Parametric Punching" by Michael Vasku, produced at

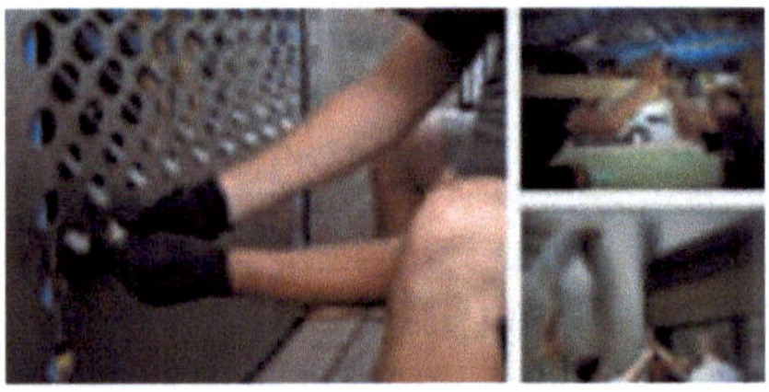

Figure 5 Manual prototyping using calculated "wedges" (Image: A. Ransom)

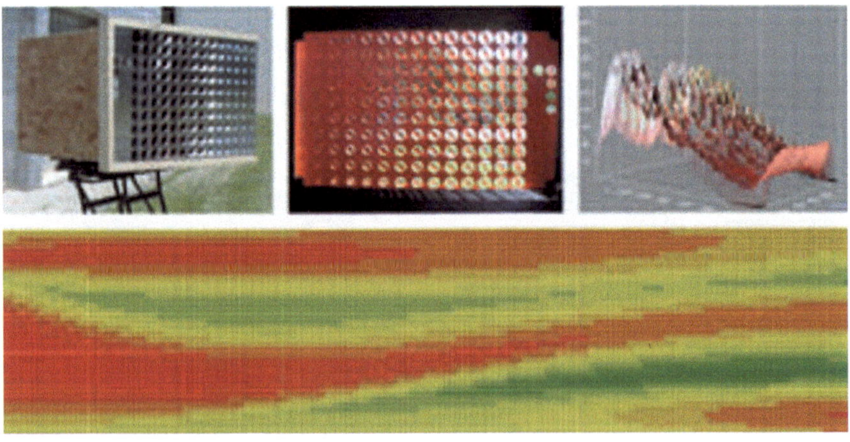

Figure 4 Performance testing of prototyped facade section. Top left: illuminance testing system; Top center: interior view of light intensity; Top right: 3D plot of pixel density histogram (diagram: Robert Schubert); Bottom: projected illuminance: red=high, yellow=moderate, green=low illuminance.

Vienna UT, presents a related, yet different, approach using the KUKA|prc automation tools to drive the manipulation of sheet metal tabs (J. Braumann and S. Brell-Çokcan 2011). Another approach suggested by Lavallee et al. utilizes a Catia-based automation strategy along with two robotic grippers to fold perforated metal sheets into individualized units using rotational movements (Lavallee et al. 2011). To facilitate robotic manufacturing, a similar integrated modular robotic programming component developed by the Harvard Design Robotics Group has been embedded into the Rhinoceros-based computational design platform used to generate the screen's final tab cut and rotation (Bechthold 2010). Using embedded model geometry this component is deployed to streamline the identification of key parameters relating to position and rotation of each perforation that are defined by the midpoint of the diameter of each part and an offset based on material thickness (Fig. 7). This planar origin is assumed to have a z-axis normal to the materials surface with a second axis tangent to the prescribed diameter, or axis of rotation. The planar orientation of each tab in relation to the screens surface is defined at a second surface mid-point on the same diameter where the associated coordinate is rotated such that the z-axis is slightly beyond normal to the final rotated tab position. Over-rotation of the tab is required to account for inherent 'spring-back' in the material and is adjusted based on model-defined material and tab dimension.

Based on the computational design model and the specific previously defined plane, Rapid Code, the c-based language used to control the tested ABB-4400 industrial robotic work cell, is generated by the integrated component. The custom modular programming interface consists of three primary modular grasshopper components. First, the project specific component isolates the geometric constraints as defined by specific model data that is used by the second component to generate Rapid Code. The third component of the modular system, an inverse kinematic solver, enables direct machine simulation within the Rhinoceros environment eliminating the need for specialized proprietary software (Fig. 7) (King, 2012).

Custom Robotic Tooling

Based on the dia-grid mentioned above a universal fixture was created to continuous-

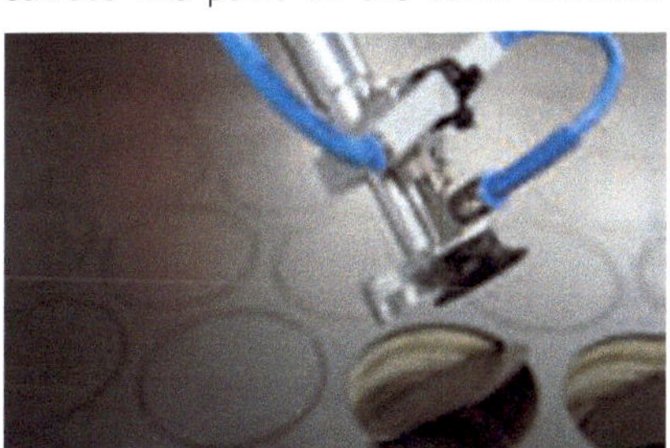

Figure 6 Initial robotic folding using custom end effector (Image A. Lee)

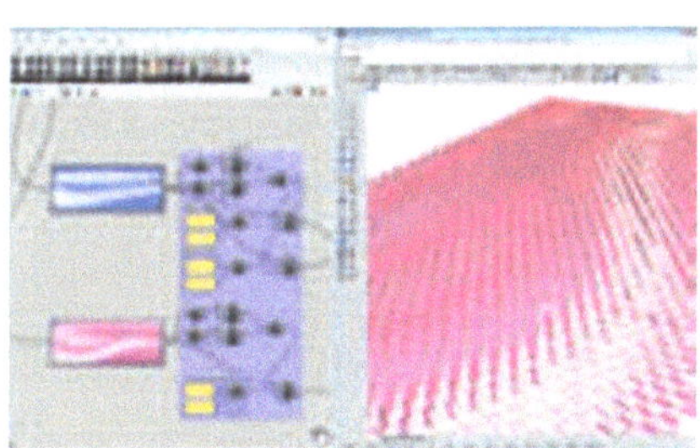

Figure 7 Grasshopper visualization of facade variation with embedded machine programming.

ly support the work object during folding. The work surface has a depth greater than the maximum tab depth and is perforated on the dia-grid with a hole diameter slightly larger than the largest tab dimension. This strategy allows the fixture to support all variations of panel based on the instantiation logic discussed previously (Fig. 8).

To facilitate the robotic actuation of tab rotation a custom end effector was designed that incorporates a pushing implement and pneumatic suction gripper assembly that enables folding based on a push-pull strategy that reduces surface deformation or 'oil canning' during tab rotation. The end effector applies an opposing bending moment on either side of the tab perpendicular to the axis of rotation The planes discussed above drive the tool-center-point (TCP) and the folding action consists of four movements. First, the tool is located directly above the surface plane and oriented according to the axis of rotation, second, the tool engages the flat surface, and third the tool rotates to a prescribed angle while maintaining a static TCP position. The fourth movement is defined by an offset plane at the same rotation as the final tab position that serves as a retract movement while maintaining to tool orientation. This positioning enables the tool to clear the work surface before returning to a normal orientation to avoid interference with final tab position.

Conclusions

The ongoing research presented in this paper is initiated through a design experiment that begins to reconnect the design idea to the associated material systems. The work represents the successful merger of a computationally driven architectural design process that links performance, both design and environmental, through an integrated design development workflow while simultaneously accounting for material and fabrication parameters. The many prototypes

Figure 8 Robot simulation using integrate inverse kinematic solver

Figure 9 Robotically actuated prototype with tab rotation from 5-80 degrees (Image A. Lee)

developed including a full-scale building integrated façade system exemplifies the architectural possibilities introduced by proposed system. Through industry collaboration with advanced fabrication consultants we have recognized a limitation in the existing industry›s ability to produce substantially differentiated architectural building components that can be addressed by the incorporation of the industrial robotic work cell. By addressing manufacturing concerns at the design level and fully developing and testing an integrated robotic fabrication scenario this research moves beyond the one-off digitally driven sculptural intervention and into the development of a truly customizable building system that can be produced industrially (Fig. 10).

Acknowledgements

Research supported by Harvard Graduate School of Design, Design Robotics Group (DRG) in collaboration with Virginia Tech School of Architecture + Design, Center for Design Research.

References

Bechthold M et al. 2011, "Integrated Environmental Design and Robotic Fabrication Workflow for Ceramic Shading Systems", *Proceedings of 28th International Symposium on Automation and Robotics in Construction ISARC2011*. Seoul.

Virginia Tech, Lumenhaus|About, http://www.lumenhaus.com/about/index.html, 2009

Grinham J, "Appliance Architecture in the Invisible College: a Pedagogical Text", MA thesis Virginia Tech, Blacksburg, 2011. Web. 1 July 2012.

Braumann J and Brell-Çokcan S, "Parametric Robot Control: Integrated CAD/CAM For Architectural Design", *Proceedings of ACADIA 11: Integration through Computation [Proceedings of the 31st Annual Conference of the Association for Computer Aided Design in Architecture (ACADIA)]* [ISBN 978-1-6136-4595-6] Banff (Alberta) 13-16 October, 2011, pp. 242-251

Lavallee J, Vroman R, and Keshet Y, "Automated Folding of Sheet Metal Components with a Six-axis Industrial Robot", *Proceedings of ACADIA 11: Integration through Computation [Proceedings of the 31st Annual Conference of the Association for Computer Aided Design in Architecture (ACADIA)]* [ISBN 978-1-6136-4595-6] Banff (Alberta) 13-16 October, 2011, pp. 144-151.

King N et al. 2012, "Robotic Tile Placement: Tools, Techniques and Feasibility", *Proceedings of the XXth ISARC*. Eindhoven.

Figure 10 Building integrated facade system on exhibit in Times Square, New York, NY. (Image: Christian Truit)

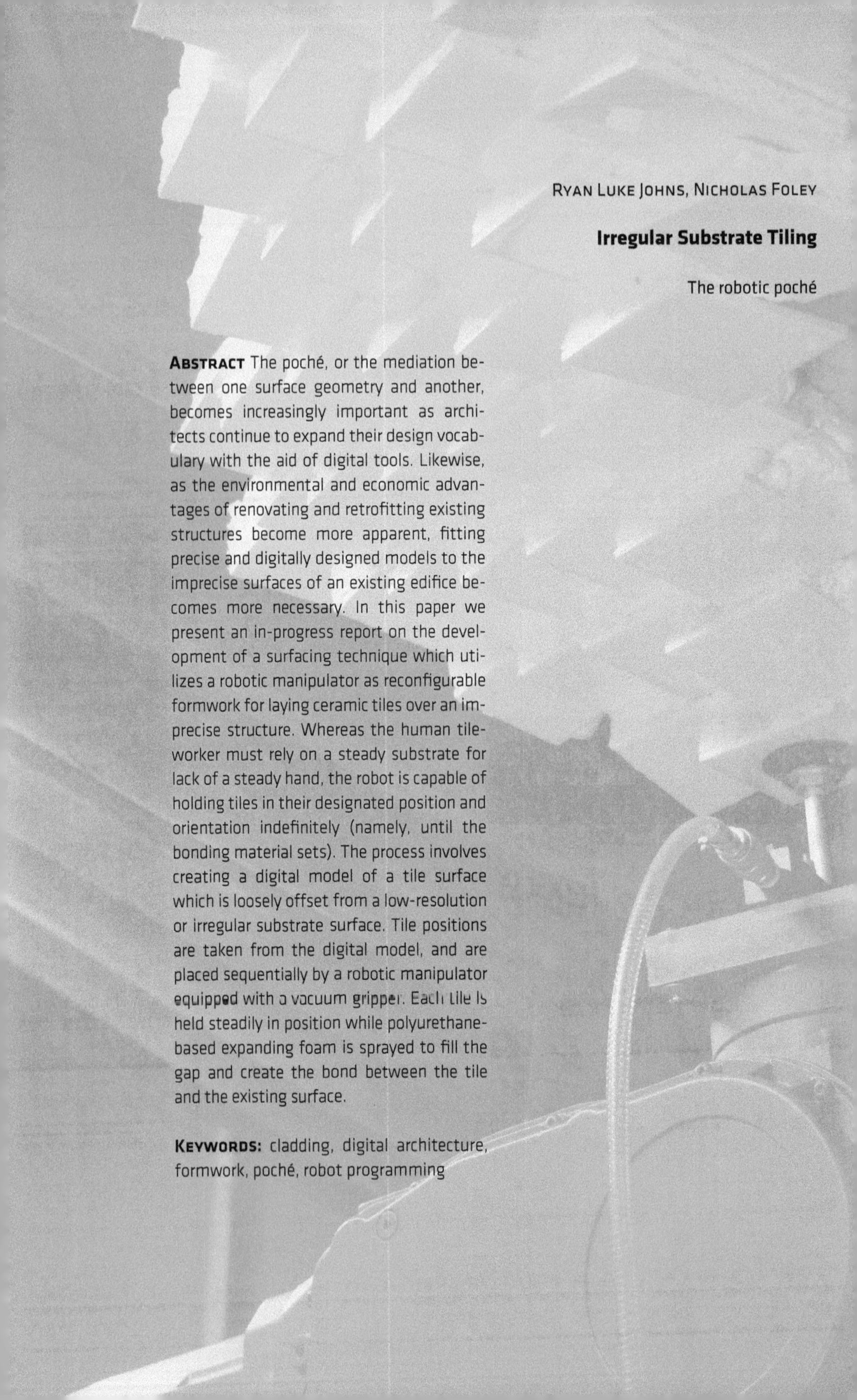

Ryan Luke Johns, Nicholas Foley

Irregular Substrate Tiling

The robotic poché

Abstract The poché, or the mediation between one surface geometry and another, becomes increasingly important as architects continue to expand their design vocabulary with the aid of digital tools. Likewise, as the environmental and economic advantages of renovating and retrofitting existing structures become more apparent, fitting precise and digitally designed models to the imprecise surfaces of an existing edifice becomes more necessary. In this paper we present an in-progress report on the development of a surfacing technique which utilizes a robotic manipulator as reconfigurable formwork for laying ceramic tiles over an imprecise structure. Whereas the human tileworker must rely on a steady substrate for lack of a steady hand, the robot is capable of holding tiles in their designated position and orientation indefinitely (namely, until the bonding material sets). The process involves creating a digital model of a tile surface which is loosely offset from a low-resolution or irregular substrate surface. Tile positions are taken from the digital model, and are placed sequentially by a robotic manipulator equipped with a vacuum gripper. Each tile is held steadily in position while polyurethane-based expanding foam is sprayed to fill the gap and create the bond between the tile and the existing surface.

Keywords: cladding, digital architecture, formwork, poché, robot programming

Introduction

Figure 1 Prototypical robotic poché surface.

The poché, or the mediation between interior and exterior surface geometries, is an inherent concern in architectural design and construction. Differentiation between "shape, position, pattern and size" (Venturi, 1966) of inside and outside surfaces enables specificity in design, but this contradiction also results in unused void space and the challenge of interfacing two dissimilar surfaces. The construction techniques used to address these design challenges can be broadly divided into two methodologies: that of creating surfaces from a structure contained within the poché (Fig. 2a) and that of casting or aggregating surfaces from a temporary external formwork or scaffold (Fig. 2b). With the advent of digital fabrication and demands for "informed" (Kolarevic, 2003) surfaces in architectural design, the former category has yielded the necessity of creating finely tuned and precisely machined framework to support cladding elements within given tolerances: materialising complexity requires a close bond between architects, data consultants, engineers, frame and cladding fabricators, contractors and surveyors (Scheurer, 2010). Likewise, formwork-based construction techniques require similar coordination, but the temporary nature of the scaffold confers the advantage of isolating production constraints: the precision of the final surface is contingent upon lost forms rather than embedded within the structure. Accordingly, the structure of the poché is given the potential of meeting other functional requirements. With these principles in mind, we propose a hybrid of these processes: one which utilizes a robotic manipulator as a reconfigurable formwork for creating precise surface connections to a loosely defined structural framework (Fig. 2c). Panels are robotically placed in an *additive* technique but also serve as a *formative* (Kolarevic, 2003; Bonwetsch et al., 2006) boundary for the cast poché which binds them to the framed substrate.

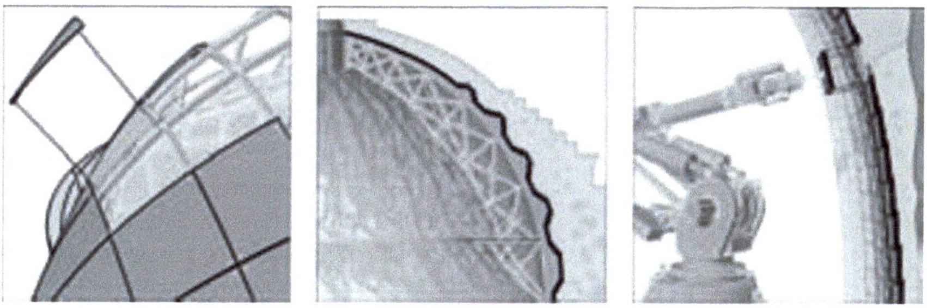
Figure 2 From left: a) Cladding relying on digitally fabricated framework; b) Pantheon section: formwork with graded aggregate fill; c) Robotic poché

Related Work

Previous work in robotic cladding in construction has focused primarily upon increasing safety and efficiency during the construction of tall buildings (Gassel et al., 2006). Robotic strength is used to lift heavy façade panels into position, generally under the guidance of a human operator (Yu et al., 2007) (Fig. 3). Rather than exploring the design potential of robotic fabrication however, prior work has been motivated by a desire to automate the construction processes of conventional designs.

In the vein of the reconfigurable formwork, the recent "Procedural Landscapes" research at the chair of Gramazio and Kohler, ETH Zurich uses both additive and formative processes to create "informed" surfaces for concrete casting by impressing sand with a robotic manipulator [1]. Use of polyurethane foam has precedence in robotic architectural fabrication in the "Foam" project, also at the ETH (Gramazio and Kohler, 2008), however these were procedural explorations which utilized the material logic of expanding foam coupled with precise paths to inform the creation of acoustic diffusers–in our research, the material (while also engaging acoustics to some extent) is used for the more traditional purpose of filling and binding.

Figure 3 The KAJIMA "Mighty Hand" panel-holding construction robot

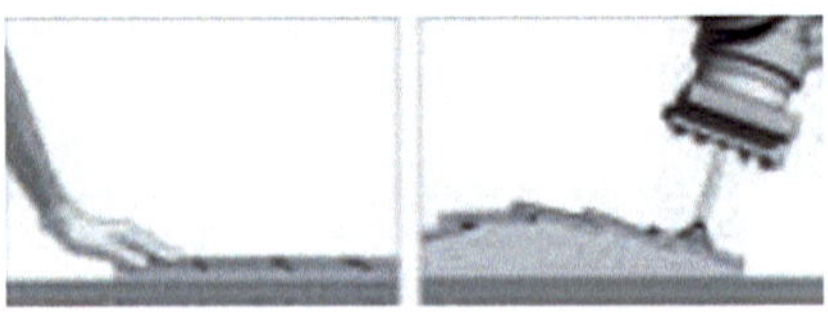

Figure 4 Human versus robotic tiling capability

Irregular Substrate Tiling

Overview

Appropriating traditional means of construction for robotic processes enables not only the possibility for aesthetic complexity (Oesterle, 2009), but also enables a re-evaluation of the potential of processes which have evolved around human capabilities and ineptitudes. For lack of a steady hand, the human tiling process requires a firm substrate to configure and support tiles during the construction process. The patience and stamina of the robotic manipulator, however, allows for an inversion of this process: the robot can hold tiles precisely in position for extended periods without fatigue. In essence, the manipulator becomes the substrate (fig 4).

The precision and variability of the robotic manipulator in this process enables "highly informed" (Bonwetsch et al., 2006) surface geometries. While most current digitally informed cladding techniques rely on a finely tuned structural frame, our process embraces and is enhanced by loose tolerances between surface panels and the structure to which they are attached. As the robot provides the calibration of the surface by holding it in the appropriate position, any ad hoc technique can be used to bridge the poché space and adhere the panel to the opposing frame/surface. Variation of these filling-and-adhering materials has the potential of increasing the utility of the void space. While this simple concept has a multitude of potential manifestations, our in-progress research explores one prototypical iteration.

Prototype Design Evolution

The evolution of our in-progress prototype was guided by the desire to explore techniques of slow robotic fabrication while responding to the practical requirements of our workspace—an imprecise existing ceiling structure with poor insulation properties. In order to decrease heat loss during winter research while reducing the noise transmission associated with our work, we focused on materials which had the potential of providing both thermal insulation and acoustic dampening. While expanding polyurethane foam met the thermal requirements, we required mass in order to reduce sound transmittance: ceramic tiles provided a cheap and easily accessible option. By utilizing expanding foam as both the means for filling the void space and adhering the tiles, we create a soft connection between the mass of our inner surface and the structure of the outer membrane—theoretically increasing the effectiveness of acoustic dampening by decoupling the surfaces [2].

As the interior surface is both robotically assembled and will remain within the context of the research facility, we elected to inform the geometry of our simple test-case by the bounding envelope of the robot's movement. The tile layout was generated in Rhino Python such that tile angles were individually oriented by the surface normal at the closest point of the robot's reach envelope and aligned to point towards its origin (Fig. 5). Tiles were placed recursively such that the distance between them was mapped based on the angle of their orientation, enabling a shingled appearance. The surface therefore serves both as an insulator for the space and a visual cue to the capabilities of the robot's movement within it.

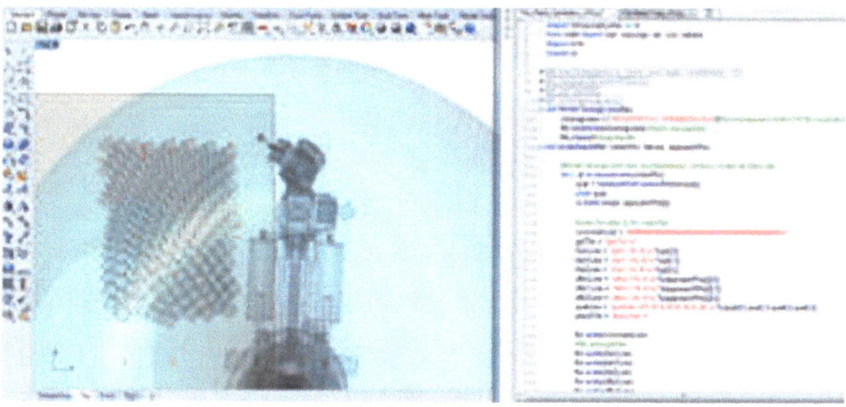

Figure 5 Tile positions and RAPID generated using Rhino Python

The large tolerances allowed by this process did not require a resolute model of the pre-existing conditions, but simply a general understanding of its key points (primarily for the purpose of avoiding collisions). We utilized our robotic manipulator as a digitizer, sending a handful of coordinate values to Rhino Python via serial interface and referencing them during the modeling process to generate our surface within a loose range (~25 cm) of the existing structure.

Fabrication

For the construction of this prototype, we created a custom vacuum gripper for our 6-axis industrial robot (an ABB IRB 6400) using a salvaged mini-fridge compressor and off-the-shelf components. The I/O system of the robot controller is wired to a relay which controls power to the compressor and to a solenoid valve which can be opened to release the vacuum.

We use 11.0 cm square ceramic tiles for an adequate balance between resolution and construction speed (Bonwetsch et al., 2006). Guided by its native RAPID language (ABB, 1997), converted from the data of the digital model using Rhino Python, the robot moves the suction cup to the loading position and turns on the vacuum pump. It then carries the tile to its designated position and raises a prompt on the controller's teach pendant notifying the user to manually apply the expanding foam. As the robot can maintain this position indefinitely (and can be shut down during a pose), the time it must wait until its next movement is determined entirely by the cure time of the filling and adhering material. In our case, we use primarily store-bought expanding polyurethane foam, which we have found to require 40-60 minutes of cure time (at 20-30 °C) before the tile can be released by the robot. In an attempt to increase production speed, we experimented with a professional two-part polyurethane

Figure 6 Prototype build process

system, and though the rapid cure time allowed the tile to be released in 2-4 minutes, it accordingly exacerbated nozzle-clogging issues to the extent that it became infeasible for continued use. The products we tested listed a suggested gap-filling ability of 7.5 cm, but we found we could fill larger volumes with careful application and the use of ad hoc filler materials (scrap styrofoam, wire mesh, dowels, etc.). The presently constructed one-square-meter section of our prototype contains 63 tiles and required approximately 70 hours of build time (Figs. 6/7).

Discussion and Conclusion

Our process effectively demonstrates a technique for using a robotic manipulator as a reconfigurable formwork, while clearly indicating that opportunities exist to streamline the foam-tile manifestation of the concept. Simple improvements to the foam delivery system, such as automated spraying, faster curing, and a self-cleaning nozzle, could improve the process speed tenfold. Further efficiency gains could be achieved through an end effector capable of orienting and placing multiple tiles in one movement. It is worth noting, however, that our primary intention is not to maximize efficiency, but to examine the design potential of this method. Indeed, the idea of maximum efficiency is in many ways at odds with the concept of the poché: as Venturi (1966) states, the residual space created by contradiction between interior and exterior geometries is "sometimes awkward" and "seldom economic."

Beyond our prototypical example which engaged thermal insulation and acoustic isolation, simple variations in material and technique present an array of available performative qualities: aesthetic complexity, light deflection, directional acoustics, and economy of material. Perhaps this process's greatest potential is its ability to produce composite surfaces which tailor the physical properties of the each element of cladding and filling to specific program requirements, creating "functional gradient materials" (Hirai, 1996).

In a production environment, accessing the full potential of this process to reduce the complexity and tight tolerances demanded by current freeform cladding systems requires mobilizing the robot for on-site construction. Mobile construction robots—like the Echord robot of ETH Zurich [3]—could be located within a working zone using not only pre-placed registration markers, but by scanning and calibrating their own previously placed tiles: effectively employing precise placement as a dynamic

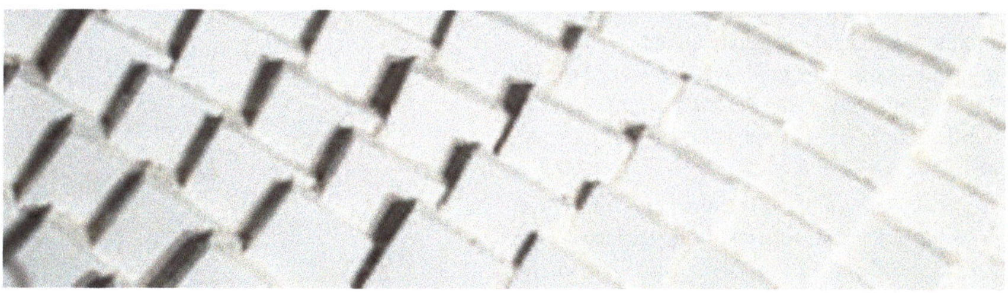

Figure 7 Prototypical irregular-substrate tile surface.

datum for growth. While the tight tolerance requirements of recent digital architecture have resulted in the need to register "the entire building... from one zero point" in place of the more traditional "ruler and tape [run] offsets from specific points" (Kolarevic, 2003), such mobile robotic paneling and re-positioning systems would afford the potential of a return to the flexibility of utilizing local origins in construction.

The precise reference system afforded by this prototypical robotic construction process encourages a reexamination of existing construction methods and implies a potential for reimagining construction order. Whether dealing with new construction or retrofitting existing structures, the loose tolerances of the architectural poché, when combined with the accuracy of industrial actuators, allow a greater liberty in producing highly informed surfaces within relatively imprecise construction constraints.

Acknowledgements

Many thanks are due to Axel Kilian for his support and advice during this research.

References

ABB Robotics AB, 1997 *RAPID Overview*, Västerås, ABB Robotics AB, Vol. 3HAC 5774-1 for BaseWare OS 3.2.

Bonwetsch T, Gramazio F, and Kohler M 2006, "The informed wall: applying additive digital fabrication techniques on architecture", *Proceedings of the 25th Annual Conference of the Association for Computer-Aided Design in Architecture*, Louisville, Kentucky pp. 489-495.

Gassel FV, Schrijver P, and Lichtenberg J 2006, "Assembling wall panels with robotic technologies", *The 23rd International Symposium on Automation and Robotics in Construction*, Tokyo, Japan, pp. 728-733.

Gramazio F and Kohler M 2008, *Digital Materiality in Architecture*, Lars Müller Publishers, Baden.

Hirai T 1996, "Functional gradient materials" in Richard et al. (eds.), *Processing of Ceramics*, pt. 2, vol. 17b, Wiley-VCH, pp. 293-341.

Kolarevic B (ed.) 2003, *Architecture in the Digital Age: Design and Manufacturing*, Spon Press, New York.

Oesterle S 2009, "Cultural Performance in Robotic Timber Construction", *Proceedings of the 29th Annual Conference of the Association for Computer Aided Design in Architecture (ACADIA)*, Chicago, pp. 194-200.

Song H 2005, *Façade-Poché*, Dissertation (PhD), University of Pennsylvania.

Venturi R 2007, *Complexity and Contradiction in Architecture*, The Museum of Modern Art, New York, pp. 70-87.

Yu S, Lee SY, Han CS, Lee KY and Lee SH 2007, "Development of the curtain wall installation robot: Performance and efficiency tests at a construction site", *Autonomous Robots*, pp. 281-291.

[1] http://www.dfab.arch.ethz.ch/web/e/lehre/211.html.

[2] http://www.tmsoundproofing.com/decoupling-explained.html.

[3] http://www.dfab.arch.ethz.ch/web/e/forschung/198.html.

DANGER

Mathew Schwartz, Jason Prasad

RoboSculpt

Unique molds for design with minimal waste

Abstract This paper describes a novel work-flow for minimal waste custom design manufacturing through the integration of traditional sculpture techniques and current manufacturing technology. This is demonstrated by using oil clay as a mold material for fiberglass layups. In addition to being reusable, the use of a malleable material allows for a non-rotating cutting tool on a 7-axis robot, addressing some current manufacturing limitations such as uniformity of the cutting tool, undercuts, and surface defects. The work-flow is demonstrated through the creation of a chair. The chair has multiple curvature and is constructed using two molds, both employing the same material. Fiberglass is used as the final products material demonstrating a selective application for minimal waste from both mold and final materials.

Keywords: minimal waste, robotics, design, sculpture, prototyping

Introduction

For thousands of years sculptors and artisans have used clay in a variety of ways. With the refinement of oil, a new type of clay was created, providing unique properties in an old material. Oil clay excels at being water tight, reusable, and maintaining its shape. The process and materials used for making oil clay allow for a highly customizable material with simple changes to material proportions. The changes in proportions can take the clay from being extremely soft and flexible to hard and brittle. The fact that it is such a flexible material has allowed sculptors to use oil clay in both the modeling and molding process. Although oil clay can be manufactured to different consistencies it retains a very unique characteristic.

The characteristics of clay have been useful in fields outside of art such as the automotive industry. In technology the extensive use of clay properties for both haptic feedback devices and virtual modeling demonstrate the versatile characteristics of this material (Cani & Angelidis, 2006; Yan, Hou, Zhang, & Kang, 2009). In the automotive industry special clay CNC machines, such as the Tarus 5-Axis[1], are used to translate computer models to physical models. These clay models can range from small scale to full size cars, and can be finished and retouched by the hands of a skilled designer.

In contrast to specialized machinery, multi-axis robots provide a machine that can be as unique or redundant as the operator decides (Gramazio & Kohler, 2008). As the number and availability of versatile multi-axis robots increases, the uses and innovations around them follows. The versatility of these robots has been exploited enormously in the last few years through novel manufacturing methods and computer-human interfaces(Payne, 2011; Willis, Xu, Wu, Levin, & Gross, 2010). This has been accompanied by new materials and ways of manufacturing(Pigram & McGee, 2011) that would not have been thought of as useful, or even possible, in the past.

Much of the new robotics work comes from the integration and invention of new technology. In contrast, this work takes processes and techniques from a long history of sculpture and integrates them in the relatively recent field of industrial manufacturing. The integration of these two fields is not direct, nor should it be. Over twenty years ago a similar research project was implemented in the GM laboratory(Chen &Gu, 1993).

Using a six-joint robot with a linear servo-track (a robot very similar to the one used for this research), researchers worked on cutting full scale car models. They named the project ROBOCUT. Details on the project are limited to the mathematics behind the inverse kinematics of the robot, which have been handled here by MasterCAM. The ROBOCUT project was highly focused on the mathematics of the robots for quick modeling. This paper demonstrates the possibilities for not just modeling but mold making as well. By taking the best of both sculpture and manufacturing, we demonstrate a novel method for low cost, reusable, and accurate mold making.

Approach

The individual elements used in this process are broken down as follows:

- 7-axis Kuka robot
- Water jet cutter
- Oil clay (200 lbs)
- Cutting head
- Foam
- Fiberglass

The 7-axis KUKA was used as the main manufacturing method with its directions created in Master CAM and the Robot Master plug-in. The water jet was used for making the robot's clay cutting tools which were designed as 2D cut paths. The oil clay was based on an in-house formula consisting of: 7.5lbs Micro-Wax, 30lbs dry ball clay, 1lb Vaseline, 3qts motor oil, and 1qt glycerin. Blue foam insulation @ 25psi was used to reduce the weight of the mold by acting as an armature. The fiberglass was 1 meter square automotive grade woven sheets.

Conceptually, the combination of elements was approached as a robotic arm imitating a sculptor, allowing for a higher transfer of knowledge for manipulating the oil clay. Although oil clay can consistently be made the same, it is by far the most dynamic, and as such least predictable, variable in the process. As with most, if not all, materials to be cut, the interaction of the material and cutting tool must be tested and understood in order to create predictable results. The appropriation of the robotic arm as a sculptor's arm circumvented many of the otherwise necessary tests to understand the material and cutting tool interaction. Using the knowledge of in-house sculptors, the human technique for cutting clay was broken into conventional manufacturing terms that could be easily transferred to machining inputs: cutting speed, resistance, approach angle, cut angle, cut depth, and retraction technique. With the exception of resistance, due to the lack of a force-feedback sensor, the human techniques were translated as inputs for MasterCAM. These inputs were later adjusted to refine the surface quality of the cut, machine capabilities, and speed.

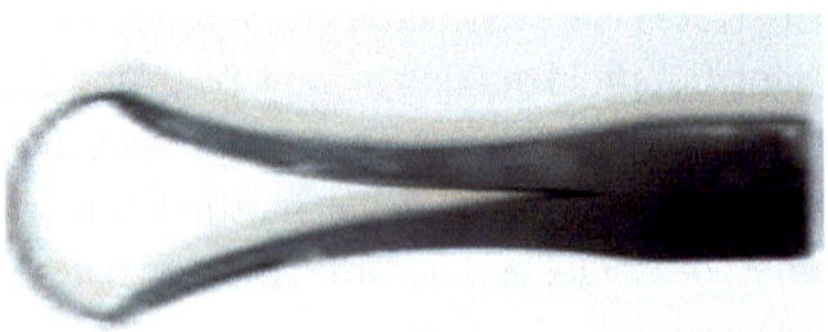

Figure 1 Final clay tool

Inline with the robotic arm mimicking a sculptor, the cutting tool was modeled on clay sculpture tools (Lucchesi & Malmstrom, 1996). The first tool was made by cutting the profile of a tool out of stainless steel on a water jet (Fig. 1). The part was then ground by hand to create a single sided beveled edge. The tool was then bolted to a generic mounting plate on the robot. This initial tool worked well in the cutting tests and served as a starting point for future customized tools. As the forms to be cut became more complex, the mounting plate of the robots end effector began to collide with the clay. Future iterations led to a redesigned interior for additional strength and a lengthened body to avoid collisions of the end effector and material. The final tool was cut from 1/4" stainless steal using a water jet, beveled on one side with a grinding wheel, and smoothed with a water stone.

The mold was created by stacking rectangular blocks of blue foam to create an armature, and laying oil clay on top. Two methods were used to compact the clay into an easily usable and diverse material; 1) Using a hydraulic press to create bricks and 2) using a roller to create sheets (Fig. 2). The hydraulic press was able to create clay bricks fairly quickly and with minimal human effort. However, the more labor intensive process of using a roller provided a more versatile clay shape that was able to drape over an armature without adding too much weight. The foam was cut into rectangular shapes independent of any form, making them reusable (Fig. 3).

Implementation

The efficacy of this process was demonstrated through the making of a chair. With the intention of making the final piece out of fiberglass, a chair with multiple locations of double curvature was designed in a CAD program (Fig. 4). In order to create the chair two molds were needed, one for the top and one for the bottom. The use of two molds demonstrates the reconfigurable and reusable approach. Both molds used the same clay, foam, and cutting tool.

As mentioned earlier, the curvature of these molds made the use of clay in brick form excessively heavy. For this application the clay was most useful in sheet form. The 200 lbs of clay were put through a clay roller. The clay sheets were placed over the foam armature and compacted by hand (Fig. 5). The oil clay was rigid enough to stay together when moving the sheet, while also being able to follow the contours of the armature. This initial setup time for the mold currently takes two people around

Figure 2 Rolled sheets of clay

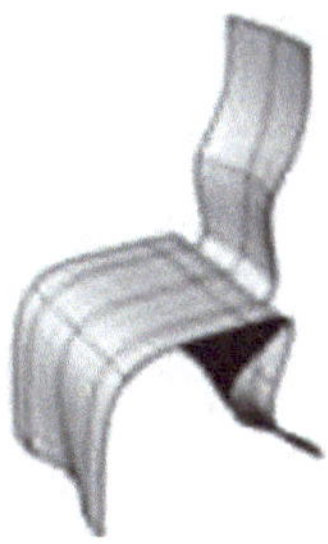

Figure 4 Computer design of chair

Figure 3 Foam armature

Figure 5 Sheets of clay compacted on armature

one hour, however, it varies greatly according to the rigidity of the clay.

Inline with the conceptual approach, the best surface quality of a cut was produced with the tool slightly more than perpendicular to the surface, with the acute angle between the surface in front of the tool's path and the tool [(Fig. 6). The angle offset from perpendicular is dependent on the thickness of the tool, the bevel of the cutting edge, and the tool's exterior surface quality. The tool used for these molds was beveled gradually from midpoint of the stock to the cutting edge. The smoothness of the bevel assists in keeping the surface smooth as the tool rides along the clay. Although we have not found an exact formula, an offset between 5 and 20 degrees produced similar results.

For the same reason that oil clay can be reused, fixing surface defects on the mold is quick and easy. The rigid but malleable properties of the clay allowed for any defects in the surface to be easily patched and re-cut. This takes the process of cutting molds from exclusively subtractive to both subtractive and additive. At present, this dynamic exists not just by machine, but as the robot subtracting material, and the operator adding material. While some materials can be added to through bonding agents, clay is unique in its ability to reattach independent of any bonding agents. Without a bonding agent there is no need for clamps or waiting, allowing clay to be re-attached without affecting other parts of the mold and ready to be cut within seconds. The ability to quickly patch defects in the clay is critical when the compacted clay has air trapped in it. These air pockets are not seen until the clay is cut, however, with little effort the air pocket was filled in by hand, and the path was recut (Fig. 7).

The firmness of the clay due to the quantity of wax helps the clay maintain its form in that, if the clay is cut, the cut piece will remain separated from the mold. This saves a tremendous amount of human labor that otherwise would be required to

Figure 6 Cutting clay

Figure 7 Air pockets found after the first pass. They were patched over and quickly recut, resulting in a smooth surface.

Figure 8 Clay either falls or is forced away on the next cut path

Figure 9 Fiberglass layup on clay mold

remove pieces of clay after each pass. While cutting the second mold we did not remove any clay during the cut. After each path the clay fell down, and any clay remaining was pushed away on the next pass (Fig. 8).

The cut speed could be varied with little to no effect in surface quality. The only case where speed was a factor was on retraction. The tradeoff to a very light mold thanks to a thin layer of clay is the lack of weight to hold the clay down. Since the oil clay bonds almost exclusively to itself, the weight of the oil clay must be sufficient to prevent the mold from sliding. A mold with multiple angles and curvature does not have a problem with sliding, however, a very thin layer of clay is prone to being pulled by the robot. We found that a slow retract rate prevented the mold from sliding, most likely due to a higher usage of static friction. Independent of retraction, and since the cutting tool does not rotate, the cut speed is determined by the load on the robotic arm servos. This eliminates many calculations previously needed to predict surface quality (Lazoglu, 2003).

After cutting the first mold a release agent was sprayed on the clay and a fiberglass layup was done directly on the clay (Fig. 9). Staying true to the aim of minimal waste, fiberglass was used for its ability to be selectively applied. With the edge of the mold clearly defined by the radius of the tool, only a few inches of fiberglass extended past the used portion of the mold (Fig. 10). Unlike viscous materials such as concrete that require twice the surface area of molds in order to control the shape, fiberglass is able to contour directly on an open mold. By using polyester resin, each layup was limited to 20 minutes. To create a fiberglass shell strong enough for a chair, without an interior spacer, eight layers were used. The entire fiberglass layup for the

Figure 10 Minimal fiberglass wasted near edge

first mold took an hour and a half, followed by a six hour wait before de-molding. The clay was easily peeled away from the fiberglass, any remaining clay was scrubbed off with soap and water. The mold release agent provides an extremely thin barrier over the mold, and is negligible when reusing the clay for another mold. The foam armature was then reconfigured and the clay was rolled again to create the second mold.

After both molds had been completed and the fiberglass parts attached, the final fiberglass form was sanded and painted (Fig. 11). The finishing of the fiberglass chair was no different to any other fiberglass product. In the end, the clay is perfectly usable for more molds and, if sculpture practices are any indication, the clay will be reusable for a long time.

Conclusions

This work presents a method for combining aspects of traditional art and industrial manufacturing to yield an independent and useful process, which informs the creation of a chair. The process of making two sequential molds for the chair produced almost no waste, as all the materials were reused.

We acknowledge the scope in which this process is applicable, as the level of practicality of this method is highly dependent on the material used. Using the oil

Figure 11 Final chair after sanding and painting

clay mold to create a physical surface pattern with plaster tiles would be an appropriate usage, however, using liquid materials like plaster and concrete to create high relief images or 3D solids creates an enormous amount of extra work. In this work we use fiberglass as the ideal material. Fiberglass easily conforms to a surface and can be selectively applied. Further usage of fiberglass can be done by the robot. In the example of a bathtub, a robot can use a cutting tool for oil clay, switch to a chopped fiberglass spray gun, and selectively apply the fiberglass with extreme accuracy.

The technique demonstrated here can be expanded in many ways, most notably by the tool design. By having the robotic arm mimic a sculptor's arm, we have opened up the manufacturing possibilities to be the same as a sculptor's. The unique shape of sculptor's tool allows for different techniques in removing material. With a single tool, the robot can cut flat surfaces, different radii, and different textures.

This research concentrates on a process for making unique designs with minimal waste, however, the combination of this and recent research on CHI could provide a new way of prototyping and making that would thrive in areas such as automotive design. This work bridges two impressive areas of research, the Haptic Control Scheme (Her, Hsu, Lan, & Karkoub, 2002) and the newly created Robotic Motion Controller (Payne, 2011). The characteristics of clay and the limitless possibilities for manipulating it make this minimal waste system not just usable, but in many situations actually a more effective system.

References

Cani MP, and Angelidis A 2006, *Towards Virtual Clay*. SIGGRAPH pp. 67-83. New York: ACM.

Chen Y, and Gu YL 1993. "Optimal path generation of a (6+1)-axis robot for auto-industrial applications." *IEEE International Conference on Robotics and Automation*, pp. 388-395.

Gramazio F, and Kohler M 2008, *Digital Materiality in Architecture*. Baden: Lars Muller Publishers.

Her MG, Hsu KS, Lan TS, and Karkoub M 2002, "Haptic Direct-Drive Robot Control Scheme in Virtual Reality." *Journal of Intelligent & Robotic Systems*, pp. 247-264.

Lazoglu I 2003, "Sculpture surface machining: a generalized model of ball-end milling force system." *International Journal of Machine Tools and Manufacture*, pp. 453-462.

Lucchesi B and Malmstrom M 1996, *Modeling the Head in Clay*. Watson-Guptill Publications.

Payne, A 2011, "A Five-axis Robotic Motion Controller for Designers." *Proceedings of the 31st Annual Conference of the Association for Computer Aided Design in Architecture (ACADIA)*, pp. 162-169.

Pigram D, and Mcgee W 2011, "Formation Embedded Design." *Proceedings of the 31st Annual Conference of the Association for Computer Aided Design in Architecture (ACADIA)*, pp. 122-131.

Willis KD, Xu C, Wu KJ, Levin G, and Gross MD 2010, "Interactive Fabrication: New Interfaces for Digital Fabrication." *Proceedings of the fifth international conference on Tangible, embedded, and embodied interaction* pp. 69-72. ACM.

Yan F, Hou Z, Zhang D, and Kang W 2009, "Virtual Clay Modeling System with 6-DOF Haptic Feedback." *Materials Science Forum*, pp. 155-160.

[1] www.tarus.com

Richard Dank, Christian Freissling

The Framed Pavilion

Modeling and producing complex systems in architectural education

Abstract According to the Encyclopedia Britannica [1] a robot is "any automatically operated machine that replaces human effort [...]". But it is much more than just another tool. It is an extremely adaptable machine open to any kind of task, when operated adequately. It is a complete new "medium", and as a result, there is a whole new "message" (Fiore and McLuhan 1967). Half a century after the introduction of robots to the manufacturing process this kinematical apparatus has finally made its way into art and architecture. Using a tangible example this paper tries to illustrate the opportunities for the contemporary building industries and the importance of teaching students the basic principles of interacting with robots. As a matter of fact, we will discuss the Design Master Studio bot/log: Parametrics/Joints constructed/designed by/in Robots/Wood.

Keywords: project and practical application of algorithmic design, computational approaches to sustainable design, 1:1 realization in wood

Robots: the Ultimate CNC Machines

In a recent lecture at Graz University of Technology (TUG) Martin Bechthold (2012), Professor of Architectural Technology at the Harvard Graduate School of Design (GSD), pointed out that the usual architectural design strategy starts with a given problem, a problem to be solved. Subsequently one develops a tool or a system that could possibly solve that problem in a suitable manner. Therefore the outcome is usually unique, but in most cases also rigid. As an example of this he refers to Japan's construction robot industry in the 1980s (Cousineau and Miura, 1998).

In contrast to this, the scientific approach of the Institute of Architecture and Media (IAM) [2] is not necessarily based on a problem. We try to take a given tool and explore its capabilities. Our goal is to get to the bottom of that tool, discover the present limits of its employment and expand the borders of what is possible. And, as we are currently working at the university, we always feel the need to focus on the process, the research and the education of future architects rather than on the pure result. In this context the head of our department, Urs Hirschberg, loves to quote Nicholas Negroponte (1994): "Don't dissect a frog, build one!"

Despite externally funded research projects in the field of geometric processing and non-standard architecture [3] the IAM has been pursuing the idea of a research-based education of master students for several years now. We canvassed promising hardware tools such as tracking systems [4], electronics prototyping platforms [5] and different rapid prototyping machines [6] [7] [8] [9] [10] in our Design Master Studios [11] with real success - including several exhibitions and publications. In 2009 the Faculty of Architecture [12] and the Faculty of Civil Engineering [13] joined forces to set up a new Robot Design Laboratory [14] in cooperation with ABB Robotics [15]. And the IAM started to delve deeply for new applications that are beyond the available industry solutions.

Apart from other obvious advantages over different CNC machinery it is the manufacturing flexibility of robots that make them rewarding for any researcher and student. Since the robotic arm can manipulate any tool that is mounted on its flange, robots already take on a large variety of tasks - some are taught online, but most of the complex ones are programmed offline (Braumann and Brell-Çokcan 2011).

A quintessential example of picking and placing with a gripper is, for instance, Gramazio/Kohler's robotically placed bricks used to form walls and columns [16] (Bärtschi et al. 2010) (Fig. 1, left). Interesting examples of stamping, drawing and even painting are offered by robotlab's *autoportrait* [17], IAM's winter semester 2010 studio *papier peint* [18] and Richard Dank's *Chinese Ink Painting Robot* at the HDA panel discussion *Should buildings grow/adapt/repair themselves? And if not, why not?* [19] (Fig. 1, center). And there are certainly a lot of welding, hot wire cutting and milling projects around, such as experiments at the smartgeometry workshops [20], designtoproduction's *SWISSBAU Pavillon 2005* [21] (Scheurer 2007) or the *ICD/ITKE research pavilion 2010* [22] (Kaltenbach 2010; Knippers and Menges 2011) and *2011* [23] (Fleischmann and Menges 2011) (Fig. 1, right) envisioned at the institutes of Achim Menges and Jan Knippers in Stuttgart, to name just a few.

So, the aim for architecture schools around the world seems pretty obvious: Teach students to model parametrically and give them full algorithmic control over the robotic arm. When they have grasped the principles and know the instrument, they will be able to produce astonishing results while experimenting.

Code: The Ultimate Way to Control the Robot

In their paper *Parametric Robot Control* Braumann and Brell-Çokcan (2011) thoroughly analyze robot on- and offline programming and the common linear workflow. Usually there are several professions and platforms involved until the design of the architect finally disembogues in produced architecture: "A designer[...] to create an aesthetic surface in CAD", "a programmer then" applies "the geometric constraints to the predefined surface, [...] followed by a technician who post-processes the geometric data output for the robot control data file." Additionally, most projects require structural engineers and people/facilities who/which are able to produce and assemble the different pieces in the end. Thanks to all these different operations the whole object will gradually evolve.

This could be a good thing in the end, but none the less the architect eventually loses control over his/her composition. What's even worse though: if the originally induced "aesthetic surface" needs to be changed for whatever reason, everything has to be done all over again. "The question arises here how to further customize the digital workflow to allow the user, i.e. the designer, to manipulate the initial CAD surface [...] and the robot control simultaneously."

We argue that all the external know-how from the collaborating partners must be incorporated in one single parametric model. This allows all the plans, figures and facts required to be directly exported. In addition even the robot code with all its parameters, from tool-data to tolerances, is written on the fly (Fig. 2, right). So the process of designing is not frozen until one presses the play-button on the robot's pendant.

As a consequence this means that future architects must be trained in designing the "aesthetic surface" as well

Figure 1 Gramazio/Kohler's brick-laying robot in action (left); Dank's Chinese ink painting robot (center); Menges/Knippers' ICD/ITKE research pavilion 2011 (right)

as being able to formalize the process. They need to be programmers (to a certain extent) and to know CNC technology with all its constraining and liberating features, so that there is no necessity to "dissect the frog"; the designer should be able to "build it" from scratch.

bot/log: Parametrics/Joints Constructed/ Designed by/in Robots/Wood

This paper presents a case study just recently finished at the IAM. It started out as a Design Master Studio in winter semester 2011/12 with a group of students and the strong Styrian woodworking industry onboard.

The task set: to design a structure and all joints solely made from timber, no glue or other fasteners or fixings allowed. For the realization use the capabilities of a 6-axis robot on an additional linear axis. Moreover the entire project must be applied parametrically! Start to analyze existing and traditional wood joints and test the possibilities of transforming them into digital and parametrical models. Next step is to improve the parametrical models with the aim of producing all joints with our robot and the milling environment. The traditional Japanese wood joint shown (Fig. 2, left) is a good example of how production with cylindrical milling tools is not possible without redesigning the joint. With this developed data start to design and simulate a walk-in structure.

The Studio concluded with 18 individual, full-scale algorithmic projects and one completely implemented and built structure –"The Framed Pavilion" (TFP). There were basically two main reasons why we ultimately decided to build this particular structure. Firstly: the erection process does not require any scaffolding at all; secondly, everything can be put together without the equipment or the hands of professional workmen. The students could assemble the frames and blocks on their own. The whole range of projects and the evolution of TFP can be found on our bot/log webpage [24].

The Evolution of The Framed Pavilion

Sabine Lehner's original design intention was to build irregular pentagonal frames mutating along an axis. The implemented algorithmic process enables the user to convert any basic surface that seems appealing. The application helps to meet the re-

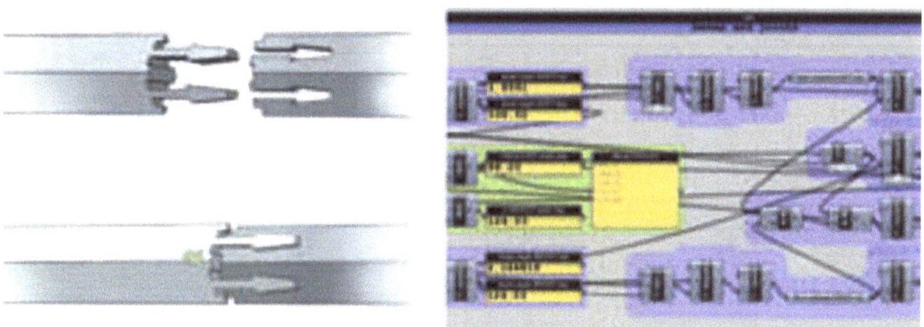

Figure 2 Traditional Japanese wood joint (left);. Visual representation of parametric model generation (right)

Figure 3 The structure as implemented and built

strictive parameters such as the positioning of the wooden dowels, the minimum and maximum beam length and joint angles. Thereby it was possible to generate a morphing shape between the interior and the exterior where height variations, gaps and openings define a special ambiance (Fig. 3).

The framework for the conceived design to production workflow was Rhinoceros [25] extended by its visual programming language Grasshopper [26]. These tools allowed us build a bridge between design, simulation and fabrication and provided all the opportunities to enlarge their functionality by specially programmed addons for our project. Due to performance and handling issues of large datasets we decided to split our parametrical process into two linked components:

1. Definition of boundary conditions and design environment for the main structure.
2. Elaboration for the joint details with building and fabrication requirements including robot code generation.

The basic setup of the realized form was defined by multiple pentagons with variable interior angles. These curves defined a lofted surface with straight sections. Afterwards we sliced the surface in user defined distances to create the square cross section wood frames (Fig. 4).

Alongside aesthetics, transport dimensions, given wood measurements and other boundary conditions, structural analysis was one of the biggest influences on constructing our rigid wood frame structure.

In collaboration with the Institute of Structural Design (ITE) [27] it was possible to define maximum beam length according to its cross section and the crease angle range between each wooden beam inside the polygonal frame where our rigid

Figure 4 Evolution from designed shape to finally defined frames.

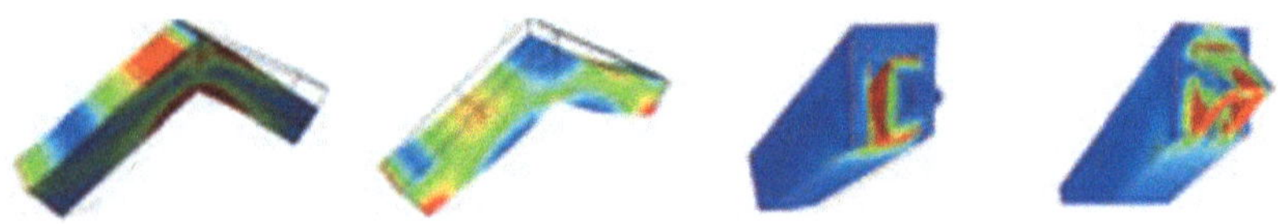

Figure 5 Simulating different load situations.

joint design carries all loads without any external fasteners and fixing. Therefore finite elements simulations with different load situations were calculated (Fig. 5).

In addition to the final shape of "The Framed Pavilion" all drawings, production lists, mounting instructions as well as material nesting results were generated on the fly with our first design component. Generated output data for each beam, as well as for each joint, defines the input for our second component where all joint information was gathered and the robot-milling code generated. The joint design was inspired by Japanese wood joints in which, after assembly, the way the joint is made is invisible (Fig. 6).

For the parametrically generation of all machining operations and tool paths following input parameters were considered:

1. Milling head for 6-axis robot with different cylindrical tool definitions.
2. Robot geometry including additional linear axis for reachability simulations.
3. Fastening structure for beams during milling.
4. Tolerance optimization between easy manual assembling and best values for friction and rigidity inside the joint.
5. Milling parameters such as cut levels, path offset distances, point step density and additional tolerances to avoid collisions.
6. Optimization of tool paths and strategies to reduce production time

Based on these conditions all necessary machining operations were specifically developed to generate automated production data. Fig. 7 shows the visual representation for different tool paths and associated tool orientations.

Keeping different robotic production environments and robot manufacturer in mind we developed two gateways to communicate with the output devices. Our component is able to export apt milling files which are standard in exchanging milling information, e.g. using robot post processors like Pi-Path for ABB robots. Pi-Path converts automatically 5 axis CNC code into multi-axis robot programs. The second output format creates the possibility of directly writing and simulating entire ABB RAPID code in real-time without intermediate steps between design environment and production. Therefore all inverse-kinematics, target information, quaternions

Figure 6 Comparison of digital parametric model (left); milled joint parts (center); assembled joint (right).

and configurations are calculated on the fly – see e.g. the Java-based simulation, code generator and live controller for ABB robots Boot The Bot for details [18].

The Production of "The Framed Pavilion"

In January 2012 over a period of three weeks our Design Master Studio students produced and assembled TFP within the production environment of the Engineering Center Wood (ECW) at the Holz Innovationszentrum in Zeltweg, Styria [28]. The prototype workshop includes an ABB IRB 6640 6-axis industrial robot on an additional 13.7 meter linear axis. This robot is equipped with a tool change unit combined with a 24.000 rpm milling head.

After nesting and cutting all wooden beams, up to 10 beams were mounted side by side on an angled supporting structure during robot milling. The workflow was designed to pick, place and mill in one process. But due to technical and temporal requirements – automatic tool changing still consumes a lot of time – we had to fix the beams manually (Fig. 8).

The final definition of the exact length for each beam and the milling process for all individual joints was done automatically afterwards. Although we could save about 48 hours of machine time thanks to milling path optimization, the production of one joint still took between 8 and 11 minutes – depending on its geometry. Several constraints had to be taken into account, as wood is obviously an anisotropic construction material.

In the next manufacturing phase, the pentagonal frames were assembled (Fig. 9). The precise production made it possible to achieve sufficient stability and stiffness in the corners by just hammering them together – no additional adhesion was needed. Eleven individual frames were then placed successively on a cradle

Figure 7 Different machining operations and milling path creation.

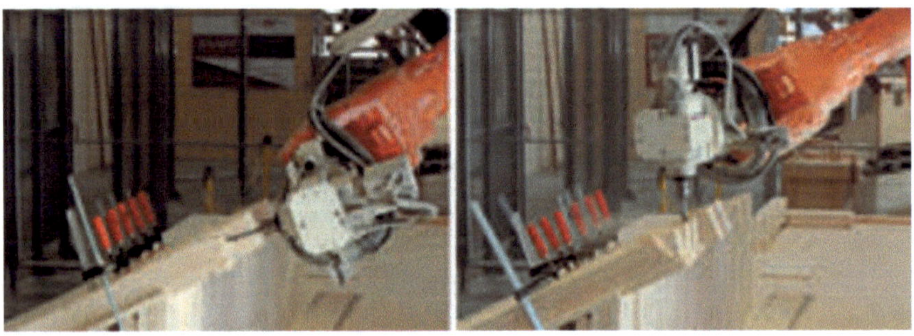

Figure 8 Positioned wooden beams and robot milling head during the production of the two joint parts.

and doweled together – with wooden plugs driven in at the predefined skew drill holes – to form transportable units (Fig. 10). During those 12 days of fabrication we were able to produce four individual units which, when positioned together, finally resulted in The Framed Pavilion (Fig. 11).

Conclusion and Outlook

Since completion TFP has been transported across Austria, exhibited in public [29], discussed in architecture magazines (Colletti et al. 2012) and has just recently been deployed in the city of Knittelfeld at its final destination. However, the steep learning curve of all students during the semester and the affirmative feedback from the woodworking industry is even more gratifying. And the experience plus the algorithmic tools we built along the way are invaluable for us teachers and researchers. 51 years after the first Unimate joined the assembly line in Ewing Township, New Jersey [30], and almost half a century after the introduction of UNISURF, the automotive industry is still the powerhouse behind robotics. They have the money and they produce the turnover. Nevertheless, they seem to have lost most of the innovative drive from the 60s (de Casteljau 1999), as they basically keep using their high-end equipment for recurring routines only – with exceptions, of course. The animating spirit of mass customization has returned to the origins of industrialization and rationalization: the textile and garment industries. But they mostly use regular CNC machines.

Figure 9 Assembling the pentagonal frames.

Figure 10 Combined transportable units (left); tilting the units into an upright position (right).

"Non-standard design in architecture is rapidly evolving, and with the designs come a need for engineering and construction methodologies to support them. [...] The most appropriate position for these new tools seems to be firmly set between the two disciplines of architecture and engineering, helping each rationalize and realize the project. The development of these digital processes not only presents the professions with a new set of tools, but also presents new challenges to traditional working methodology. Perhaps the biggest challenge for the non-standard designer will be to accept that, in order to optimize the processes, the designer will no longer detail the form of a design, but will design the process which generates the details." (Scheurer 2007) Perhaps the new generation of architects that is now graduating from universities worldwide can close the gap between robotics and parametric design – or at least make a substantial contribution to achieving this.

Figure 11 Interior view of the four units forming "The Framed Pavilion"

Acknowledgements

The Design Master Studio *bot/log* would not have been possible without the generous support of Jörg Koppelhuber and Michael Lackner from the ECW and all partners of the Holzcluster Steiermark [31]. We would also like to thank the House of Architecture Graz (HDA) [32] for the opportunity to exhibit the findings to a large audience, our colleagues at IAM and ITE for their stimulating collaboration, and, last but not least, all of our students for their wonderful commitment and their inspiring projects.

References

Bärtschi R, Knauss M, Bonwetsch T, Gramazio F and Kohler M 2010, "Wiggled Brick Bond" in Ceccato C, Hesselgren L, Pauly M, Pottmann H and Wallner J (eds.), *Advances in Architectural Geometry 2010*, Springer, Wien, pp. 137-148.

Bechtold M 2012, *Design Robotics*, Graz Master Lectures, 10.

Braumann J and Brell-Çokcan S 2011. "Parametric Robot Control: Integrated CAD/CAM for Architectural Design", *Proceedings of the 31st Annual Conference of the Association for Computer Aided Design in Architecture*, ACADIA, Calgary, pp. 242-251.

Cambridge Scholars 2010 (eds.), *Computer Graphics Professionals: Ivan Sutherland, James H. Clark, Pierre Bezier, Edwin Catmull, John Warnock, Ronald Fedkiw,* James D. Foley, General Books LLC, Memphis.

Cousineau L and Miura N 1998, *Construction Robots: The Search for New Building Technology in Japan*, American Society of Civil Engineers, Reston.

Colletti M, Dank R, Jobst S and Tschapeller W 2012, Mit allen Mitteln – analoge/digitale Versuchsanordnungen, Konstruktiv: Digitalisierung, no. 286, pp. 12-16.

de Casteljau P 1999, "De Casteljau's autobiography: My time at Citroën", *Computer Aided Geometric Design*, no. 16, pp. 583-586.

Fiore Q and McLuhan M 1967, *The Medium is the Massage: An Inventory of Effects*, Bantam, New York City.

Fleischmann M and Menges A 2011, "ICD/ITKE Research Pavilion: A case study of multi-disciplinary computational design" in Gengnagel, C, Kilian, A, Palz, N and Scheurer, F (eds.), *Computational Design Modeling, Proceedings of the Design Modeling Symposium Berlin*, Springer, Berlin/Heidelberg, pp. 239-248.

Kaltenbach F 2010, "Teaching by Doing - Forschungspavillon in Stuttgart", *Detail*, 10/2010, vol. 50, pp. 994-995.

Knippers J and Menges A 2011, "ICD/ITKE Research Pavilion 2010", *A+U: Timber Innovation*, 07/2011, no. 490, pp. 10-15.

Negroponte N 1994, "Learning by doing: don't dissect the frog, build it", *Wired Magazine*, no. 2.07.

Scheurer F 2007, "Getting complexity organized: Using self-organisation in architectural construction", *Automation in Construction*, vol. 16, no. 1, pp 78-85.

Tarn TJ, Chen SB and Zhou C (eds.) 2007, *Robotic Welding, Intelligence and Automation*, Springer, Berlin/Heidelberg.

[1] Encyclopedia Britannica, http://www.britannica.com/search?query=robot

[2] Institute of Architecture and Media, http://iam.tugraz.at/

[3] IAM Research, http://iam.tugraz.at/?&0&1&3&0

[4] Wacom, http://wacom.com/

[5] Arduino, http://arduino.cc/

[6] Universal Laser Systems, http://www.ulsinc.com/

[7] Epilog Laser, http://www.epiloglaser.com/

[8] Dimension Printers, http://www.dimension-printing.com/

[9] SCM Group, http://www.scmgroup.com/

[10] STEP-FOUR, http://www.step-four.at/

[11] IAM Studios, http://iam.tugraz.at/?&0&1&2&10

[12] Faculty of Architecture, http://dekarch.tugraz.at/

[13] Faculty of Civil Engineering, http://bau.tugraz.at/

[14] Robot Design Laboratory, http://rdl.tugraz.at/

[15] ABB Robotics, http://www.abb.com/robotics/

[16] Architektur und Digitale Fabrikation, http://dfab.arch.ethz.ch/

[17] autoportrait, http://robotlab.de/auto/portrait.htm

[18] papier peint, http://iam2.tugraz.at/studio/w10/

[19] Should buildings grow/adapt/repair themselves? And if so, why not?, http://www.hda-graz.at/galery.php?item=501

[20] smartgeometry, http://smartgeometry.org/

[21] SWISSBAU Pavillon 2005, http://www.design-toproduction.ch/content/view/11/43/

[22] ICD/ITKE Research Pavilion 2010, http://icd.uni-stuttgart.de/?p=4458

[23] ICD/ITKE Research Pavilion 2011, http://icd.uni-stuttgart.de/?p=6553

[24] bot/log, http://iam2.tugraz.at/studio/w11/

[25] Rhinoceros, http://www.rhino3d.com

[26] Grasshopper, http://www.grasshopper3d.com

[27] Institute of Structural Design, http://ite.tugraz.at

[28] Engineering Center Wood, http://www.hiz.at

[29] Mit allen Mitteln, http://www.hda-graz.at/event.php?item=6961&lang_id=en

[30] The Automation Hall of Fame, http://www.pearcedesign.com/ahof/honorees.html

[31] Holzcluster Steiermark, http://holzcluster-steiermark.at/

[32] Haus der Architektur Graz, http://www.hda-graz.at/

Ryan Luke Johns

Augmented Reality and the Fabrication of Gestural Form

Abstract Architectural design is developed in conjunction with technological innovations. These developments are not merely informed by new tools and techniques of production, but also by technologies of representation and dissemination (Carpo 2001) . The last decade has seen a marked increase in both realms: parametric design, CAAD (Computer Aided Architectural Design) and CAM (Computer Aided Manufacturing) on one side, and networked mobile visualizations on the other (augmented reality, smart phones, Microsoft's Kinect technology, Web 2.0, etc.). In this paper we utilize a combination of these technologies to explore the design potential of using robotic fabrication tools in conjunction with a specially developed low-cost augmented reality system. We propose and implement a work-flow in which forms are (1) generated using skeleton-tracking and human gesture, (2) visualized, explored and modified in 3D first-person-view in situ with a head-tracked see-through augmented reality headset, and (3) fabricated in position using a robotic manipulator. We will discuss the communication protocol behind several variations of this procedure and their architectural implications upon design scale, on-site design, and the modular.

Keywords: augmented reality, digital architecture, robot programming

Introduction

The inherent link between technological development and architectural design innovation is one of both empowerment and restriction. As William Mitchell poignantly observed, "architects tend to draw what they can build and build what they can draw" (2001, cited Kolarevic 2003). While industrial robotic manipulators have recently provided the potential of *highly informed* (Bonwetsch et al., 2006) design fabrication, a coupling of these technologies with developments in accessible representational techniques would enable another means of informing design for mass customization (Piller 2004). Mario Carpo (2001) illustrates that, while design and construction technologies are clearly linked to the development of architectural styles (trabeation for the Ancient Greek, the arch for the Romans, stereotomy for the Gothic, reinforced concrete in modernism, and more recently, digital fabrication), they can also be influenced by technologies of representation and the dissemination of media (notably, the effect of the printing press upon the Renaissance). With the prominence of social networking, Web 2.0, and highly-capable smart phones, new forms of representational media have become more fluid and, in turn, accessible to designers.

In this paper, we examine a series of experiments which utilize a combination of representational and fabrication techniques with potential utility in on-site architectural design and mass customization. Namely, we develop a low-cost augmented reality (AR) system using widely available commercial products for use in a workflow in which forms are generated using skeleton-tracking and human gesture, previewed using a see-through AR headset, and fabricated *in situ* via robotic manipulator.

Related Work

There have been numerous research projects involving gestural form-finding (Greenwold, 2003) and many more that suggest the potential application of augmented reality systems in architectural design (Feiner et al., 1996).

The intent of this research is not to develop or dwell upon technology in skeletal tracking or augmented reality, but rather to implement them as simply and as cheaply as possible in order to explore their ability to inform architectural design, robotic fabrication, and mass customization. In this sense, the project contains some of the same ideas behind the cell-phone-designed mTable of 2002 (Gramazio and Kohler, 2008)—by empowering non-designers with software that turns their own off-the-shelf hardware into highly capable and often clumsily-controlled design tools, architects are forced to rethink their role in a world where digital fabrication technologies have enabled the potential of mass-customization.

Initial Research

This project naturally evolved from research begun at the Gramazio & Kohler Professorship for Architecture and Digital Fabrication, ETH Zurich, which explored the on-site potential of robotic fabrication through the use of laser scanning technologies and a robotic manipulator mounted on a movable platform. Initial tests utilized a robot-mounted *Kinect* (Kean et al., 2011) scanner connected to a nearby PC, which was in-turn connected to the robot controller via Ethernet connection. Using the *simpleOpenNI* [1] library for *processing* [2], we were able to track the 3D hand coordinates of the human user in real-time. The program was written such that hand movements could be interpreted to generate a virtual brick wall along the gestured path as it was drawn. By reading the orientation and position of the robotic manipulator, the *Kinect's* local coordinates could be transformed to match the coordinate system of the robot, and the processing code was written such that the generated brick positions and orientations could be translated into the native language or the robot (ABB, 1997) and sent directly to the controller. Using a vacuum gripper attached to the same end-effector as the *Kinect*, the robot could then proceed to construct the brick wall along the designated path (Fig. 1).

For more information on this research project, see "In-situ robotic fabrication" (Project leader: Volker Helm; Collaborators: Dr. Ralph Bärtschi, Tobias Bonwetsch, Selen Ercan, Ryan Luke Johns, Dominik Weber), Professorship for Architecture and Digital Fabrication, ETH Zurich [3].

Augmented Reality System

Overview

While the potential of coupling the gesture recognition of the *Kinect* with robotic manipulators has been explored on numerous occasions, there is generally a gravitation towards human mimicry via telerobotics (De Luca and Flacco, 2012; Itauma et al., 2012) rather than utilizing gesture as guiding factor for more complex processes (i.e. brick laying). By combining the *highly informed* detailing made possible by computer scripting and industrial robotics with gestural inputs, defining complex structures intuitively on-site becomes more feasible.

In order to experiment with the potential of shaping, interacting with, and approving the parameters of gesturally-based forms *in situ* prior to robotic fabrication, we opted to utilize a see-through, head-mounted augmented reality system.

Figure 1 Brick wall robotically fabricated along gestured path, Gramazio & Kohler, ETH Zurich

To maximize options for expanded functionality and to avoid the cost of AR-specific commercial products, we developed our own device using cheap, off-the-shelf components.

Hardware

In searching for the components necessary for an augmented reality system–position tracking, orientation sensors (electronic inclinometer/accelerometer and compass), networked communications, portable power, resolute screen, and an operating system that supports localized software (Feiner et al., 1997) – it quickly became clear that all of these elements were available inside the majority of today's smartphones. Repurposing such a widely available product ensured low cost, compact form-factor and the potential of making any developed applications accessible to a mass audience. For its existing integration with the *processing* environment [4], the *Android* OS was selected and a used *Motorola Droid X* became the core of the augmented reality system.

The headset was assembled from the hardware of a scrap headlamp and laser-cut acrylic parts, with the Motorola *Droid X* mounted above the eyes with downwards-facing screen, reflecting onto an angled sheet of transparent, mirrored acrylic (Fig. 2).

Multi-device Interface

By creating a custom interface between three mass market electronic devices (*Kinect*, personal computer, and smartphone), we are able to create a robust gestural interface using components that exist within millions of homes worldwide [5]. The interconnectivity of the devices functions in the following manner (Fig. 3):

1. Both the PC and *Droid X* (headset) are running custom applications written in *processing* which are constantly communicating with one another wirelessly over the internet using OSC protocol [6].
2. The *Kinect* is connected via USB to the PC and provides the data used for 3D tracking of the user's joint coordinates.
3. Head pan, tilt and roll are calculated using the mobile phone's accelerometer and geomagnetic sensor [7] , while head position is read from the *Kinect* data.

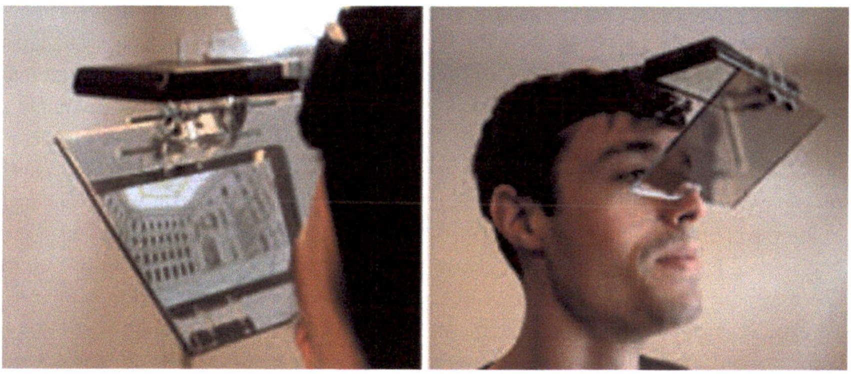

Figure 2 Smartphone based augmented reality headset; Equirectangular image credit: Ilja van de Pavert

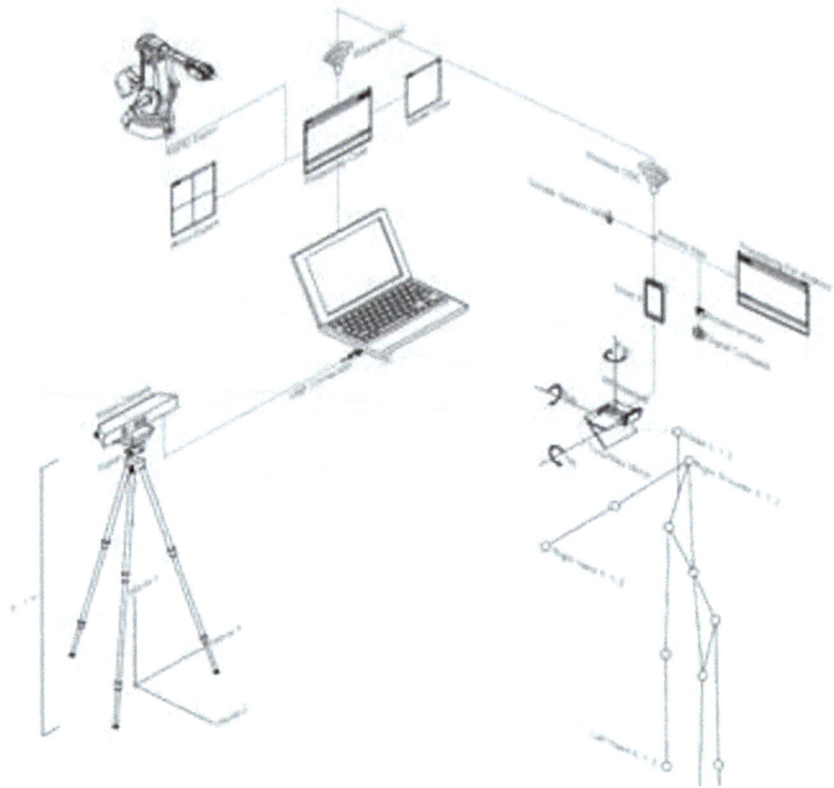

Figure 3 Interface of hardware and software for gestural AR system

4. The touch screen of the *Droid X* activates *Google's Speech Recognizer* [8] to listen for voice commands which operate the program.

Software Development and Capabilities

In order to gain familiarity with *processing for android* and an understanding of how to utilize the phone's sensor readings, we first implemented a simple panoramic viewer which rotated the viewing direction within a textured sphere [9] (mapped with an equirectangular image) based on the user's head orientation (Kwiatek, 2005) (Fig. 4).

Once we were familiar with the settings required to create a fixed position, orientation-based viewer, these techniques were combined with the skeleton-tracking capabilities of the *Kinect* order to enable a fully navigable and modifiable AR environment. The software on the PC side reads skeleton data, and is constantly sending the head and hand positions wirelessly to the headset. The *processing* 'scenes' in both versions of the software are fundamentally the same—sharing a common world origin directly below the *Kinect* on the ground plane. The headset software merely places its virtual camera at the received head-XYZ coordinate and orients the camera frustum based on the values read from the accelerometer and compass. As head-rotation (in plan) is based on world azimuth angles and the *Kinect* is not always placed due-north of the viewer, each session begins by the user facing the *Kinect* and "calibrating" the scene such that the angle between the calibrated azimuth and true-north is factored into future camera rotations.

The program on each device is equipped with the same expandable set of gestural form-finding techniques: at the current state of this prototype, the primary functions are "loft" surfaces and "brick" surfaces. All commands are accessed by tapping the touch-screen to initialize voice recognition, and then speaking the command (which is registered by the *Droid* and immediately sent to the PC). In example, the spoken command "loft" initializes the generation of a surface that is lofted between the paths of the right and left hand, while the command "brick," initializes a brick wall which follows the path of the right hand in plan and is built to the height of the hand in elevation.

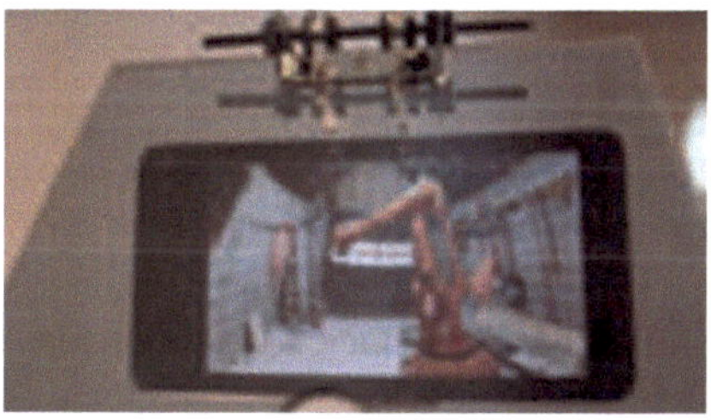

Figure 4 Orientation-responsive equirectangular image viewer developed with processing for Android

Multiple functions can be run simultaneously (Fig. 5a), forms can be added to or erased, and multiple objects can be generated within the same program. The user can walk around and explore the scene before speaking the command "Rhino" to open the exported geometry in the 3d modeling software (Fig. 5b) on the PC for prototyping (Fig. 5c) or can export RAPID for direct use with the robotic manipulator (Fig. 6).

Discussion

While augmented reality systems and gestural form-finding are certainly not new topics, we propose that their architectural potential is reinvigorated through integration with industrial robotics and the question of design scale. If we regress to the time of the primitive hut, we find an architecture that is both designed and constructed at the human scale (Fig. 7a)–one

Figure 5 From left: a) simultaneous gestural generation of brick wall and loft surface using hand coordinates; b) geometry exported to Rhino 3d; c) lasercut scale model

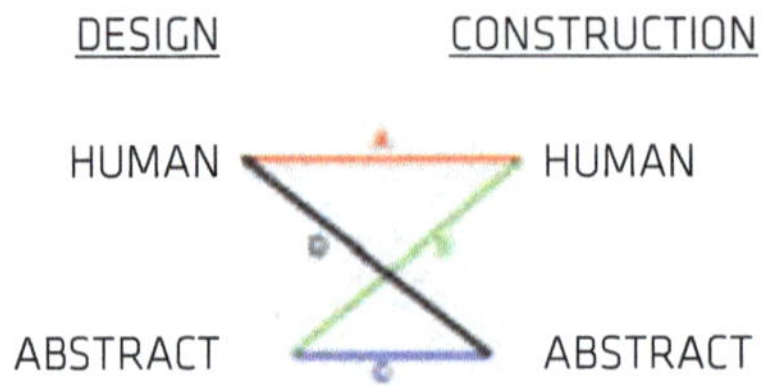

Figure 7 Relationships between scales of design and construction in architecture.

that is informed by the modularity of man and restricted by the strength and reach of his body. As drawing technology entered the picture, however, architects began to design at abstract scales while construction detailing remained limited by the capabilities of the worker (Fig. 7b). With the advent of computer modeling and industrial fabrication, both architecture and construction have lost their association to the human body: digital models are generated in abstract scales and fabricated using machines whose scale, strength and precision go far beyond human potential (Fig. 7c). We suggest that a coupling of gestural form-finding with highly capable industrial robotics enables an exploration of the last remaining trajectory: one in which design is done at a human scale and construction is performed with a level of strength and complexity that is entirely inhuman (Fig. 7d).

Conclusion

In this paper, we implement a workflow in which architectural forms can be generated based on bodily movement, previewed and altered using an augmented reality system, and translated back into the physical world through means of digital fabrication. Using a prototypical software interface, we present a method for adding informed complexity to spontaneous forms. In this instance, we generate a brick wall or a loft surface along the path of the hand, but foresee a potential future in which design functions could be added by other developers and architects much like "apps" are added to smartphones. In this way, the software could be expanded to enable a wide array of modeling techniques which are tailored to consumer demands or specific developments in computational design and fabrication technologies. By providing individuals with intuitive means for roughing out architectural forms at the human scale, and then equipping them with easy techniques for exploring, editing, detailing and fabricating those forms, such interfaces make the design process more accessible to non-architects. The potential implications of mass customization, therefore, can only be realized when the technology for repre-

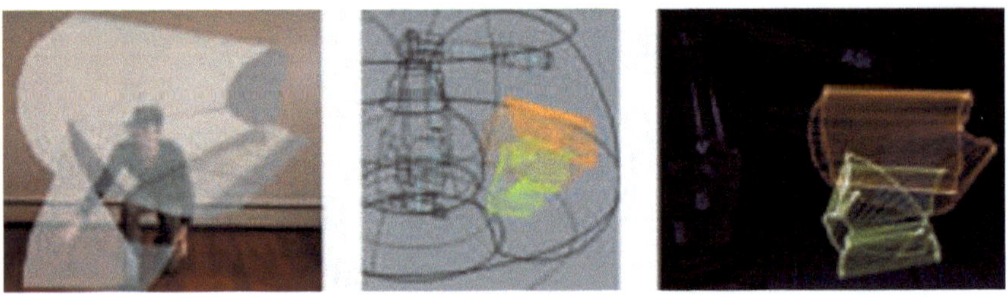

Figure 6 From left: a) loft surfaces generated using AR headset and hand coordinates from the Kinect; b) surfaces exported to Rhino for viewing; c) RAPID code generated for robotically produced light painting of surface

senting and disseminating design options is given as much attention as the tools for fabricating them.

Acknowledgements

This research owes much to the Gramazio & Kohler Professorship for Architecture and Digital Fabrication, ETH Zurich for their kind support during my research stay, and to the ThinkSwiss Research Scholarship for their funding during that period. I would like to thank Axel Kilian for being supportive of my research, Alejandro Zaera-Polo and Ryan Welch for their valuable criticism and insights regarding the augmented-reality headset and interface, and Nicholas Foley for his partnership in developing *greyshed*.

References

ABB Robotics AB, 1997 *RAPID Overview*, Västerås, ABB Robotics AB, Vol. 3HAC 5774-1 for BaseWare OS 3.2.

Bonwetsch, T, Gramazio, F and Kohler, M 2006, "The informed wall: applying additive digital fabrication techniques on architecture", *Proceedings of the 25th Annual Conference of the Association for Computer-Aided Design in Architecture*, Louisville, Kentucky pp: 489-495.

Carpo, M 2001, *Architecture in the Age of Printing: Orality, Writing, Typography, and Printed Images in the History of Architectural Theory*, Benson, S. (trans.) , The MIT Press, Boston.

De Luca, A and Flacco, F 2012, "Integrated control for pHRI: Collision avoidance, detection, reaction and collaboration", *The Fourth IEEE RAS/EMBS International Conference on Biomedical Robotics and Biomechatronics*, Rome, Italy, pp. 288-295.

Feiner, S, Macintyre, B, Höllerer, T, and Webster, A 1997, "A Touring Machine: Prototyping 3D Mobile Augmented Reality Systems for Exploring the Urban Environment." *Personal Technologies*, Vol. 1. , pp. 208-217

Gramazio, F and Kohler, M 2008, *Digital Materiality in Architecture*, Lars Müller Publishers, Baden.

Greenwold, S 2003, "Spatial Computing", Graduate School of Science in Media Arts and Sciences, Massachusetts Institute of Technology, Massachusetts.

Helm, V, Ercan, S, Gramazio, F and Kohler, M, 2012, "In-Situ Robotic Construction: Extending the Digital Fabrication Chain in Architecture", *Proceedings of the 32nd Annual Conference of the Association for Computer-Aided Design in Architecture*, San Francisco.

Itauma, II, Kivrak, H and Kose, H 2012, "Gesture Imitation Using Machine Learning Techniques", *Proceedings of the Signal Processing and Communications Applications Conference (SIU), pp.* 1-4.

Kean, S, Hall, J and Perry, P 2011, *Meet the Kinect: Programming and Scripting Natural User Interfaces*, Apress, New York.

Kolarevic, B (ed.) 2003, *Architecture in the Digital Age: Design and Manufacturing*, Spon Press, New York.

Kwiatek, K 2005, "Generation of a virtual tour in the 3D space applying panoramas, exercised on the sites of Dresden and Cracow", Thesis (Diploma), AGH University of Science and Technology.

Piller, F 2004, "Mass Customization: Reflections on the State of the Concept", *The International Journal of Flexible Manufacturing Systems*, Springer, pp. 313-334.

[1] http://www.processing.org.

[2] http://code.google.com/p/simple-openni/.

[3] http://www.dfab.arch.ethz.ch/index.php?lang=e&this_page=forschung&this_page_old=forschung&this_type=&this_year=&this_id=198

[4] http://wiki.processing.org/w/Android.

[5] http://community.guinnessworldrecords.com/_Kinect-Confirmed-As-Fastest-Selling-Consumer-Electronics-Device/blog/3376939/7691.html.

[6] http://www.sojamo.de/libraries/oscP5/.

[7] http://developer.android.com/reference/android/hardware/SensorListener.html

[8] http://developer.android.com/reference/android/speech/package-summary.html.

[9] http://processing.org/learning/library/texturedsphere.html.

Steven Keating, Neri Oxman

Robotic Immaterial Fabrication

Abstract In this work a KUKA KR5 sixx R850 robotic arm was transformed into a novel multi-fabrication platform capable of additive, subtractive, formative, and immaterial fabrication processes. We define immaterial fabrication as a novel class of fabrication category where material properties are manipulated without direct mechanical forces to create design environments and objects. Design studies discussed in this paper include real-time light renders generated by dynamic control of light sources and annealed patterns created by manipulating heat fields. The paper focuses on the immaterial sensing and fabrication processes developed, including volumetric scanning measurements of optical, thermal, magnetic, and electromagnetic fields and methods of spatial data output. In addition, the concept of informed fabrication utilizing robotically-controlled environmental sensing to influence and inform fabrication is discussed, explored, and demonstrated.

Keywords: digital fabrication, robotics, informed fabrication, immaterial fabrication, light painting

Introduction

Industrial robotic arms are used in fabrication applications apart from assembly lines with increasing frequency. From robotic bricklayers to graffiti robots,robot arms are expanding into new roles that challenge our current view of robotics and redefine fabrication techniques (Gramazio and Kohler, 2008; Robots in Architecture, 2011).

As the significance of digital fabrication continues to grow in digital design and fabrication, the definition of fabrication becomes increasingly useful both as an organizational and a generative tool. Fabrication is classically defined as a process of "construction from parts" and is traditionally broken down into categories based on how the "parts", or raw materials, are mechanically manipulated to construct an object. The three widely accepted fabrication categories include additive, subtractive, and formative processes (Chua, Leong, & Lim, 2010).

Additive processes are construction methods that add material to produce an object. Most 3D printing technologies (such as fused-deposition, stereolithograthy, and laser sintering processes) are included in this category. In contrast, subtractive fabrication techniques remove material to produce the manufactured object. Most machining processes are subtractive fabrication methods and include milling, turning, and grinding. Finally, fabrication methods that mechanically shape a set amount of material are known as formative processes such as bending, forging, and forming. Manufacturing methods that combine additive, subtractive and formative techniques are referred to as composite or hybrid processes. With these definitions in mind, we aim to explore two new classes of fabrication that use robotic arms: immaterial fabrication and informed fabrication. Robotic arms have the benefits of speed, agility, and flexibility, and can be used as both inputs (sensing) and outputs (modifying the physical environment). In addition to mechanical outputs, elements of an environment can be transformed as an output without the movement of physical material. Instead of physical matter, properties and fields can be made into spatial outputs of the system, such as light, sound, heat, radiation, and radio waves. Heat, for example,can be applied to a metallic object in varying quantities to impart an annealing pattern. While a digital fabrication method is implemented here, the medium has altered from relocating physical matter in a specific design to repositioning a heat source in an intended design. The design process is still a process constructed from parts—in this case it is the alteration of the crystal structure—but not by manipulating material with direct mechanical force, as is characteristic of additive, subtractive, and formative processes. To facilitates the characterization of this type of environmental fabrication and distinguish it from physical construction, we use the term immaterial fabrication. In this paper we will explore this definition and demonstrate different examples of immaterial robotic fabrication distinct from conventional additive, subtractive, and formative processes.

Robotic arms are considered advanced compared with traditional fabrication methods (Pires, 2007). As mentioned, robotic arms can be used as input or output devices. When implemented as an input device, sensors are coupled with the arm to allow spatial measurements of the environment. For example, an optical scanning system can be combined with the robotic arm to automatically generate 3D data of objects in an environment (Callieri, 2004). As an output device, end effectors are coupled with the arm allowing the robot to modify its environment. Such environmental modification can be made useful for a variety of digital and physical automation purposes such as fabrication, entertainment, or organization. For example, a milling robot cuts foam to create a sculpture, a dancing robot moves its limbs to entertain an audience, and a cleaning robot tidies up a mess.

The coupling of input and output fabrication capabilities of a robotic arm allows for a system capable of producing objects that incorporate environmental data. This use of environmental feedback to directly inform and influence fabrication offers many potential new avenues for design and manufacturing which will be discussed in this paper. We use the term *informed fabrication* to refer to combinations of environmental sensing and fabrication.

The Multi-functional Robotic Fabrication Platform

To explore the concepts of robotic fabrication, we set out to build a robotic arm platform capable of each type of fabrication category: additive, subtract, formative, and immaterial. This paper focuses on immaterial fabrication processes and environmental sensing, but a brief description of all of fabrication capabilities is provided.

For all experiments, a KUKA KR5 sixx R850 robotic arm was used. The KR5 sixx R850 is lightweight (29 kg), fast (maximum speed of 2.0 m/s), and has a reach of 850 mm with a repeatability of +/- 0.03mm[1]. A KUKA KR C2 sr controller was used for communication with the robotic arm. All programming was executed implementing Python scripts to generate KUKA Robotic Language (KRL) code, with the exception of 3D printing and milling tool

Figure 1 The multi-functional robotic arm platform configured as an ABS 3D printer (left). Printed objects have a layer height of 0.3mm (right).

paths. For this purpose, a G-Code to KRL Python script was written to facilitate the use of commercial printing and machining CNC software. Custom end effectors were made to facilitate the various fabrication processes and controlled through the programmable outputs of the arm.

Additive fabrication was demonstrated using an *acrylonitrile butadiene styrene (*ABS) extruder attachment. By extruding layers of molten plastic on top of previous layers, the robotic arm platform was able to 3D print objects with a layer resolution of 0.3 mm from a computer aided design (CAD) file (Fig. 1).

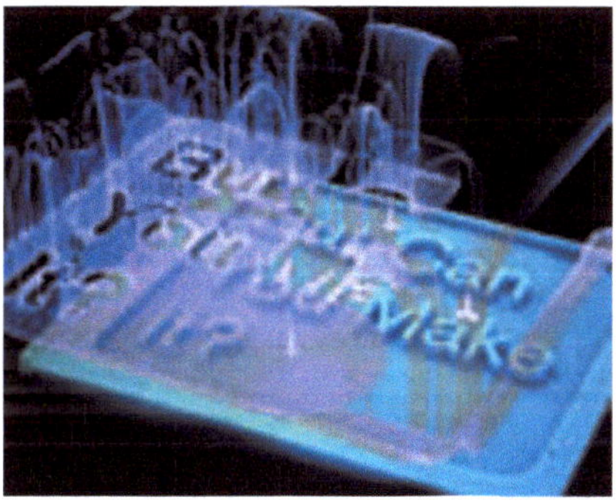

Figure 2 A polystyrene sign milled using the robotic arm platform. The tool paths are seen using long-exposure photography.

Figure 3 A clay mold is sculpted using the robotic arm platform according to a CAD file.

Using a milling attachment, the robotic arm successfully milled various polyurethane and wooden panels employing a CAD file, demonstrating subtractive fabrication (Fig. 2). In addition, the combination of 3D printing ABS followed by a facing milling operation created a hybrid process capable of achieving a finer surface finish than 3D printing alone.

Formative fabrication was explored through clay sculpting using a modeling end effector. By mechanically depressing the clay as informed by CAD data, various relief patterns were formed which served as molds for casting objects (Fig. 3).

By demonstrating the use of a single robotic arm as a multi-functional fabrication platform capable of additive, subtractive, and formative processes, the flexibility of robot arms in digital fabrication is made clear. Moving past traditional fabrication techniques, we believe robotic arms are also well suited for novel fabrication possibilities; namely immaterial fabrication, sensing, and informed fabrication.

Immaterial Fabrication

The conventional fabrication categories are defined by the interaction of mechanical forces with the raw stock material; additive processes build structures up, subtractive processes carve structures out, and formative processes reshape material into the final structure. Such categorization is sufficient when dealing with homogenous physical matter, however these definitions become problematic when fabricated parts cease to be based on mass and physical matter alone. What happens when designs are fabricated out of material properties rather than mass, such as crystal structure, elasticity, and density? How should designs that are fabricated with fields other than mass be regarded, for instance designs that are informed by electromagnetic and thermal fields? Can the definition of fabrication be extended beyond purely mechanical movements of mass?

Driven by the necessity to design and deliver highly complex material parts, new technologies and applications are increasingly focusing on material properties and behavior. Materials with gradient properties are an ideal example. Functionally graded materials—materials designed with spatially varying properties—offer many advantages over conventional homogenous structures. The ability to tailor structural and material properties spatially can improve functionality and material efficiency. For example, annealed metals are often designed with heat treatments that impart gradient material properties suitable for structure applications. By imparting heat the crystalline features of a metal structure can be changed to produce and control various properties, such as hardness. The application of heat in a pre-designed spatial pattern to produce a desired structure, postulates a new class or category of digital fabrication that we term immaterial fabrication.

Immaterial fabrication processes are based on non-mechanical forces and fields, such as electromagnetic, thermal, radioactive, and acoustic fields. In the case of annealing, as previously discussed, both thermal (conductive and convective energy transfer) and electromagnetic forces (radiative energy transfer) are used to affect the material and create the designed structure. This definition is still based on the formal definition of fabrication proposing the construction of parts, yet it allows for controlled designed manipulation of non-physical parts, such as photons.

Many possibilities for immaterial fabrication exist with regard to media such as light, sound, heat, and material properties. Light was selected and explored as an example medium and several methods of light design fabrication were executed.

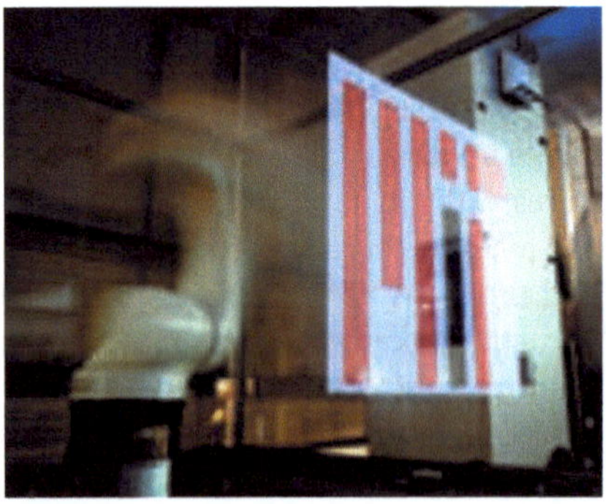

Figure 4 An image is fabricated in physical space and captured using long-exposure photography. This technique of capturing light paths is known as light painting.

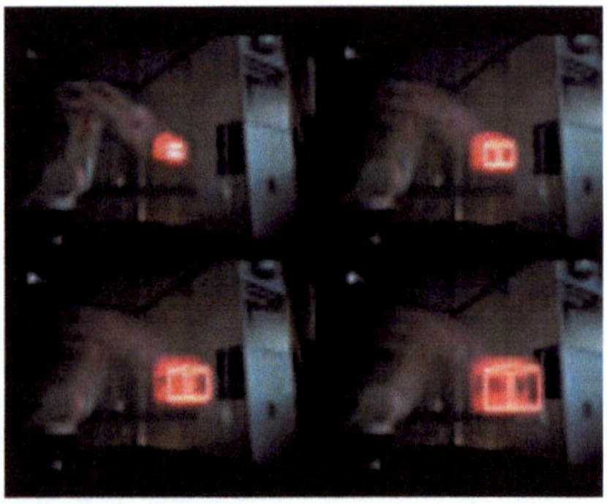

Figure 5 A series of 3D light structures are fabricated and captured as long-exposure photographs, as frames of an animation sequence of a growing cube.

Light painting is a photographic technique where a long-exposure image is used to show motion paths of lights within the image. Light painting has been used for decades by photography artists and recently explored in several robotic installations including Outrace (Weisshaar&Kram, 2010) and Halo: Remember Reach [2]. We classify light painting as a type of immaterial fabrication since the designed structure is defined through a manipulation of the electromagnetic field (i.e. light generation).

Light painting was initially explored by moving a controlled light source in a designated spatial design from a CAD file. 2D color images were generated using this method and were captured using long exposure photography (Fig. 4). 3D structures and animations were also generated, where the robotic arm rendered each frame of the animation in real space (Fig. 5).

The light painting examples explored in this paper constitute a slow volumetric display. Combining this technique with a static camera to visualize the designs, a new form of digitally controlled animation is made possible in which each frame of the animation is rendered in the real environment. While this application is primarily artistic, the use of immaterial fabrication may promote and contribute to industrial purposes such as localized heat treatments, specific curing designs, or magnetic patterning. Instead of outputting light, an effector can produce complex heat treatments, electromagnetic fields, and magnetic designs for target structures.

Environmental Sensing

The opposite effect, where data is captured instead of being exported, is achieved through the use of sensors. By applying a sensor as an end effector, volumetric sensor arrays can be simulated quickly and cheaply. Any type of environmental sensor that can be mounted to a robotic arm may be used to simulate a sensor array. This sensor array can have a programmable scanning structure to allow for custom spatial resolution. The simulated array has a reduction in temporal resolution due to the serial nature of scanning and this temporal resolution is dictated by the scanning speed, distance of the scanning paths, and the total volume scanned.

As a first example, the reverse setup to the light painting experiments was explored. Instead of moving the light source and keeping the camera static, the camera's location is dynamically controlled by the robotic arm and static environmental light was used (Fig. 6). Termed *inverse light painting*, a controlled light source was placed in the environment and the camera was moved in designed paths to generate an arbitrary image in the form of a long-exposure. This is seen in Fig. 7, where the desired image is centered and the background is a blurred combination of the light from the rest of the environment.

This setup can be taken one step further to create a camera with a synthetic aperture of any given size within the reach of the robotic arm. By translating a camera with the shutter open in the desired shape and size of the synthetic aperture, an effective synthetic aperture camera is created. Fig. 8 compares the result of a scene captured with the regular camera aperture and the same scene captured with a very large synthetic aperture with a robotic arm.

Figure 6 The setup used for inverse light painting has a robotically controlled camera and a fixed environmental light source.

Figure 7 By moving the camera in specific paths according to a CAD file, this long-exposure photograph displays an image painted using environmental light. The background is a result of the blurred environment.

Synthetic apertures can be utilized for a number of applications in computational photography. For example, as seen in Fig. 8, larger apertures can see through occlusions if the aperture is larger than the occlusion. Compared to building a complex and expensive array of hundreds of cameras, simulating a large camera array with a robotic arm offers the benefits of simplicity, costs, and flexibility at the cost of temporal resolution. For steady-state environments, the reduced temporal resolution does not affect the data.

Volumetric measurements can be taken as well, where a sensor is moved through a 3D space to collect spatial measurements. Based on the previous light exploration, a volumetric reading was obtained by moving a photodiode through a 500 mm cube. The measurements were taken in the dark with a single light emitting diode positioned at the bottom of the robotic arm to provide an example light field. By sampling the light intensity, a 3D map can be generated showing the spatial light intensity corresponding to the scene (Fig. 9).

The potential applications for robotic sensing are vast, ranging from pure scientific applications to applied analytical ones. Examples include optical scanning, acoustical mapping for sound reduction, spatial chemical analysis, heat transfer data acquisition, structural inspections, X-ray analysis, tomography, and much more. The inherent flexibility of robotic arms is ideal for such scanning applications, as arms can unobtrusively explore spaces, be easily reconfigured, and have a small physical footprint.

Figure 8 An environment (left) and the resulting inverse light painting from the same scene (right) shows the effects of a large synthetic aperture. Note that occlusions smaller than the synthetic aperture, like the windows on the right of the scene, can be imaged through if the focal plane is tuned past the occlusions.

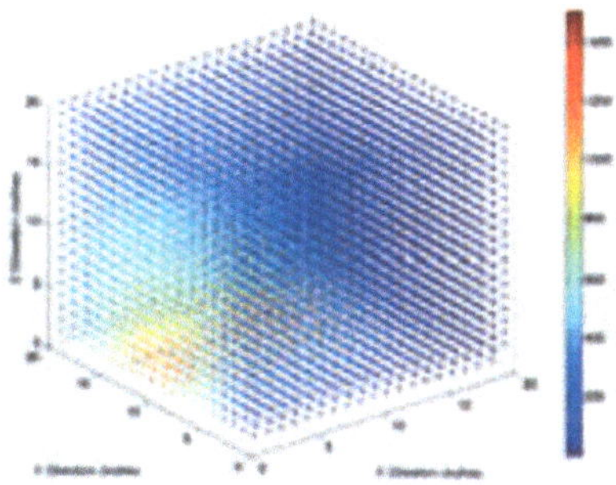

Figure 9 Volumetric light intensity measurements using the robotic arm show a single light source near the bottom left.

Informed Fabrication

The combination of immaterial sensing and physical fabrication is here referred to as informed fabrication, where environmental feedback contributes to the finished design product. Using sensing equipment as an effector, the robotic platform can map out an environmental field or material property and use such information to control the fabrication process. For example, using an X-ray imaging system as a scanning sensor for crack detection on an aircraft part (Xu et al. 2010). Using the information from the sensor effector, a welding effector can then be used to apply a repair weld to the precise area required. This method is made fast and efficient by combining operations and it facilitates a secondary X-ray scan to evaluate the repaired weld seam. Informed fabrication can involve real-time feedback to enable process control. This allows for subtle corrections to ensure proper fabrication, such as correcting for observed thermal warping in 3D printing or chip removal in milling. Informed fabrication can be applied to any CNC manufacturing method, but is especially suited for robotic arm systems that have the required flexibility, internal space freedom, and agility.

Using our previous examples of light painting as immaterial fabrication, adding a sensing input to light painting creates an informed process. We set up several different sensors on the robotic arm to inform the light painting process including microwave and magnetic fields (Figs. 10 and 11). Using a scanning pattern, hidden fields were visualized using the light painting technique. The intensity and color of light was informed in real-time by the spatial field strength, producing images captured in long-exposure photographs. This was accomplished using a tri-color light emitting diode controlled by a microcontroller attached to various sensors.

By setting a threshold sensor value to turn on the light and mapping sensor values higher than the threshold to a color chart, the environmental data is represented in the light painting. As seen in Figure 10, a microwave oven produces microwave radiation that leaks outside of the oven. The magnetic field strength around a laptop is seen in Figure 11, indicating the location of the hard drive. This method of field visualization is very useful for analysis, as it allows data to be directly matched to an environment.

Conclusion

The research work presented in this paper proposes the term immaterial fabrication as a novel category of digital fabrication and construction. By altering the design medium (or substrate) from physical matter to physical properties or force fields, it promotes design processes informed by invisible forces such as heat, light and load. Though fabrication is traditionally defined as the process of constructing wholes from parts, such"parts"need not be limited, as we propose, to homogeneous physical solids. The photons used to fabricate designs implementing the light painting method serve as the fabrication medium used to

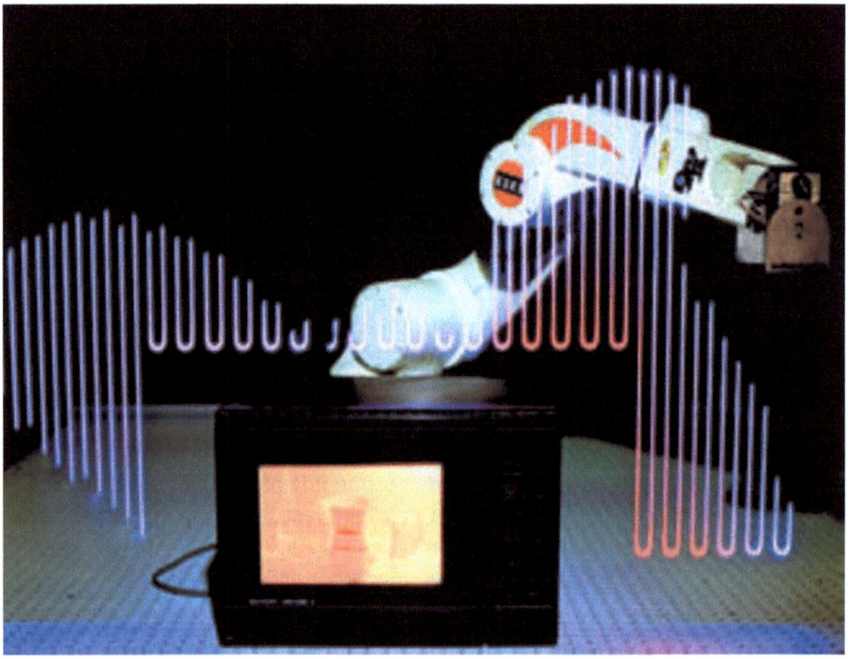

Figure 10 The microwave field around a microwave oven is seen using a scanning probe sensor and real-time light output. Note the higher field strength in the right corner indicates the location of the magnetron. Also, the sharp corners leak higher amounts of radiation.

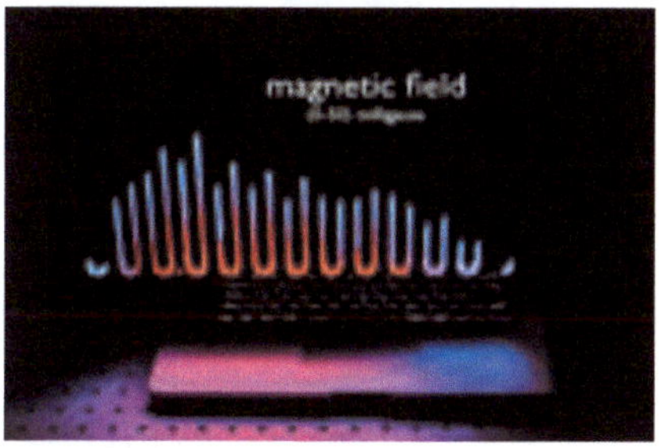

Figure 11 The magnetic field around a laptop is seen using a scanning probe and real-time light output. Note the higher field strength on the left side indicates the location of the hard drive.

construct immaterial designs. Other immaterial properties can be utilized to embody a design, such as magnetic or thermal radiation. Based on this definition, the multi-fabrication platform can be utilized in a range of interesting applications that are largely unexplored. Extending the use of robotic arms from fabrication to sensing allows valuable sensor arrays to be simulated using single sensors and scanning motion paths. This facilitates volumetric environmental data acquisition that can then be used for a variety of applied applications.

Informed fabrication combines fabrication and environmental sensing. With sensor data informing the fabrication process, the manufacturing process can start to take on roles both for process control and for design itself. Finally, this paper demonstrates the potential of immaterial and informed fabrication to transcend the utilitarian automation-centric role of robotic fabrication by proposing novel research areas where such platforms may not only execute but also inform the design process from its earliest stages to its complete and fully integrated physical manifestation.

Acknowledgements

This work was supported by an NSF EAGER grant award entitled "Bio-Beams: Functionally Graded Rapid Design & Fabrication", award number: 1152550. We would like to thank the Mediated Matter research group at the MIT Media Lab for their support. Specifically, we would like to acknowledge Thomas Lipoma, Julian Merrick, and Ali AlShehab for their assistance with robotic programming.

References

Callieri M 2004, "RoboScan: an automatic system for accurate and unattended 3D scanning", *Proceedings of 2nd International Symposium on 3D Data Processing, Visualization and Transmission, pp. 805-812.*

Chua CK, Leong KF, and Lim CS 2010, *Rapid Prototyping: Principles and Applications*, World Scientific

Gramazio, F and Kohler M 2008, *Digital Materiality in Architecture*, Lars Müller Publishers.

Kram R and Weisshaar C 2010, Outrace: The Paper, online, accessed July 1, 2012, www.outrace.org

Pires JN 2007, *Industrial Robots Programming: Building Applications for the Factories of the Future*, Springer.

Robots in Architecture 2011, Robot Masterclass @ Design Modelling Symposium Berlin 2011, online video, accessed July 1, 2012, <www.robotsinarchitecture.org/712/master-class-berli>

Xu J, Liu T, Yin X, Wong BS, and Hassan SB 2010, "Automatic X-ray Crack Inspection of Aircraft Wing Fastener Holes", 4th International Symposium on NDT in Aerospace, Mo.5.A.4

[1]. www.kukarobotics.com/en/*products/industrial_robots/small_robots/kr5_sixx_r850/*

[2]. http://www.rememberreach.com/

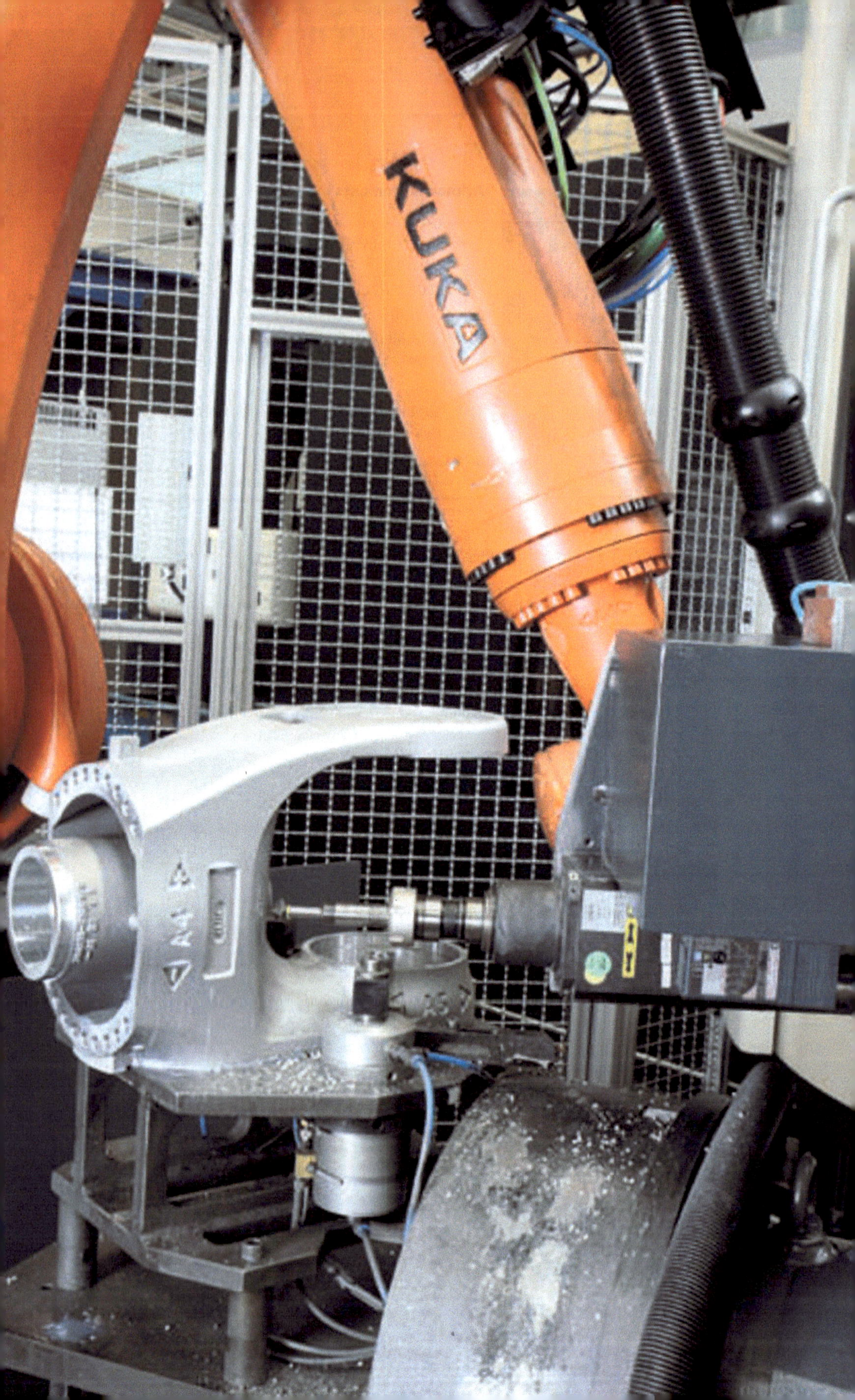
KUKA

Alois Buchstab

KUKA: Innovations in Industrial Robotics

Automation with Industrial Robots

The use of industrial robots has increased steadily in recent years. Automation is the decisive key to higher productivity and greater cost-effectiveness. It improves product quality, reduces cost-intensive use of materials and minimizes the consumption of dwindling energy resources. Robots replace the rigid and expensive special machines that were still customary fifteen years ago with highly flexible automation solutions.

In the past industrial robots were used almost exclusively in the automotive sector and in series production. Thanks to the systematic ongoing development of robot and control technology, industrial robots have now become established in many other sectors besides the automotive industry. The primary objective here is the development of applications in new markets – in the fields of plastics, metalworking, foundry, electronics, medical technology, but also in the creative industries, from entertainment to art and architecture.

New Robot Technologies

In 1996, KUKA presented the first PC-based robot controller, marking the dawn of a new era of "real" mechatronics, characterized by the precise interaction of software, controller and mechanical systems. Within the past two years, KUKA has introduced two completely new robot families, the KR QUANTEC series of high-payload robots, and the KR AGILUS, a small robot with extreme speed and high precision. Another recent development is the KUKA LWR (light-weight-robot), a small robot originating from the aerospace industry, whose seven axes allow extremely human-like motions. These advances in new hardware are supported by newly developed software and interfaces such as the KR C4 robot controller, the SafeOperation package for safe human-robot interaction, and the KUKA CNC package, which allows robot to be used as CNC machines, without requiring external software.

KUKA KR QUANTEC: Efficient High-Payload Robots

KUKA KR QUANTEC robots are characterised by their extremely high power density, offering up to 160 kg less weight and 25% less volume, compared to the Series 2000 robots, while keeping reach and payload unchanged. The KR QUANTEC robot family covers the entire high payload range from 90 to 300 kg, with reaches from 2,500 to 3,100 mm. As the most compact robots in their class, the reduced space requirements open up new fields of potential applications for production in confined areas.

The series has been designed by implementing a concept based on shared parts with just four motor and gear unit variants. All models have the same hole pattern for the mounting base, the same as that for the previous series, and an identical tool flange on the wrist, making the KR QUANTEC series thus fully compatible with existing cell layouts based on the Series 2000. The design of the KR QUANTEC series is distinguished by minimized disruptive contours and a compact wrist for accessibility, even in confined spaces.

As such, the KR QUANTEC robots can be used in virtually all branches of industry: the palletizers in the logistics sector, the foundry variants in foundry settings, the shelf-mounted robots in the plastics industry and the press-linking robots in the metalworking industry.

KUKA KR AGILUS and LWR: Innovative Small Robots

The most recent developments by KUKA in the area of small robots are the KR AGILUS and the LWR robots. While the new KR AGILUS small robot is characterized by precision and speed, the 7-axis LWR, a sensitive lightweight robot with integrated sensors, opens up a large range-of new possibilities both for innovative research and industrial applications.

The KR AGILUS is characterized by its extreme speed, short cycle times and high level of precision and safety. For handling tasks, especially Pick&Place, it offers minimized cycle times while at the same time working with great precision, enabling manufacturing quality of the highest standard. Its speed and accuracy make the performance of the KR AGILUS unique in its payload category. The basic model, KR 6 R900 sixx, weighing 51 kg, can carry a maxi-

Figure 1 KUKA KR QUANTEC robot in a tooling application

mum payload of 6 kg. The energy supply system of the KR AGILUS is integrated into the robot to save space and includes a 100 Mbit Ethernet cable, three 5/2-way valves for compressed air, a direct air hose, six digital inputs and two digital outputs. The KR 6 R900 sixx can reach points both near the robot base and also in the overhead area, performing its tasks as a floor, ceiling or wall-mounted robot.

The KR AGILUS is especially suited for operation in general industry, wherever automation with low payloads is required. In the context of universities or even architectural offices, robots such as the KR AGILUS can be used to quickly prototype robotic tasks, before moving on to a heavy-payload robot. KUKA SafeOperation makes the KR AGILUS especially suitable for such environments, as it greatly simplifies safe human-robot interaction. The software and hardware components of KUKA.SafeOperation monitor velocities and workspaces of both robot and external axes. This dispenses with the need for mechanical axis range monitoring systems and opens up new, cost-effective options for cell configuration and human-robot interaction. Similarly, the LWR has been specifically designed to share its workspace with the human operator in the future. A sensitive, lightweight robot, the LWR comes very close to the motion sequences of the human arm. The operator can manually guide the robot to different positions in the workspace and control and teach it using the very simple user interface. The LWR is able to perform demanding tasks that require high precision and a sensitive but powerful touch. With its in-built sensitivity, achieved by means of the integrated sensors, it is ideally suited to handling and assembly tasks. Its low weight of just 16 kg makes the robot energy-efficient and portable.

New Robotic Software and Interfaces

The robot control unit is the brain of every

Figure 2 KUKA LWR light weight robot

robotic arm. In addition to robot, motion, sequence and process control, safety control has also been seamlessly integrated into the new KR C4 control system. The KR C4 thus not only ensures the simple implementation of dedicated monitoring functions, but even more importantly the control technology ensures that the motion and velocity of the robot can be influenced safely.

By replacing limiting hardware with commonly-used, open industry standards, such as multi-core and Ethernet technology, the KR C4 offers a large performance and development potential. Based on these technologies Ethernet-based field bus systems, such as ProfiNet or Ethernet/IP, can be simply integrated as software functions. In this way the KR C4 concept will automatically benefit from future leaps in development and performance increases.

Figure 3 KUKA KR AGILUS robot

Furthermore, expanded software packages for a wide range of applications can be implemented at the controller. Especially interesting in the context of architectural fabrication is the KUKA CNC application, which enables the robot and machine tool to work together more efficiently as a system and simplifies production.

The smartPAD controller acts as the interface between user and robot control unit. Weighing only about 1,000 grams, the KUKA smartPAD offers a wide range of new user-friendly features, such as a USB port for convenient saving and loading of data directly on the control panel. It is operated using a large, high-resolution, 8.4" antireflection touch screen and a small number of keys, making it possible to control eight axes switching. When working with the KUKA smartPAD, the user is always offered precisely the operator control elements that are actually needed at any given moment.

Advances in Safe Human-Robot Cooperation

According to the International Federation of Robotics (IFR), there are around 234 industrial robots in Germany for every 10,000 employees in the manufacturing industries. As the number of robots increases, so, too, does their proximity to humans with the associated potential hazards. The objective is cooperation between the robot and the human, without endangering the latter.

Protection against injury takes the highest priority. Since it is impossible to preclude entirely the possibility of collisions between robots and humans in collaboration spaces, the minimum objective is to reduce the risk of injury to a tolerable level. International norms and standards help manufacturers and integrators to implement safe systems.

Where humans and machines work together in close proximity and physical safeguards impede work sequences, other measures must be taken to ensure the safety and protection of the human workers. What is required is a safe, "intelligent" robot that reacts immediately in the event of danger.

With the integration of Safe Technology into the KR C4 controller software, KUKA has taken a step towards the concept of a safe and intelligent robot for which there is increasing need in order to perform collaborative tasks. Such a robot must be equipped with safety controllers, permanently sense the motions of the human worker, determine the risk of collision and adapt its own Cartesian motions.

Research towards Green Robotics

KUKA is committed to conserving environmental resources and developing robots that offer maximum energy efficiency. The specifications regarding energy savings and increased energy efficiency must therefore be taken into consideration early in the process. Drives and components are optimally rated, for example, in order to avoid overdimensioning the robot. Furthermore, research by KUKA has recognized that programming and control are increasingly relevant in terms of energy-efficient robot operation. The latest studies show that there is significant potential for savings in path programming. In the past, the goal was always to program the shortest path. However, energy-efficient robot paths often differ greatly from direct point-to-point motions. The task is to generate a robot path in which the interplay of the axes consisting of repeated acceleration and deceleration and is coordinated so as to create the most energy-efficient motion profile.

The solution is sophisticated software tools in which the user merely defines the start and end coordinates. The software then calculates the energy-optimal path within the specified workspace quickly and with minimum effort.

Robot Design

Functionality and aesthetic appearance are not mutually exclusive. In the development of new KUKA products, industrial designer Mario Selic is consulted at a very early stage. The development engineers implement the

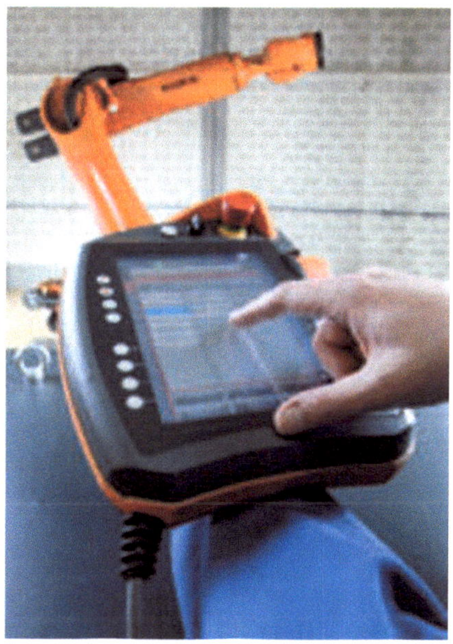

Figure 4 KUKA smartPAD with touch interface

latest robot-specific technology. The close interplay of design expertise and engineering prowess results in robots that not only meet the highest mechanical requirements, but also look good, as evidenced by the KR 270 R2700 ultra which won this year's red dot "best of the best".

For the company, the design reduces costs and energy consumption and increases the service life of the products, while also enabling the customer to benefit from ergonomic and intuitive operator guidance. For example, organically designed components with smooth transitions between structural shapes improve the mechanical force transmission and increase component strength. The nature of the design gives the robots a high degree of stability and stiffness.

Designing with Robots

Today, industrial robots are much more than robotic arms that replace manual labor in the automotive industry – they are present in nearly all industries and are used for a wide variety of tasks – even in various creative industries. KUKA robots have been present in movies such as *The DaVinci Code*, *James Bond: Die Another Day*, and *Tomb Raider*. They can also be found on the other side of the camera, as part of a system that uses LWRs as camera platforms for repeatable, complex camera movements.

With the development of new interfaces such as KUKA|prc, and projects such as the Red Bull arch, where a group of artists themselves fabricated 83 2x3 m foam molds with a KUKA robot, the creative industry has proven to be a serious user of industrial robots, capable of dealing with complex problems themselves. KUKA's open architecture supports these advances by providing users with the opportunity to delve deep into the software and customize the machines to individual needs.

www.kuka-robotics.com

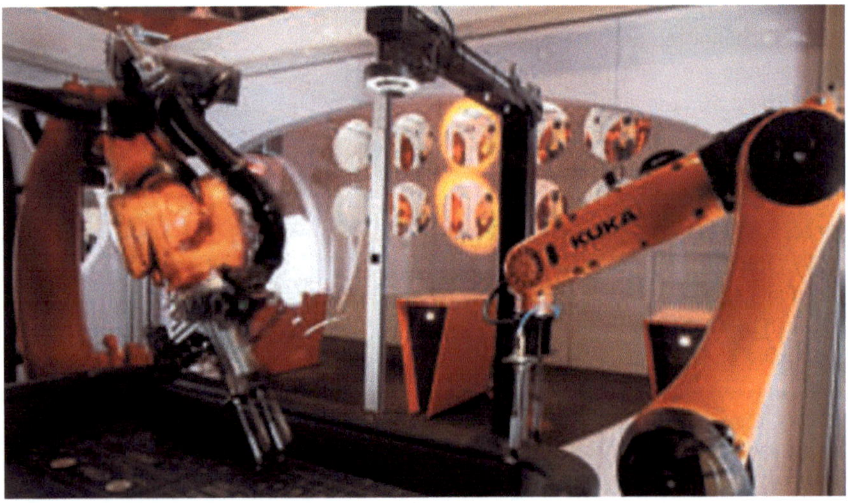

Figure 5 Cooperating KUKA robots at Automatica

KUKA
KR QUANTEC F
KR QUANT
automation
becomes easy
automation
becomes easy
Montageprüfung

Valentine Troi

splineTEX

Architectural composite materials

Introduction

The rapid development in CAD (Computer Aided Design) software in recent years has made new, previously unimaginable shapes for architecture and design possible. Very often, however, these ideas remain in virtual space. While elaborate freeform shapes and organically-morphed geometries have become state-of-the-art in architectural design software, the 1:1 conversion of free-form structures into reality – i.e. into systems with material properties, where a thickness of 0 does not exist – is still highly complicated and therefore expensive.

This – often unattainable – aesthetic appeal of a CAD-wireframe informed the work of Valentine Troi, at that time lecturer and researcher at the Institute of Experimental Architecture and Construction at the University of Innsbruck, which finally led to the development of splineTEX.

splineTEX

splineTEX is a multi-phase composite material that can be simply formed into the desired shape in the soft state, before it is then cured, rendering the elaborate and costly production of molds currently required for the preparation of free-formed structural components superfluous. Using splineTEX fibre composite profiles, complex geometric shaped, structural elements can be produced across a range of sizes at a considerably reduced cost. splineTEX structural elements are available in diameters from 5 mm up to 50 mm. Carbon and glass fibers are processed as standard. If desired, natural fibers (hemp, linen, flax) and special fibers (hybrids, aramid, basalt) also provide unique advantages. Carbon fiber structural elements suitable for weight-sensitive applications can achieve a flexural modulus of 75,000, corresponding to the bending modulus of elasticity of aluminium, at a

Figure 1 SplineTEX forming (left and center); hardened splineTEX element (right)

density of 1, 45 g / cm ³ (aluminium density 3 g / cc), i.e. the same performance at half the weight.

New splineTEX Markets

The technology has been optimised for the production of structural elements with a length up to 3 meters, although in special cases lengths up to 10 meters can be produced. After three years of development work on splineTEX the client base is no longer made up solely of architects and designers. The automotive industry, as well as the aerospace and boat construction sectors are also showing a keen interest in "flexible composite profiles", and the use of this material, originally developed for architectural and design applications as an ultra-light alternative to formed metal components (e.g. aluminium), has been widely discussed at materials technology trade fairs and symposia.

Robotic Forming of splineTEX

Prototypes for detailing a composite-based car of the future are currently at the planning stage. Working together with the Association for Robots in Architecture, an automated production concept for a CFRP body structure is being produced under contract for an automotive OEM. Instead of constructing a large number of individual molds, the flexibility of the material allows a large-scale winding of splineTEX material blanks around a single mold, producing multiple splineTEX parts in a single robotic process. The large workspace required, high precision and repeatability of the winding in

Figure 2 superTEX tower: Vertical shaped net structure made with splineTEX carbon.

a short cycle time is achieved by cooperating KUKA robots, with one robot rotating the tool, while the other robot holds the splineTEX blank, Using KUKA|prc, the robotic fabrication process can be simulated beforehand in the parametric design environment Grasshopper. Similarly, material simulation allows the quick verification of the winding process. It is expected that the combination of flexible, modular molds with fully parametrized robot programming using KUKA|prc will enable the rapid shaping of a high number of individual splineTEX parts.

Further research to discover new ways in which robotic arms, as inherently multifunctional machines, can be used to shape the similarly adaptable and multifunctional material splineTEX is currently ongoing.

www.supertex.at

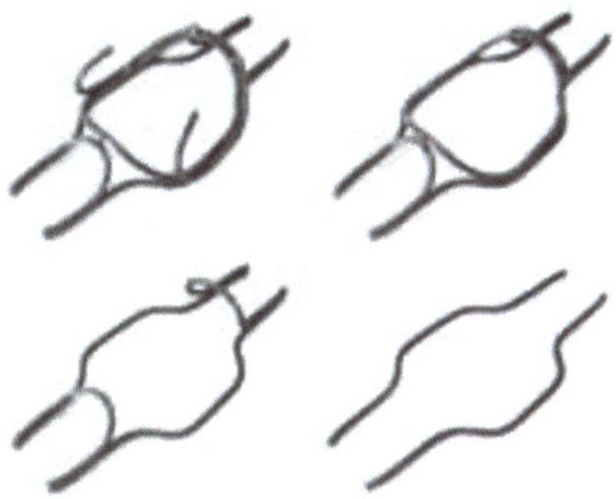

Figure 3 Automotive chassis assembly via parametric, modular tube structures

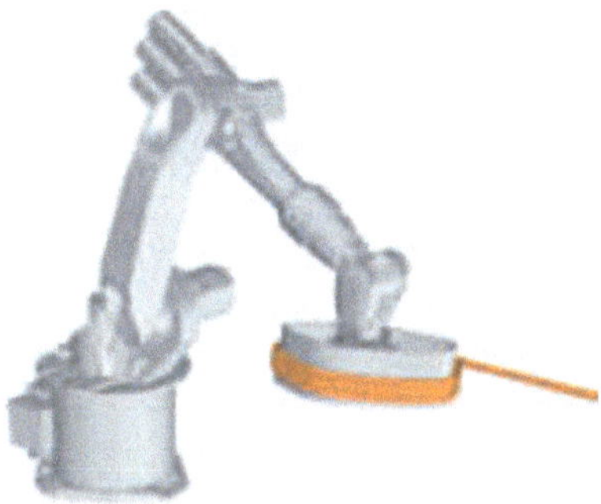

Figure 4 Robotic winding process fully simulated in a CAD environment

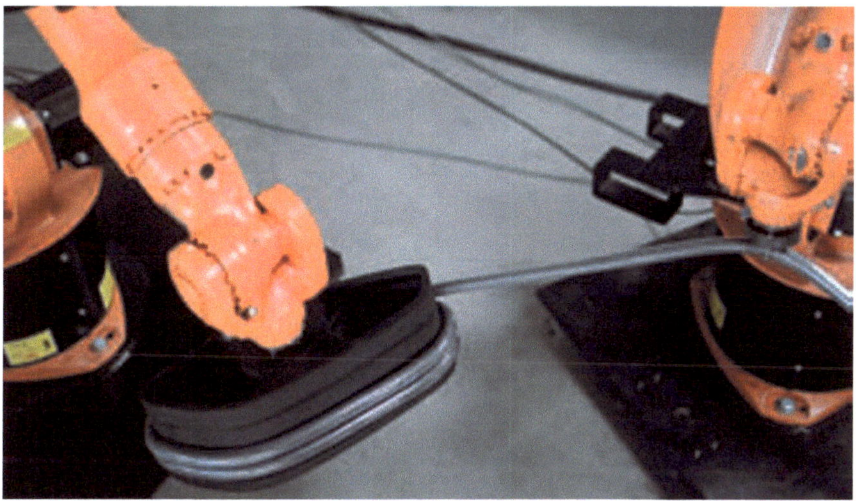

Figure 5 KUKA robots forming a 1:5 scale chassis part, parametrically programmed in KUKA|prc

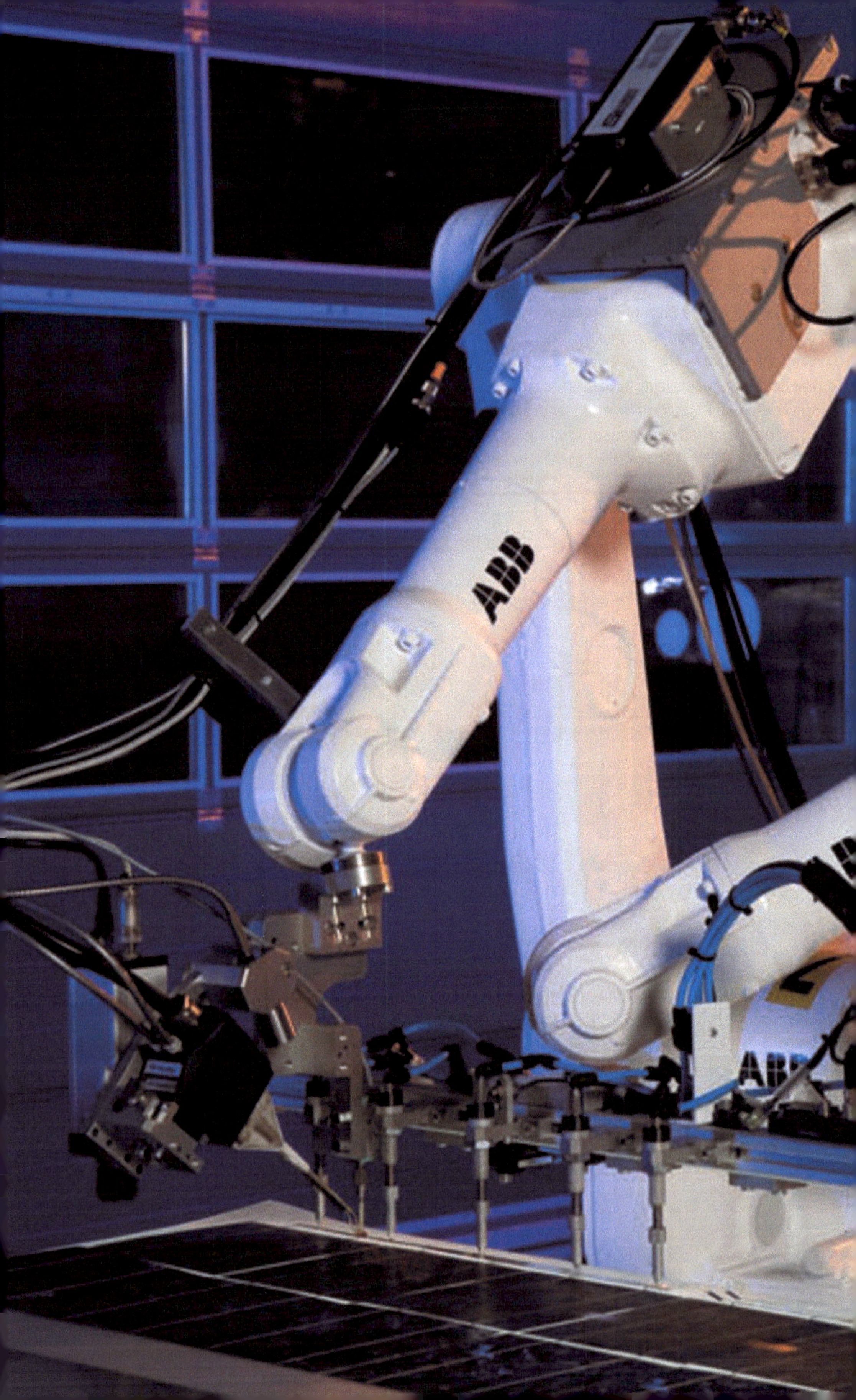
ABB

David Kittl, Martin Kohlmaier

Industrial Robots in Architecture

Trends and innovations from ABB

Robots have been proven to deliver significant benefits in a wide variety of applications. Introducing robots to production processes typically leads to a significant transformation in productivity and efficiency, with higher levels of output, product quality and especially flexibility.

Innovations – Fuel for Differentiation

Innovations such as ABB's Force Control technology for surface finishing processes, true offline programming with ABB's RobotStudio software, frequently in combination with dedicated CAD/CAM-solutions, MultiMove robot systems with up to four coordinated manipulators and revolutionary manipulator models like the compact IRB 2600 with sharp accuracy and short cycle times or the powerful IRB 6660 for tough machining processes enable undreamed-of possibilities for new applications in architecture and the construction industry. In today's industrial production factors like uptime, energy efficiency and increased workplace safety are the keys to success when it comes to automation solutions with robots.

Force Control Technology

One key development that promotes the more efficient use of robots in the processing field is a new technology that regulates force and position. ABB's new RobotWare Machining FC (force control) simplifies the previously complex track programming during processing. One aspect of the technology, FC Pressure Process, provides improved process quality and programming through force-controlled motion perpendicular to the surface. Force in the sensor-controlled direction and speed along the surface is constant. The path is adapted to the curvature of the surface, using a controlled material-removal rate. The programming is done in process parameters instead of positions, thereby taking significantly less machining and operating time. With FC Speed Change, another aspect of the technology, cycle times on a predefined path can be improved. A controlled material removal rate is based on the force acting on the tool. By using the force acting on the tool, a maximum process speed can be maintained, while automatically slowing down when the process forces are too high. The force in path direction is constant, while the speed is variable. The results are increased path accuracy and minimized risk of damage to work objects, tools and the robot. As a benefit of using the new force control technology, the system becomes significantly more flexible and can also be used for smaller lot sizes and a larger number of variations.

Offline Programming & Virtual Robot Technology

RobotStudio is ABB's premier solution for offline programming. Offline programming reduces the risk by visualizing and confirming solutions and layouts before the actual robot is installed, and generates higher part quality through the creation of more accurate paths.

The most timesaving feature of RobotStudio is the so-called AutoPath routine. By using a CAD model of the part to be processed it is possible to automatically generate the robot positions needed to follow geometry, providing a significant advantage for CAD-based workflows as used in architecture.

MultiMove & Motion Control

The IRC5 industrial robot controller builds upon more than four decades of robotics experience. With the control of up to four robots by only one controller, with a compact drive module added for each additional robot, MultiMove opens up previously unimaginable operations, thanks to the perfect coordination of complex motion patterns. Based on advanced dynamic modeling, the IRC5 optimizes the performance of the robots for the physically shortest possible cycle time and precise path accuracy. Together with a speed-independent path, predictable and high-performance behavior is delivered automatically, with no tuning required by the programmer.

Figure 1 Cooperating ABB robots

Safety Solutions

Operator safety is a central quality of the IRC5 robot controller, fulfilling all relevant regulations with good measure, as certified by third party inspections. Electronic position switches (EPS) and SafeMove represent a new generation of safety, enabling more flexible cell safety concepts, e.g. involving collaboration between robot and operator. These innovations mark a major step in removing the bonds placed on heavily regulated industrial robots that toil in isolated settings. "Safe Stand-Still" – as a key function – supervises the stand-still of all robot axes without having to switch the robot to "Motors Off". It enables operators to perform tasks in the immediate vicinity of the robot, saving cycle time and wear to contactors and brakes. Developed and tested to comply with international safety standards, SafeMove is a software and electronics based safety solution that ensures safe and predictable robot motion.

Follow the Trend

Robots offer speed and accuracy that cannot be achieved with human labor, they reduce operating costs and scrap and, most importantly, are extremely flexible. The new technologies outlined above, from CAD-based offline programming to the simultaneous control of cooperating robots, allow robots to be used in entirely new fields such as architecture, where a fluent CAD workflow is imperative and multiple robots may be required to process large workpieces.

www.abb.com

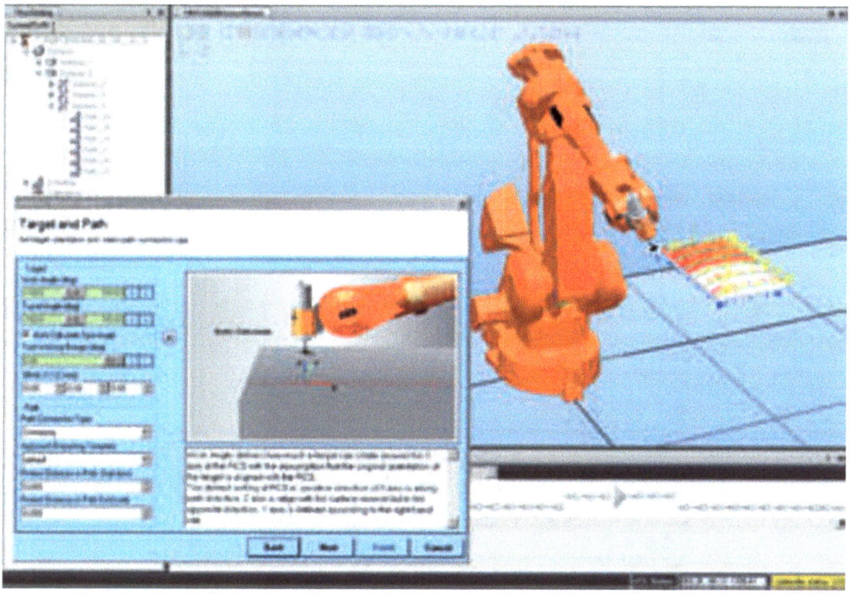

Figure 2 Robot programming using ABB RobotStudio

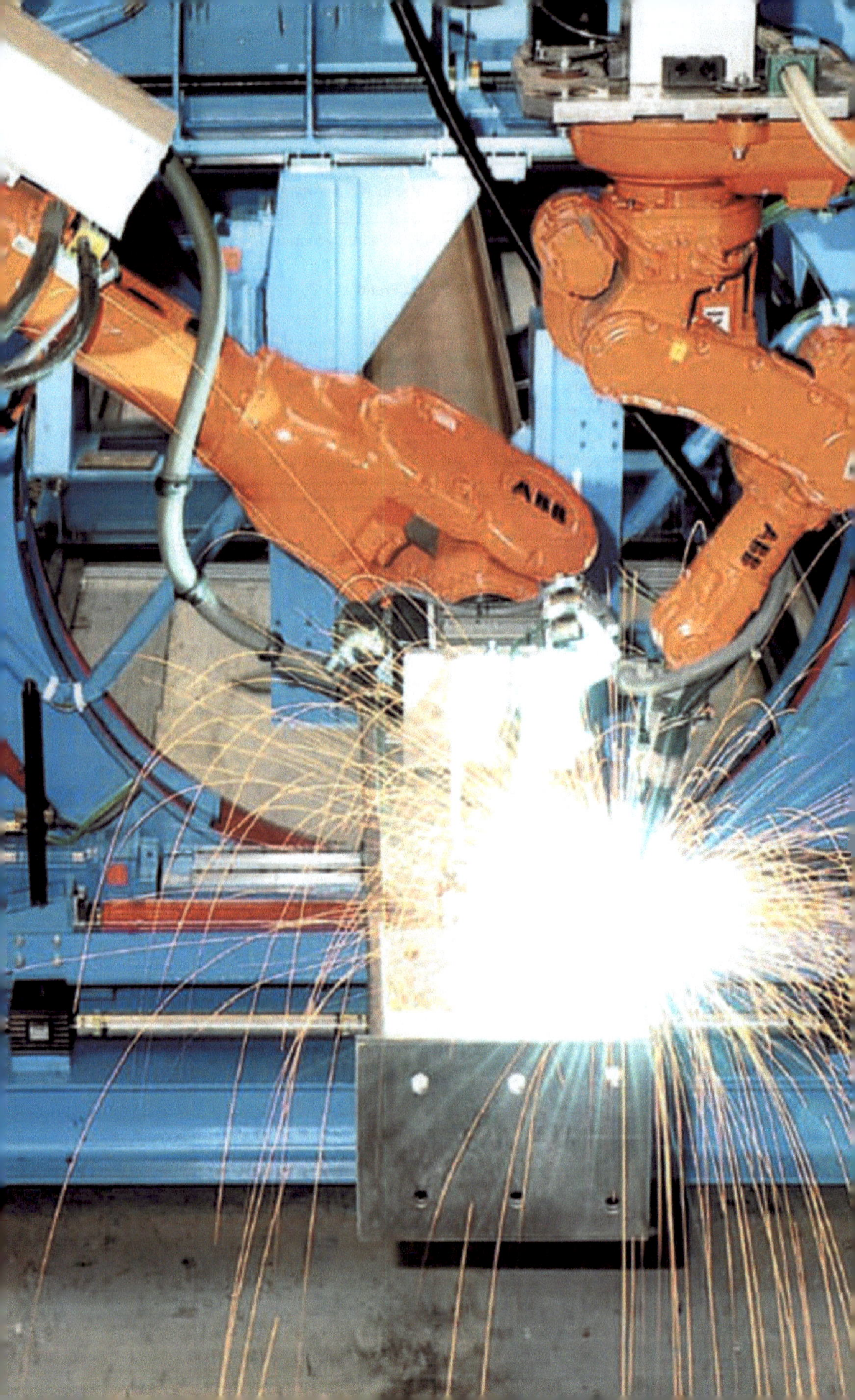
ABB
ABB

Andreas Hofer

Revolution in Steel Beam Fabrication

Introduction

Based in Styria in southeastern Austria, ZEMAN has developed a fully automated production line for practically any type of steel beam that includes add-on parts. The individual components arrive on a conveyor at one end, are processed by industrial robots, and emerge a short time later at the other end as an assembled steel beam that then only needs to be given a protective coating and loaded onto a truck.

Current Steel Beam Fabrication

Until recently, the only way fabricators could put their steel beams together was by hand. Traditionally this meant assembling structural steel elements by manually welding on the headplates, baseplates, connection plates and whatever stiffeners might be needed to meet structural design requirements. There are several drawbacks to this. As this approach is no longer economically viable in Western countries, more and more firms have relocated their production to low-wage countries, only to find that problems with technical standards, meeting deadlines and quality control often eat up much of the cost advantage. As a result, structural steel fabricators in Western Europe have increasingly lost ground to competitors from further east. These, in turn, have to contend with a constant shortage of qualified personnel, and frequently with an inability to meet the required quality standards.

The Robot-based Steel Beam Assembler

Zeman has now developed a fully automated production line, the Steel Beam Assembler (SBA), capable of performing all the steps involved in fabricating a steel member. Instead of manual labor, computer-controlled robots execute the CAD engineering drawings 1:1 – with no errors, and in a fraction of the time that would otherwise be needed. The first step is that a conveyor system feeds all the add-on parts into the production line past a high-powered scanner, which not only detects the position of each of the randomly placed parts, but also captures its actual dimensions and compares these with the target values given in the CAD drawings. This ensures compliace with the specified tolerances. The scanner relays all these data to the positioning robot, in real-time. This robot now has the job of picking up the parts one by one, and positioning them correctly on the steel beam. To do so, it uses several different magnetic grippers with which it can safely pick up the various add-on parts that can be of differing size and weight. If any parts come in from the infeed table facing the wrong way, the positioning robot can briefly set them down on a holding device on the robot table, and then pick them up again with the right orientation. Again, the positioning robot handles these tasks much more quickly and accurately than a human being ever could. There is no longer any need for laborious measuring up, or for the comparatively "fiddly" job of manually attaching the add-on parts. With diagonally arranged parts, too, the task of positioning these dimensionally accurately, and fitting them in the correct positions, is also carried out smoothly and flawlessly. Angles, tappets and frame corners are also no problem. In the case of thick add-on parts and large weld-seams, it is important to pre-heat the parts to be joined. Otherwise the metal may be distorted, or stress cracking may occur. This pre-heating is also carried out by the positioning robot, using a heat torch.

Once the plates are correctly placed on the steel beam, one of the two welding robots comes into action, tack-welding the plates to the main member in the order dictated by the positioning robot. If higher capacity and output are needed, the production line can be designed with two parallel lines. In this dual-line configuration, while the two welding robots finish welding the workpiece on one line, the positioning robot can carry on working on the other. The welding robots are also equipped with all the necessary tools: welding torch, plasma cutting device and laser measuring system. Changeover to whichever tool is needed is also performed fully automatically. The two welding robots are mounted on a shared longitudinal trackway, but can be separately controlled. For the transverse and vertical strokes, there are two further traverse-axes for each robot. The vertical axis also has a telescoping unit which enables the line to be installed in production buildings of the usual height. Each of the welding robots is also equipped with laser measuring heads. This enables them to recognise the actual fit-up situation and compensate for any weldment tolerances. And with the plasma cutting device the robots can trim the edges ready for large-volume welding seams, and make any openings needed in the web, or cut-outs on the members. When the steel beam is finished,

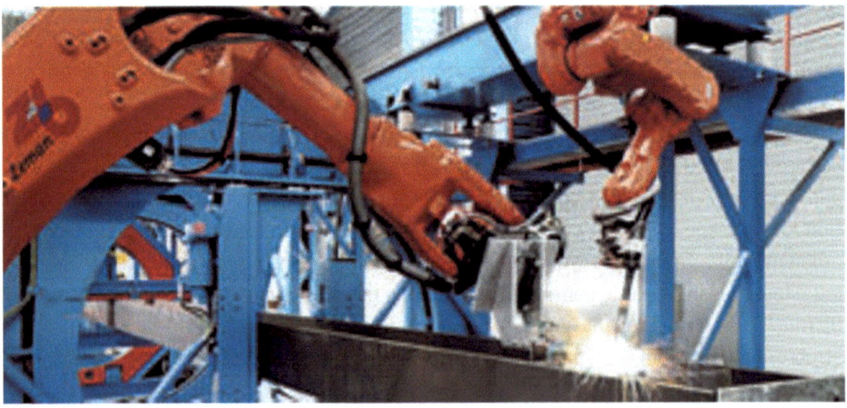

Figure 1 Robot welding at Zeman

it emerges from the line via the unloading device. The line is designed in such a way that not even a power cut has any damaging consequences; once power has been restored, work can resume unhindered.

Conclusion

SBA's complete automation means that high wage and ancillary labor costs etc. now play only a minor role, making industrial production in e.g. Western Europe much more competitive against rivals in Eastern Europe or Asia. For western producers, the substantial reductions in labor, energy and materials costs mean a shorter payback period. For low-wage countries, on the other hand, the machine is attractive because it enables them to supply top-quality goods even without well-qualified personnel.

Reductions in production time of up to 85 percent can be achieved: Instead of at least eight hours for a tonne of steel, it now takes just under two hours for the beam to be completely finished. In addition to advances in efficiency, the SBA's automation facilitates the production of highly individual steel members. This enables planners to employ custom-tailored structural elements, thus saving costs and allowing for new geometrical solutions.

Acknowledgements

This innovative technology is the outcome of nearly five years of research and development, and 4.5 million euros, partly funded by the Austrian Research Promotion Agency (FFG). One of the main challenges was the development of the software, which was headed by Zeman itself with the help of four programming subcontractors. Various firms were involved in the design of the hardware, with Plasmo and Arsenal Research creating a highly accurate scanning system that is robust enough for the shopfloor, Fronius supplying the welding technology, Güdel the linear guideway system, and ABB acting as robot supplier and principal hardware developer.

www.zeman-stahl.com

Figure 2 Robotic steel beam fabrication.

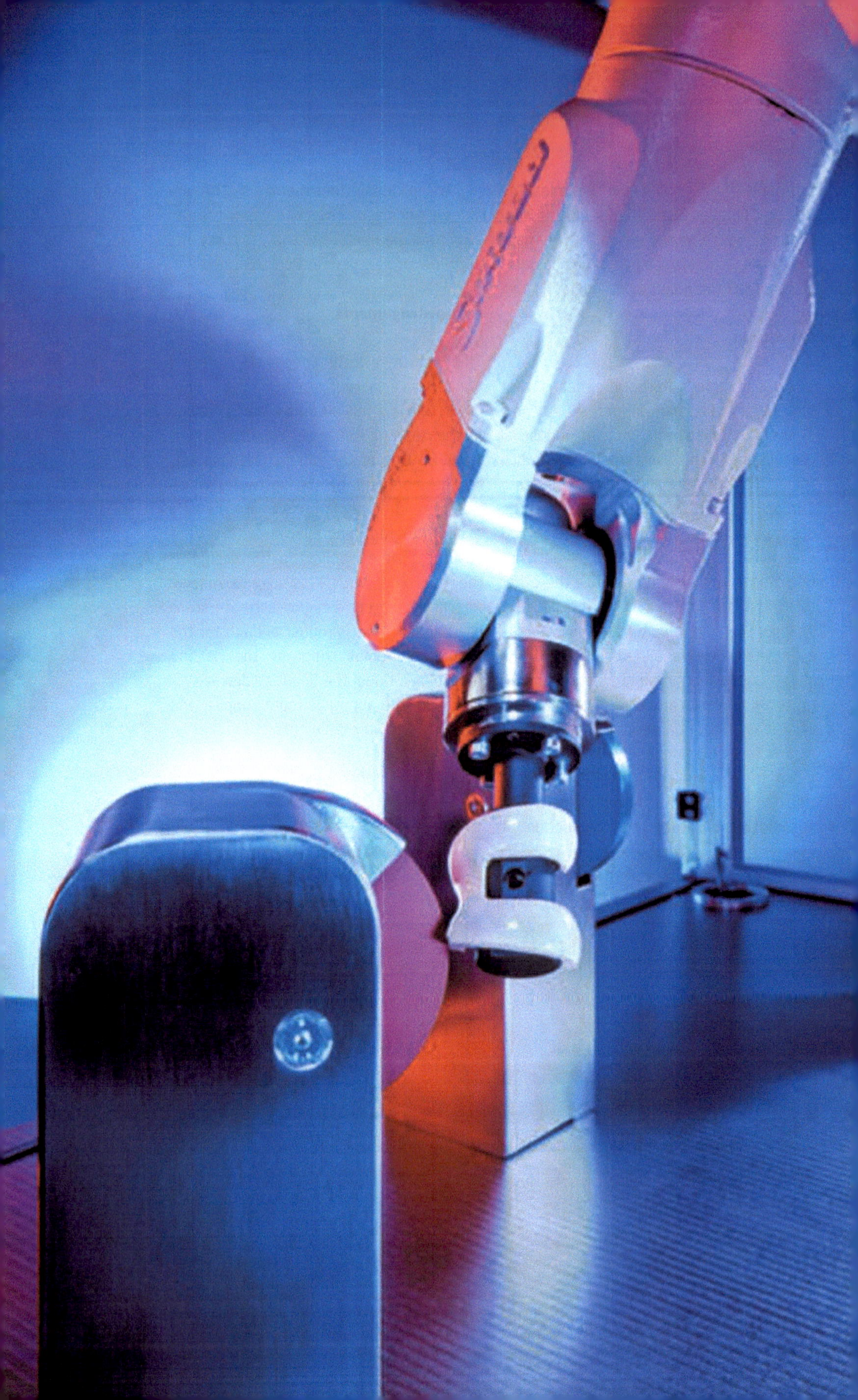

Manfred Hübschmann, David Arceneaux

21st Century Art: The Marriage of Inspiration and Innovation

Robots evolve to become the artist's high-performance tool

Introduction

The precision to capture and accurately realize an artist's vision; the speed to produce and reproduce objects en masse with flawless consistency; the strength to lift, move and build with stone and brick, quickly and efficiently. On the frontiers of design and architecture, robots are empowering the creative human element to challenge the old "1% inspiration, 99% perspiration" adage, freeing up more space for creativity by taking over labor-intensive tasks – and performing them more quickly, accurately and flexibly than a human being ever could.

This is not the basic pick-and-place industrial machine of the 1960s. When the Swiss mechatronics company Stäubli introduced its RX Series 6-axis robotic arm in 1992, it was unlike any robot that had ever been seen. It featured a unique fully enclosed structure armor that enabled it to withstand the rigors of harsh industrial environments, made it easy to clean, and protected its inner workings, while allowing it to move with remarkable dexterity and precision.

Stäubli Robots in Industry: the Possibilities Multiply

The RX Series provided the basis for what is now a wide range of 4- and 6-axis robots from Stäubli, including the TS, TX and TP Series low, medium and heavy payload robots, with handling capability from 1kg to 250kg. The range is highlighted by the TX40, TX60 and TX90 6-axis robots, launched in 2004. The most widely used of all Stäubli robots, the TX60 and TX90 (low and medium payload robots respectively) belong to the fastest and most precise robots on the market.

Both feature the enclosed structure of the original RX Series, along with a unique spherical work envelope that allows maximum utilization of cell workspace. The flexibility of multiple mounting configurations, including floor, wall and ceiling options, makes them easy to integrate.

Stäubli robots have been adapted to a wide array of industrial applications. They provide safe, sterile handling for repetitive tasks in the pharmaceutical, life science and food industries. Their speed, reliability and accuracy have enabled new operations in the automotive, plastics, machining, semiconductor and solar industries, optimizing efficiency, safety, productivity and product quality, even where very delicate handling is required.

TX60 and TX90 robots are also available in specialized versions adapted to the specific requirements of various industries. These more application-specific automation solutions include:

- Painting robots: spraying and painting robots controlled using PAINTIXEN Advanced Control Software, which allows the user to control all paint parameters.
- Plastics robots: 6-axis robots perform operations such as insert loading, assembling and fitting components, precision stacking and packaging, as well as trimming, despruing, reorienting and loading of molded parts.
- Cleanroom and Stericlean robots: these consist of ISO Class 2/3 and Class 4 certified robots for use in semiconductor, biotechnology, pharmaceutical, medical and other industries. Stericlean robots are designed specifically for the VHP (Vapor Hydrogen Peroxide) decontamination process used in the pharmaceutical/life science industries.
- Humid environment robots: the TX HE robot is adapted for applications in very humid environments, such as waterjet cutting, cleaning and food processing.

Beyond the Factory Walls

In the latest phase of their evolution, Stäubli robots are being adapted to innovative uses far beyond the traditional factory setting. Their high speed, precision and flexibility are sparking the imaginations of those looking for alternative fabrication and production techniques as well as more

Figure 1 Range of Stäubli robots

freedom to create. The elegantly designed robots are even popping up in postmodern art and high-tech entertainment.

Sculpting in Stone

A supplier of natural stone products in the U.K. needed a new approach to machining stone when it was offered the opportunity to provide artistically sculptured sandstone planters and seating units for its city's main square, a popular theater district experiencing a renaissance. The stonemason chose Stäubli 6-axis robots for their ability to machine stone in a wide variety of standard and non-standard shapes with a high degree of precision. Originally designed for high speed machining, the robots allow the user to integrate a variety of different spindles directly into the robot forearm. User-friendly software is used to control the robot and manage all of its functions. The robots proved capable of machining not only natural stone but also a range of other materials, from alloy and stainless steels to Inconel 600 and aluminum, with consistent accuracy. They also opened up new possibilities for surface texturing of stone, significantly adding to the aesthetics – through automation, rather than perspiration – allowing the artisan to achieve textures that would be difficult, if not impossible, by any other production process. The endeavor also proved more cost-effective than traditional CNC machining.

Designing the Future of Architecture

In 2012 a prestigious architectural school in Los Angeles, California opened a design and construction studio featuring six Stäubli robots. The robot-centered environment serves as a platform for hands-on collaboration and experimentation with the robots, enabling students and researchers to delve deeper into the physical realities of design concepts in real time.

The studio is comprised of two adjoining spaces. The main area features five large Stäubli robots configured into a multi-robot work cell. Adjacent to this is a lab equipped with a compact Stäubli TX40 6-axis robot. In this pioneering workspace, Stäubli robots support a wide range of applications, including on-site construction. The 6-axis robots can move and cut in the X-Y-Z planes, like more traditional 3-axis CNC mills, but they can also rotate 360 degrees around an object. The finished product is more exact, so there is no need to use sandpaper to smooth out edges and surfaces. With fewer constraints and greater independence, the architects are able to venture beyond the standard design, programming, fabrication and construction sequence.

Those leading the robotics program at the school are investigating techniques of networked manufacturing in architectural design – in essence blazing a trail for what could well be the future of architecture. Various projects explore movement as well as free-form fabrication with advanced composite materials.

Robots in Art and Entertainment

Stäubli robots have been featured in a number of movies over the years. Their innovative design and fluid, anthropomorphic movements make them a natural fit for films set in an imagined future. Here in the 21st century, the role of robots in entertainment is being expanded by a French company that stages high-tech multimedia productions starring Stäubli robots.

As they are capable of a range of subtle movements, the robots are programmed to "dance" to music in live events as well as in video and movie projects. Along with music and lighting, the productions incorporate props, wherein the robots hold and swing various end-of-arm objects, such as light sticks, which can be switched quickly. The robots' six degrees of freedom make an endless variety of staged scenes possible. Choreography is done either live, by a musician, or programmed using software that controls the robots as well as music and other elements of the performance. The robots can accelerate rapidly, with 6D spline trajectories allowing smooth or jerky movements. The result is a compelling spectacle of light, sound and motion.

Stäubli robots also serve as pieces of moving modern art. In Switzerland, a Stäubli TX60 is mounted on the lobby wall of an exclusive 5-star hotel. Grasping and moving light fixtures, the robot is an ultra-modern design element, blended intriguingly with a restored historic ceiling and antique fireplace. The hotel earned SpaFinder's International Best Interior Design award, in addition to numerous other national and international awards.

www.staeubli.com

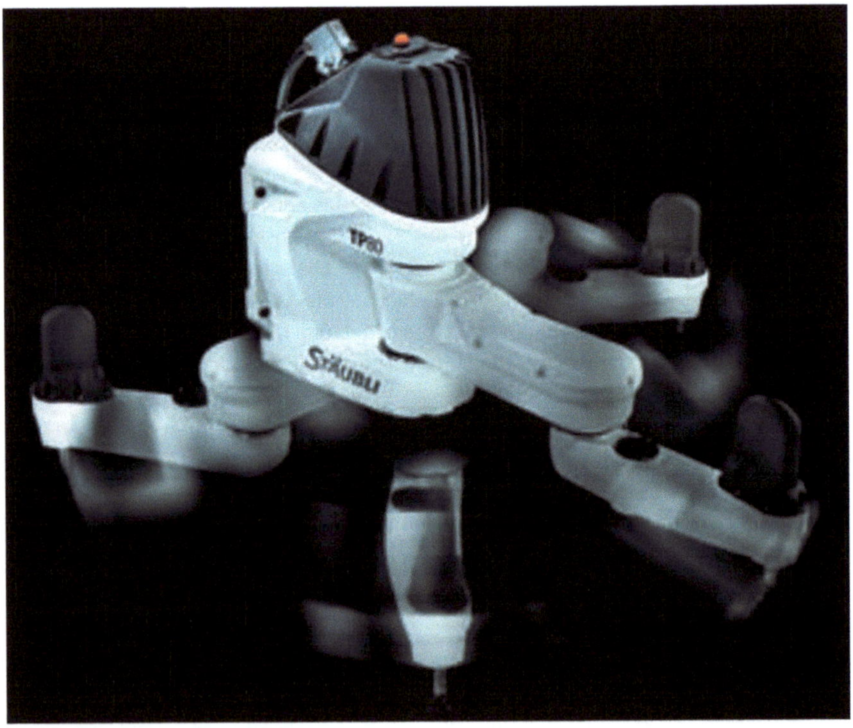

Figure 2 Stäubli TP-80 Fast Picker robot

STÄUBLI
UNIMATION

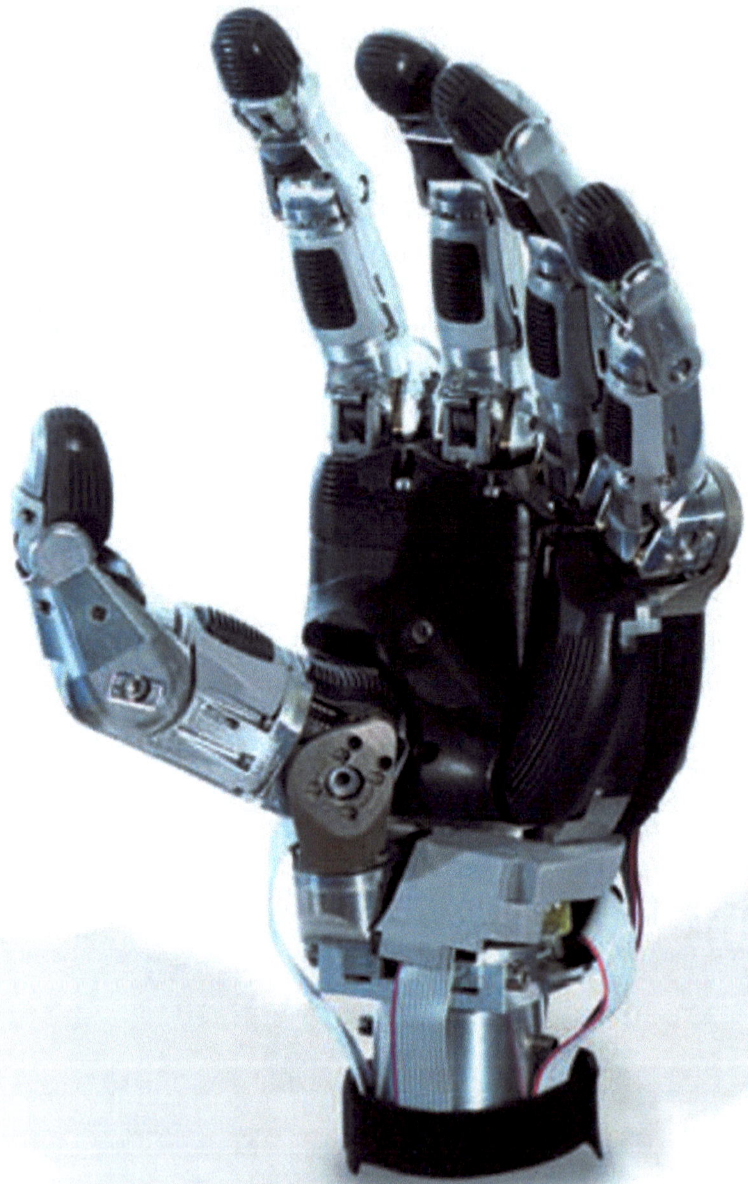

Christian Binder

Modular Robotics

From individual modules to complex robotic structures

Introduction

Robotic arms are generally designed as universal machines capable of performing a multitude of different tasks, similar to a human arm.

However, there are also very specific applications, which cannot be covered by commonly available industrial robots, as e.g. non-standard kinematic layouts with 6 or more degrees of freedom, or cordless operation using batteries may be required.

Modular Robotics

Schunk, best known for its clamping technology and gripping systems, has been developing robot modules as independent functional units for years. These units can be joined to complex systems via compatible and standardized interfaces. Compact and flexibly combinable rotary actuators, lightweight manipulators and servo-electrically actuated grippers, are used to create unique and modular-designed special robotic solutions. These custom-built robots

Figure 1 Rail-mounted POLAB robot

are used mostly for lab automation, service robotics, and research and development.

Exemplary service robotics applications are AMaRob, a semi-autonomous robot designed to help disabled and elderly people with their daily life activities, and the Care-O-bot, a robot that assists people in home environments. These special systems require mobile, light-weight, battery-powered robotic arms that can be safely operated next to their users. A research and development application is the POLAB Shuttle system, which consists of a modular robotic arm that is mounted on a mobile platform (Fig. 1). Moving along rails, the robot can work in a shared human-machine workspace and e.g. deliver samples to analysis equipment.

Innovative Robot Modules

As an expert in gripping technology, Schunk has developed many innovative gripper modules for robotic arms. Among the most sophisticated is the 5-finger hand, which uses nine drives to carry out various humanoid gripping operations. The seemingly simple 2-finger parallel gripper WSG also features high-end technology inside, such as its own web-server for configuration and diagnosis, as well as a microSD slot for storing and exchanging programs. Even the gripper fingers can be highly customized, as the polyamide fingers are laser-formed, a rapid prototyping process that allows highly complex geometries and a delivery time of just three days. Another recent development is the Powerball ERB – a single robotic module which contains two axes with minimum space requirements (Fig. 2).

Powerball Light Weight Arm

Three of these Powerball components make up the core of the Schunk Powerball Lightweight Arm LWA 4.6 (Fig. 3). Despite its low weight of 12 kg, it can dynamically handle loads of up to 6 kg, and has a gripping radius of more than 700 mm, while using only an average of 80 W of energy. The Powerballs contain all supply lines for gripper and tools, and can even be equipped

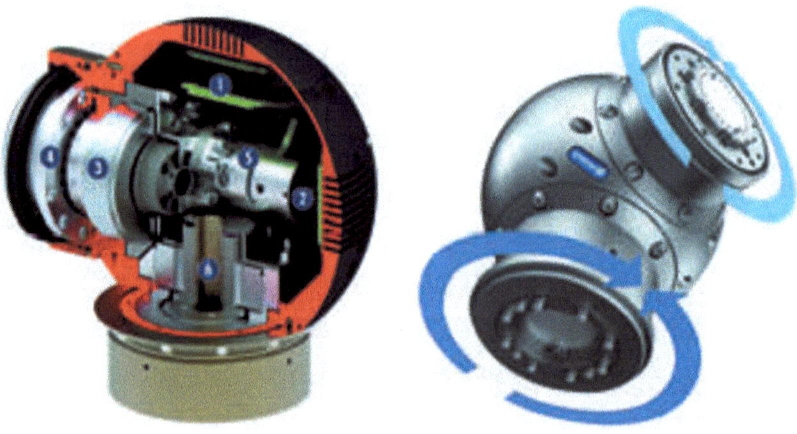

Figure 2 Powerball robot component

with an integrated force-torque sensors that allows sensible interaction with the environment. Even without the sensors, the LWA's design prevents dangerous crushing and shearing movements, making it safe to use around people.

Conclusion

The use of a modular robotic system allows the creation of highly optimized robotic arms for a wide variety of applications. Through open interfaces and even open-source hardware drivers, these robotic arms can then be easily integrated into a wide range of software systems, matching modular robotic hardware with customized software. These solutions are already used for research and service robotics applications, but have the potential to lead to entirely new types of robotic applications in the hands of the creative industry.

www.schunk.com

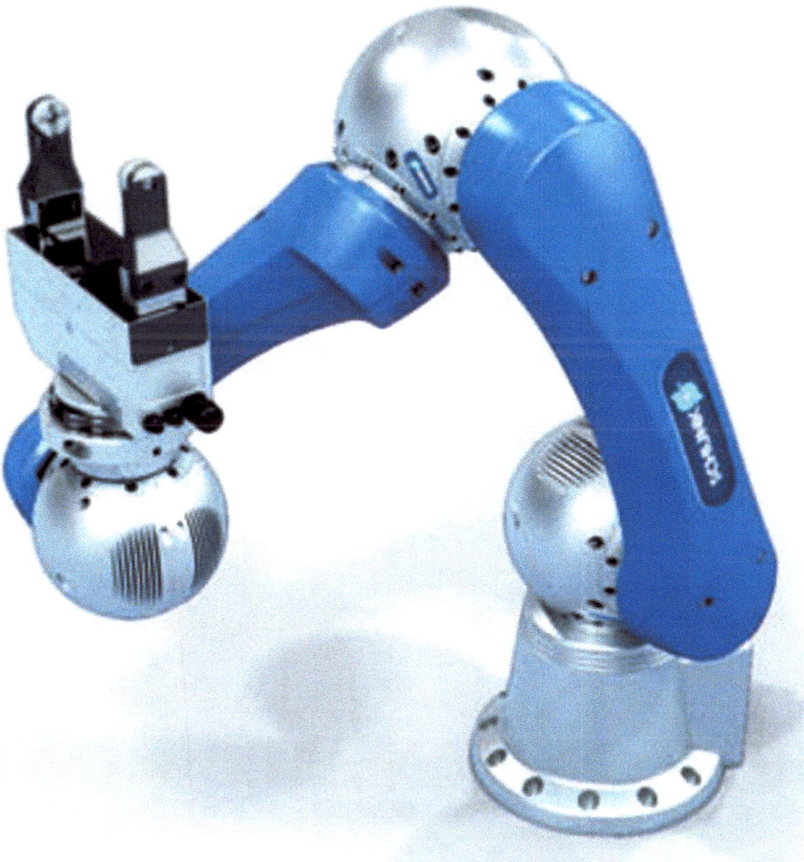

Figure 3 Powerball-based light weight arm LWA 4.6

Alfred Kaser

New Perspectives for Architecture and Design

Fabrication using robotic machining centers

Introduction

One of the main challenges – be it in engineering, architecture, or design – is to combine form and function. With regard to both of these aspects, the choice of material is a deciding factor when designing elaborate three-dimensional structures, as it has an impact not only on the structural properties, but also on an object's visual and haptic appeal. Fabricating such structures in an economic, energy- and resource-efficient way requires in-depth knowledge about the properties of materials, as well as innovative approaches to how they can be processed.

Robotic Machining Centers

A² Anlagentechnik & Automation GmbH is based in Seligenstadt/Germany and develops and produces highly flexible robotic machining centers (RMCs). These devices offer repeatability of up to 0.2mm making them suitable for applications that until recently required expensive 5-axis CNC machining centerss. In comparison to these machines, RMCs offer the following advantages:

- Reduced costs of purchase and operational expenses
- Greater availability
- Low maintenance effort
- Faster tool changes
- Greater flexibility
- Easier programming and handling

RMCs are capable of processing the same materials as 5-axis CNC machining centerss. This enables architects and designers to use a full range of materials for their designs, from aluminium, steel, wood, and composite materials to aerated concrete, ceramics, and even natural stone. Due to the fast and easy tool changes, the robot can perform multiple, different operations such as milling, drilling, cutting, sawing, deburring, grinding, or polishing.

Additionally, robot solutions feature a combination of high flexibility and great range. A robot with six axes covers in its base configuration a range of 2800mm, which translates into a work envelope of 68 m^3. By placing the robot on a linear axis, this workspace can be enlarged at will, thereby enabling an economic processing of large parts.

Robot Programming

Another advantage of industrial robots is their comparably easy programming. The very first robots demanded profound knowledge of their proprietary programming language, as robot positions had to be programmed manually, point by point. This changed with the rise of robot programming strategies such as teaching, which is still used frequently today. Teaching requires the user to manually guide the robot's tool center point (TCP) to each position. The coordinates are then saved, and can later be combined to a single robot trajectory. Robot simulation programs marked another significant innovation in the area of industrial robotics, as they allow the programming of robots in virtual space. Through this so-called offline programming, complex applications can be generated without having to stop the production, thereby considerably raising the machine's uptime. Furthermore, intelligent tools for collision avoidance allow the early detection of collisions with tools, workparts, or other technical equipment in the robotic cell before they happen, allowing to user to act on this information and to change the programming. Recently, CAD/CAM systems acquired the ability to translate three-dimensional CAD data into robot control data. Nowadays, it is even possible to control industrial robots via industry-standard G-code (DIN/ISO), which is translated by the robot itself into robotic movements. This enables CNC programmers to work in their standard programming environments, without requiring additional robotic knowledge.

Outlook

Robotic material processing using industrial robots has become state of the art in the metal, plastic, and wood industries. The experiences gained therein can now also be applied to the processing of ceramics and natural stones. This leads to highly inter-

Figure 1 Machining of aluminium in a RMC

esting perspectives for architecture, art, and design, as robotic processes allow machining centers to realize concepts faster, and cheaper – with a lot size of one, but for the price of serial production. Even in an artistic context, the use of robotic labour is no longer out of the ordinary, with robots being used to support sculptors by roughly shaping the stone for the sculptor, who then follows up by manually processing the surface and working out the intricate details. The RMC's flexibility, combined with its large workspace and easy programmability, enable new fabrication strategies and make them uniquely suited for the non-standard requirements of the creative industry.

www.a-quadrat.eu

Figure 2 High-powered milling spindle mounted on a KUKA KR500 robot

Eric Dokulil

Parametric Robot Control without CAD/CAM

Dynamically generated parametric robot commands for the fabrication of pneumatic cylinders

Introduction

Grasl Pneumatics is a medium-size enterprise specialized in the fabrication of pneumatic cylinders for light-domes. The core of this product is the patented profile of the tubes, which allows the free placement of the pneumatic supply, as well as the pivot point of the cylinder according to the customer's specifications. However, the custom CNC machine's variable insertion depth made manual loading impossible, instead requiring a machine that would automatically take the tube from a pallet, position it in the CNC machine, and rotate the part so that both ends can be machined. Due to their large workspace and agility, the use of an industrial robot for this process was proposed, finally settling on a KUKA KR60L45 robot with 45 kg of carrying weight.

At the time of the robot programming, the sequence of commands was already fixed:

- Take a tube from a pallet fed by an automatic conveyor belt
- Insert the tube into the CNC machine
- After side 1 is finished, remove the tube and insert it into a fixture
- Take the tube from the other side of the fixture.
- Insert it into the machine for the processing of side 2
- Place the tube back into the pallet

The robot interacts with the tubes via a custom-made gripper, which holds the elements from inside. For each tube diameter, the robot can automatically retrieve a fitting gripping tool from the gripper shelf. Each pneumatic cylinder therefore requires individual manufacturing parameters, but still has to be processed automatically, one box of elements of various lengths at a time. A common way of dealing with such a matter would be to create one job for each individual element, and then just execute the fitting file. However, a quick calculation shows that, taking a tube length of 150 to 2030 mm and at least 5 tube diameters into account, such an approach would result in 94,000 variants, or nearly 5,000,000 robot positions. This quickly showed that the only proper way of dealing with such a problem is to develop a parametric model that can accommodate all these variants in one shared definition with individual parameters. As Grasl Pneumatics required an integrated solution, the parametric model was generated entirely in KUKA Robot Language (KRL).

Advances Robot Strategies

Robotic movements are always defined in relation to a tool coordinate system and a base coordinate system. This approach has the advantage, that e.g. the same job can be applied to various objects at different positions just by shifting the tool or base coordinate system. The robotic movement programming strategy that was used for the pneumatic tubes utilizes this concept. When comparing a gripper without a tube, and a gripper holding a tube, one notices that the tool centerpoint (TCP) is simply shifted along the tool's main axis. The different length of the tube can therefore be easily compensated by editing the properties of the tool instead (Fig. 2). In KRL this can be done by adding the position vectors of the tools. Of course, such an approach requires a perfectly defined tool axis, something that cannot be achieved with the robot's on-board tools, as even a slight deviation can lead to inaccuracies in the centimeter range when dealing with 2 m long elements. Therefore, a laser was used to exactly calibrate the orientation of the tool. In

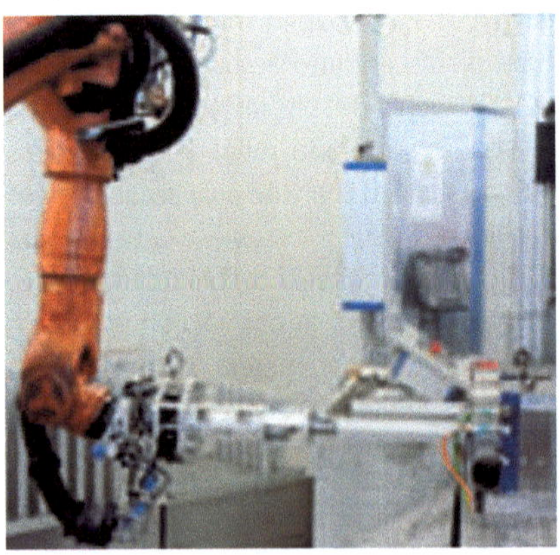

Figure 1 Robot inserting tube into CNC machine

theory this concept would make it possible to deal with all tube lengths, just by programming the movement once with a reference tube and then adjusting the length value of the tool. However, the large differences in tube length lead to very different robot postures for the same sequence of movements. In general, there is more than one way for a robotic arm with six degrees of freedom to approach a give point. While one strategy may work for the reference tube, it may not be ideal for an especially long or short tube. Therefore, the program evaluates all possible movement strategies from one point to another, and chooses the strategy that leads to the least amount of axis rotation.

All these motion programs are then linked together to be accessed by a central program structure. The CNC milling machine and the industrial robot are tightly linked, but retain a certain amount of independence. The robot acts similarly to a state machine and decides what to do based on the state of the gripper, the milling machine and any additional periphery. Instead of using a fixed cycle, the robot can react to instructions from the CNC machine immediately, and once it has inserted the tube, instruct the CNC machine itself to start the according program.

Evaluating KUKA|prc

KUKA|prc (parametric robot control) is a plugin developed by the Association for Robots in Architecture that allows accessible and intuitive robot programming, based on the parametric modelling environment Grasshopper, whose approach towards visual programming has made it a popular tool in the creative industry. The advantage of parametric design for fabrication is that processes – such as the handling of the pneumatic tubes – can be programmed once, but still allow the variation of parameters – in this case the length of the pneumatic tubes.

At the time of the project, KUKA|prc was not yet available and therefore not an option. However, recent evaluation of the software has shown, that even complex handling tasks can be programmed in KUKA|prc. As KUKA|prc immediately visualizes and kinematically simulates parametric toolpaths, the user can fluently and intuitively interact with the virtual robot and optimize its trajectory. It has to be noted, though, that such software requires basic CAD literacy, which is sometimes not available on the shop floor. However, projects like the Red Bull arch show that even new robot users can quickly acquire CAD, KUKA|prc, and robot skills.

Conclusion

Parametric modelling allows the reduction of 94,000 individual movement programs to a single parametric definition that adapts itself to the current requirements. Instead of placing the fabrication logic in an external computer and using the robot as a simple positioning unit, all programming is contained within the robot's own control unit, reducing cycle time, while retaining full flexibility. The described robot program has proven itself over many years, and clearly shows that parametric logic can not only be contained in design tools, but also in the robot code itself. However, for future projects, the use of CAD-based parametric control such as KUKA|prc will provide an alternative for the quick and intuitive programming and simulation of parametric robot programs.

www.dokulil.com

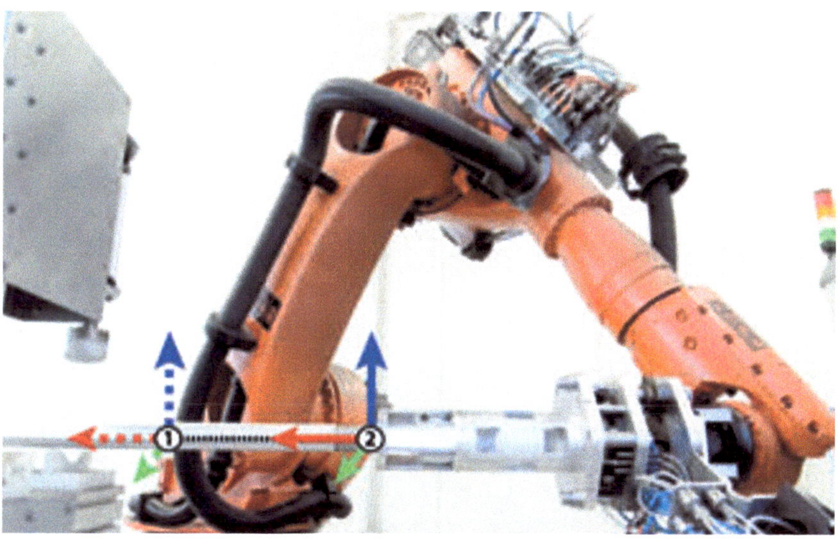

Figure 2 Tool coordinate system for reference tube (1) and longer tube (2), using the same programming

MICHAEL BAUER

A Custom Robotic Trimmer for Modern Timber Constructions

Introduction

Graf-Holztechnik, a timber construction company located in Horn, in the province of Lower Austria, has established itself as a leading trimming specialist in the wood industry.

Combining traditional craftsmanship knowhow with state-of-the-art technology, Graf uses up-to-date computer technology in all stages of production, from static and constructive planning to manufacturing. This ensures a maximum of reliability, speed and efficiency and gives free rein to creativity and the realization of extraordinary ideas.

CNC in the Timber Construction Industry

In the 1990s the call for ever shorter building times in timber construction gave rise to new ideas and developments in trimming, from the traditional drawing yard to trimming machines, in order to allow for a higher degree of prefabrication. CNC trimmers offer a number of decisive advantages for timber construction:

1. production of 3D components with free shapes and three-dimensional curvatures,
2. high precision and repeat accuracy (tenth-of-a-millimetre tolerances),
3. fast and efficient trimming,
4. direct input of planning and design data,
5. high degree of prefabrication.

This allows for the production of parts and shapes for modern timber structures that would otherwise have been impossible or too imprecise, or which could only have been produced at prohibitive cost. To meet these requirements, Graf-Holztechnik has added two CNC trimming units to its line of production machinery, combining the advantages of an established Hundegger trimming/joining machine, with an experimental, robotic, multi-axis trimming machine. The Hundegger K2i is a highly-advanced trimmer/joinery machine for solid-wood processing and timber trimming. The machine allows the trimming of workpieces of variable length without requiring time for measuring, marking or set-up. Designed for the rapid trimming of large series, the unit operated by a single worker can achieve an annual average output of 1.5 to 2.0 m^3 per hour.

A Custom Multi-axis Robotic Trimmer

In cooperation with Hage, a special-purpose machine builder in Obdach, Styria, Graf developed a prototype robotic trimmer with an overall length of 45 m capable of precisely trimming workpieces up to 20 m in length and with a width of up to 2.4 m.

The production process requires a workflow consisting of several tightly interlocked programs:

1. With the aid of a CAD (Computer Aided Design) program the production planning department prepares three-dimensional design plans which are then sent to a BTL unit by via a custom post-processor.
2. The Easywood CAD/CAM (Computer Aided Manufacturing) software developed by the Italian software firm DDX issues the commands for the individual processing steps to the trimmer
3. The DDX files are loaded at the robotic trimmer
4. The clamping carriages automatically asume the requisite positions.
5. The operator mounts the workpieces on the clamping carriages and aligns them along the zero laser line.

Subsequently the trimming process is started and runs automatically. Depending on the number and scope of the individual processing steps, drill-holes and milling operations, this can take between 10 and 60 minutes, with the carriages moving at a rate of up to 80m/min. The unit can handle workpieces weighing up to 9 tons. An essential feature is that all work-piece sides can be processed without re-clamping. This is an indispensable prerequisite for achieving a precision in the millimetre range needed to meet specifications.

The heart of the machine is a spindle with 5 degrees of freedom, operating at a speed of 12,000 rpm which can be adjusted in four directions and turned along

Figure 1 Custom robotic trimmer

two axes.

Its spindle arm moves along the workpiece over a distance of 2.8 m horizontally and 2.4 m vertically and is infinitely adjustable; this means that in combination with the adjustable carriages the machine can process all six sides of the workpiece at full spindle output without re-clamping, allowing for the trimming of all conceivable free shapes.

Milling and drilling tools are stored in a drum-type holder for 20 tools up to 15 cm diameter and 40 cm length. Two additional tool boxes above the spindle hold circular saw blades and planing heads. The circular saw blades have a maximum diameter of 800 mm and are driven by water-cooled 30 kW spindles with a torque of 70 Nm.

Conclusion

Advances in the development of CNC machines for up-to-date timber applications open up new perspectives for wood as a building material, an ecological product with a high sustainability potential that offers architects a large measure of freedom and versatility. Compared to articulated robotic arms, as used in the automotive industry, custom robotic machines such as the trimmer by Hage, are highly optimized for their particular task and offer e.g. superior accuracy and force. However, once ongoing research into robotic hard- and software allows industrial robots to achieve the precision of today's timber trimming machines, robotic arms may provide an affordable alternative to specialized machines.

www.graf-holztechnik.at

Figure 2 Freeformed wooden elements

Peter Mehrtens

Rapid On-site Fabrication of Customized Freeform Metal Cladding Panels

Robotic devices in shipping containers shifting from research to state-of-the-art

Motivation: Research in the Lab, the Factory and On-site

Within contemporary architecture, freeform surfaces have gained increasing popularity over the last few years. Forming surfaces on a workstation has become easy due to the powerful cad packages and digital toolsets available to the designer of today. However, on the road to turning conceptual design models into real buildings, some interesting and challenging work often lies ahead. This includes the detailed computational design of building envelopes, precise part fabrication and the act of building such structures.

A number of highly interesting research projects related to robotics and automated digital fabrication processes within the architectural, engineering and construction industries are being or have been carried out at various research institutions and university faculties around the world. For a number of years, German façade systems developer BEMO SYSTEMS has endeavoured independently into this topic with a very strong focus on turning concepts to reality – both in terms of innovative technology as well as built work.

Conventional Roll Forming

The metal forming process of cold roll forming is well established in the field of building materials, for example to produce corrugated sheets, trapezoid sheets, standing seam profiles.

Roll forming is a bending technology with rotating tool motion, by which an initially flat metal sheet is transported through a series of roll forming-stands that gradually change the sheets shape (DIN 8586:2003-09). Successively each forming stand defines an intermediate stage of the final cross section, into which the sheet is pressed. Minor over-bending by tool design accounts for the spring back of the bent profile, when it leaves the last forming stand.

The key to conventional roll forming is a carefully engineered layout and tooling of the forming rolls. The arrangement and shape of the upper and lower forming rolls is designed with the aid of a so-called flower pattern, which is the sequence of profile cross-sections at each stand of rolls. The number of forming steps required to form a sheet into a profile is influenced by the profile shape to be roll-formed, the properties and characteristic of the raw material and tolerances within to produce the panel.

A full production line also encompasses the processes of unwinding material from a coil, cutting the strip to the right length, and feeding the forming station.

Mobility of Production Units

When constructing large roof areas with standing seam profiles, it is beneficial to maximize the area of the joint- and penetration-free water-bearing-layer. In order not restrict panel sizes to specific lengths required for transport, the production units were designed to be mobile.

The MONRO roll forming technology was developed to fit into a customized 40-foot intermodal container (Fig. 1). The complementary machinery for bending standing seam panels is built into a 20-foot container (Fig. 2). For a streamlined workflow, as many in-line work stages as possible are integrated in the two portable production units.

Advanced Roll Forming Technology for the MONRO Standing Seam Profile

A mill for unwinding the coil is the first process, which the line of machines takes care of. A straightening device is located at the feeder. For cutting panels to length, the mobile unit is set up to use a pre-cut die. This means only a single blank runs through the machine. Also near the start of the production line an optional set of forming rolls offers shaping linear ribs into the panel – primarily used when making panels with symmetric edges or when forming straight (developable) panels.

The cutting-edge technology literally sets-in where the panel is cut to shape: On both sides moveable edge cutters truncate excessive metal, which is separated and transported to a tray above. The main roll forming work is carried out by two rows of forming stations. The 3D roll forming process requires not only transporting sheet material through a set of forming stations, but especially demands that these tools move to the right position concordantly with the feeding speed. The pairs of rollers actively altering the shape must gradually move in synchronous manner to satisfy geometric constrains, such as perpendicularity to the varying panel edge (Fig. 3).

A computerized node control, responsible for changing the cross-section profile within a single sheet, enables fast and precise mass customization of panels. As the bending process involves movement

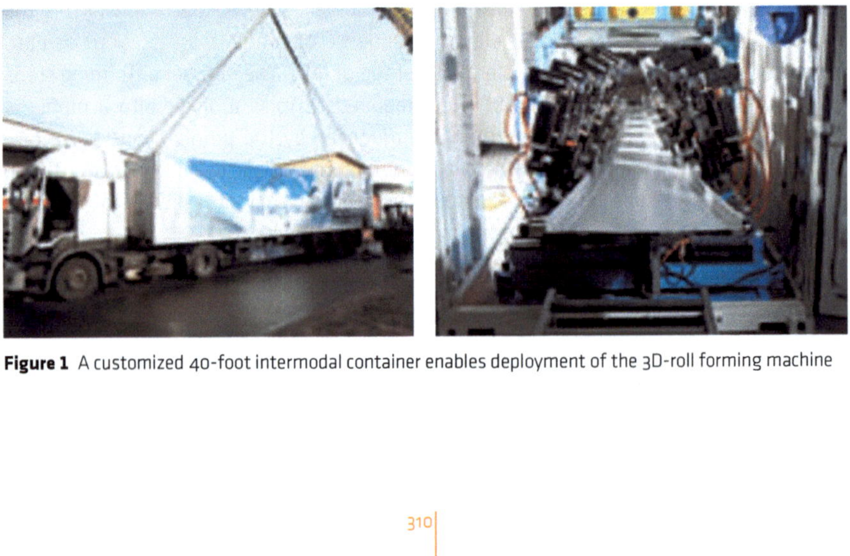

Figure 1 A customized 40-foot intermodal container enables deployment of the 3D-roll forming machine

and friction, lubrication is used to create a thin barrier between the roll dies and the panel surface, reducing wear-off and resulting alteration of the end-effectors size.

The tooling of the rolls that form the large and small seams on the left and right sides of the panel can be designed with a conventional flower pattern. However, due to the fact that 3D roll forming has varying cross sections, the conventional flower pattern design is replaced by software calculating the intermediate cross sections of each panel. The 12 left and 12 right forming stations can individually move horizontally (along the Y-Axis) and vertically (along the Z-Axis) within planes perpendicular to the material transport direction and they can rotate (B-Axis). These 3 degrees-of-freedom combined with the common (X-axis) for all robotic devices within the panel shaping system, which is given by the linear sheet transportation, adds up to 24 end-effectors, each addressing 4 axes.

As opposite sided stations work in pairs, it is even possible to tilt the seam by lifting only one row of forming tools.

Bending

Six pairs of forming rolls in active and passive positions bend the standing seam profiles as the profiled sheet metal is transported through the bending machine. A proprietary numeric control software correlates tool positioning to the overall panel shape and required radii. Important was that the inventors not only considered driving separate radii per edge, but to simultaneously tilt and shift active rolls sideways to match asymmetric panel edge tangents and changing width as the sheet progressively travels. Between the two work stages of roll forming and bending, the sheet is manually rotated by 90°, so that the final workpiece can be taken from the machine easily. Convex and concave bending is possible – even within a single panel.

Underlying Geometric Concept and Resulting Panel Shapes

Clastic and anti-clastic surfaces can be approximated by models of developable strips. The MONRO technology is an abstracted application of developable strips, in which

Figure 2 The 3D-bending machine is built into a 20-foot intermodal container

Figure 3 Each MONRO standing seam profile is gradually formed by 24 individually moving stations

the alignment curves of the metal panels coincide with the reference surface. In the abstract model these alignment curves are the edges of the strips. The ribs and the seams of the panel are offset normal to the reference surface.

In cases, where the panel geometry is rationalized, the panels may resemble straight developable strips. For architectural applications, the exclusive use of such panels restricts the designer to orienting closely to principle curvature lines (Fig. 4), or at least to designing the alignment pattern as a family of geodesic curves – if twisting the panels (or strips) is permitted. This is likely to be acceptable for interior applications and certain surface shapes, however does not suffice for exterior application on buildings – i.e. the water-bearing-layer, as in this case study. When applying purely straight developable strip models to building envelopes, the surface first needs to be segmented into patches that can be covered with parallel geodesics. This segmentation can cause joinery work at undesirable locations. The second problem may be that in some cases the pattern of parallel geodesics can hinder proper water run-off.

The technical ability to produce panels that resemble developable strips with non-parallel edges, and even with curved edges gives architects and engineers greater freedom in designing building envelopes as general double curved surfaces. Reducing constraints in the shape of developed strips can be beneficial to the design

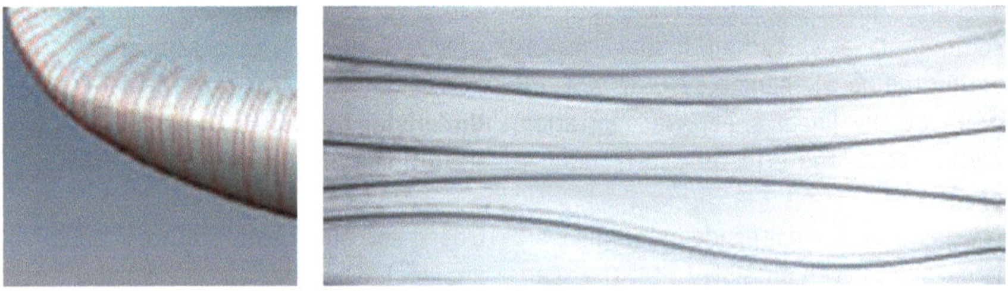

Figure 4 Principle Curvature Lines indicate directions for families of geodesics that form the centerlines and/or the edges of strips which represent almost straight developable surfaces with low torsion (left). Sample surface displaying metal standing seam panels with independently curved edges (right).

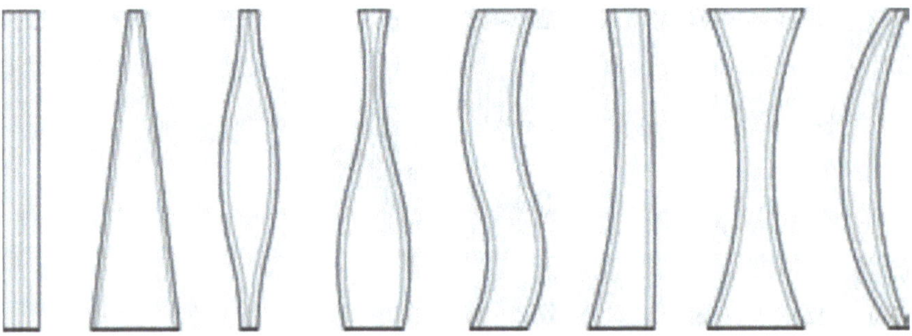

Figure 5 Shape examples of developed strips, producible as facade and roof panels via roll forming and bending

in terms of aesthetics (visually continuous patterns are feasible), and for constructive reasons, i.e. when long, joint-free panels are preferred or even required. Possible configurations of shapes and manufacturing possibilities are shown in Figures 4 and 5.

The width of standing seam panels may vary between 100 mm to 1000 mm, as the raw material is on coils of up to 1250 mm width. Panel length may exceed 100 meters when panels are fabricated on site, because their length is theoretically restricted only by coil length. Seam alignment curves and surfaces are modeled as geometric entities with curvature continuous property (as NURBS). When it comes to driving the production line, a numeric approximation, consisting of coordinates, tangent vectors and feeding speeds, is extracted from the curves. Files storing CNC data for cutting, roll forming and bending procedures are generated per panel. Cutting and forming tools, shaping the panels, move according to the instructions read from the machining file.

Examples of Built Work

The initially developed machines were in fact working prototypes, and the doubly-curved structures realized in the meantime can be seen as proof-of-concept. The Budapest Sports Arena (Fig. 6) was the first envelope clad with MONRO panels (27000 m^2). The complete workflow has evolved into a fully digital process. Covering large roof and facade areas, e.g. of entire concert halls, stadia or airports, has become feasible in terms of design, quality, time and cost. Digital stages of work include the modelling of curvature continuous (smooth) surface patches, design of panel layout and optional optimization of geometry, 3D laser scanning of as-built load-bearing structures, responsive generation and dimensioning of substructure, extraction of production data, computerized numerically controlled fabrication of parts, photogrammetric quality verification of building elements and tacheometric surveying on site. Mass customized skins of

Figure 6 Papp László Budapest Sportsarena, Budapest, Hungary (completion Feb 2003); architecture: KÖZTI (Skardelli György, Pottyondy Péter), Hungary; photos: KÖZTI

have also been realized with this technology on large public venues such as the multipurpose hall "ISS Dome" in Germany and "Le Tarmac" concert hall in France (Fig. 7) – covering areas of 8000 m^2 and 6200 m^2. In both cases the areas are assembled largely of individual parts. The shape of the pebble-like domes emerged from the functional arrangement of the interior spaces. Lightweight façades can be built of various metallic materials and finished with numerous coatings to match the conceived design intent. The roof of the Main Station Local Transport Hub in Graz, currently under construction, is treated with an eloxal coating, giving it a distinctive appearance in terms of color and reflection, as well as improving corrosion resistance.

Acknowledgements

The ideas, research and development of the inventors Wolfgang Maas and Dr. Lars Ingvarsson, combined with the entrepreneurship of BEMO SYSTEMS GmbH in close collaboration with their machine building partner ORTIC AB, have made the technology for on-site fabrication of customized freeform metal cladding panels available. Their work covers in particular engineering the process of roll forming curved standing seam panels – from concept via flower design and roll tool development to machine construction and application of this technology in the construction industry.

www.bemo.com

References

Photography as mentioned, except: Fig. 1 left BEMO; Figs. 1 right - 4 Roman Benz, Atelier Busche.

DIN 8586:2003-09 Manufacturing processes forming by bending - Classification, terms.

Figure 7 Concert Hall "Le Tarmac", Déols Châteauroux, France (completion Aug 2007); architecture: blond&roux architects, Paris, France; Photos: Pauline Turmel

List of Contributors

Felix Amtsberg
Institute for Structural Design, TU Graz
felix.amtsberg@tugraz.at

David Arceneaux
Stäubli Robotics U.S.
d.arceneaux@staubli.com

Ralph Bärtschi
ROB Technologies, ETH Zurich
baertschi@rob-technologies.com

Joshua Bard
Carnegie Mellon University
jdbard@gmail.com

Michael Bauer
Graf-Holztechnik GmbH
office@graf-holztechnik.at

Gabriel-Bello Diaz
IAAC
gabrielbellodiaz@googlemail.com

Martin Bechthold
Harvard Graduate School of Design
mbechthold@gsd.harvard.edu

Christian Binder
SCHUNK Intec GmbH
christian.binder@at.schunk.com

Tobias Bonwetsch
ROB Technologies, ETH Zurich
bonwetsch@rob-technologies.com

Johannes Braumann
Association for Robots in Architecture
johannes@robotsinarchitecture.org

Sigrid Brell-Çokcan
Association for Robots in Architecture
sigrid@robotsinarchitecture.org

Jan Brüninghaus
Chair of Industrial Robotics and Production Automation, TU Dortmund
jan.brueninghaus@tu-dortmund.de

Alois Buchstab
KUKA Roboter GmbH
AloisBuchstab@kuka-roboter.de

Baris Çokcan
IIArchitects-Int
cokcan@2amimarlik.com

Jens Cortsen
Maersk Mc-Kinney Moller Institute, University of Southern Denmark
cortsen@mmmi.sdu.dk

Richard Dank
Institute of Architecture and Media, TU Graz
dank@tugraz.at

Federico Diaz
Artist
federico@fediaz.com

Karola Dierichs
Institute for Computational Design, University of Stuttgart
karola.dierichs@icd.uni-stuttgart.de

Eric Dokulil
Association for Robots in Architecture
eric@dokulil.com

Kathrin Dörfler
Institute of Arts and Design, TU Vienna
doerfler.kathrin@gmail.com

Alexandre Dubor
IAAC
alexandre.dubor@gmail.com

Yaron Elyasi
Etto Studio
Yaron@ettostudio.com

Jelle Feringa
TU Delft
jelleferinga@gmail.com

Michal Firstenberg
Mediated Matter Group, MIT Media Lab
michal10@mit.edu

Christian Freissling
Institute of Architecture and Media, TU Graz
freissling@tugraz.at

Nicholas Foley
Greyshed
nhfoley@gmail.com

Fabio Gramazio
Professorship for Architecture
and Digital Fabrication, ETH Zurich
gramazio@arch.ethz.ch

Jonathan Grinham
Virginia Tech Center for Design Research
jgrinham@gmail.com

Ben Hagendorfer-Daniell
University of Michigan
ben.hagenhofer.daniell@gmail.com

Matthias Helmreich
Professorship for Architecture
and Digital Fabrication, ETH Zurich
helmreich@gramaziokohler.com

Andreas Hofer
Zeman GmbH
hofer@zeco.at

Manfred Hübschmann
Stäubli Robotics Germany
m.huebschmann@staubli.com

Ryan Luke Johns
Greyshed / Princeton University
rlj@ryanlukejohns.com

Alfred Kaser
A² Anlagentechnik & Automation GmbH
A.Kaser@a-quadrat.eu

Steven Keating
Massachusetts Institute of Technology
stevenk@mit.edu

Nathan King
Harvard Graduate School of Design
Virginia Tech Center for Design Research
nathanking.king@gmail.com

David Kittl
ABB AG Österreich
david.kittl@at.abb.com

Matthias Kohler
Professorship for Architecture
and Digital Fabrication, ETH Zurich
kohler@arch.ethz.ch

Martin Kohlmaier
ABB AG Österreich
martin.kohlmaier@at.abb.com

Martin Kölldorfer
Artist
martin.koelldorfer@hotmail.com

Carsten Krewet
Chair of Industrial Robotics and
Production Automation, TU Dortmund
carsten.krewet@tu-dortmund.de

Oliver David Krieg
Institute for Computational Design,
University of Stuttgart
oliver.krieg@icd.uni-stuttgart.de

Brandon Kruysman
Southern California Institute of Architecture
kruysman@gmail.com

Bernd Kuhlenkötter
Chair of Industrial Robotics and
Production Automation, TU Dortmund
bernd.kuhlenkoetter@tu-dortmund.de

Lukás Kurilla
Faculty of Architecture, CTU Prague
kurilluk@fa.cvut.cz

Silke Langenberg
Professorship for Architecture
and Digital Fabrication, ETH Zurich
langenberg@arch.ethz.ch

Jared Laucks
Mediated Matter Group, MIT Media Lab
jlaucks@mit.edu

Benjamin Lehnert
Mediated Matter Group, MIT Media Lab
blehnert@mit.edu

Steven Mankouche
University of Michigan
mankouch@umich.edu

Iain Maxwell
University of Technology Sydney
maxi@supermanoeuvre.com

Wes McGee
University of Michigan
wesmcgee@umich.edu

Peter Mehrtens
Bemo Systems
mail@petermehrtens.de

Achim Menges
Institute for Computational Design,
University of Stuttgart
achim.menges@icd.uni-stuttgart.de

Clemens Neugebauer
Artist
neugebauer@tmo.at

Silvan Oesterle
Professorship for Architecture
and Digital Fabrication, ETH Zurich
oesterle@arch.ethz.ch

Neri Oxman
Massachusetts Institute of Technology
neri@mit.edu

Tom Pawlofsky
Kunstgiesserei St. Gallen
tom.pawlofsky@gmx.de

Stefan Peters
Institute for Structural Design, TU Graz
stefan.peters@tugraz.at

David Pigram
University of Technology Sydney
dave@supermanoeuvre.com

Jason Prasad
University of Michigan
jasonpr@umich.edu

Jonathan Proto
Southern California Institute of Architecture
jonathan.proto@gmail.com

Florian Rist
Institute of Arts and Design, TU Vienna
frist@fs.tum.de

Romana Rust
Institute for Digital Media,
Technical University Graz
romana.rust@gmail.com

Matthew Schulte
University of Michigan
schultem@umich.edu

Mathew Schwartz
University of Michigan
cadop@umich.edu

Thibault Schwartz
ECZT, Robots in Architecture
ts@thibaultschwartz.com

Tobias Schwinn
Institute for Computational Design,
University of Stuttgart
tobias.schwinn@icd.uni-stuttgart.de

Asbjørn Søndergaard
Aarhus School of Architecture
Asbjorn.Sondergaard@aarch.dk

Dorthe Sølvason
Maersk Mc-Kinney Moller Institute,
University of Southern Denmark
dorthe@mmmi.sdu.dk

Hanno Stehling
designtoproduction GmbH
stehling@designtoproduction.com

Yoav Sterman
Mediated Matter Group, MIT Media Lab
sterman@mit.edu

Ladislav Svoboda
Faculty of Architecture, CTU Prague
ladislav.svoboda@fsv.cvut.cz

Valentine Troi
superTEX composites GmbH
valentine.troi@supertex.at

Andreas Trummer
Institute for Structural Design, TU Graz
andreas.trummer@tugraz.at

Elizabeth Tsai
Mediated Matter Group, MIT Media Lab
elyt@mit.edu

Lauren Vasey
University of Michigan
lauren.vasey@gmail.com

Jan Willmann
Professorship for Architecture
and Digital Fabrication, ETH Zurich
willmann@arch.ethz.ch

Scientific Committee

Kristy Balliet
Ohio State University

Martin Bechthold
Harvard GSD

Thomas Bock
TU Munich

Baris Çokcan
II Architects int

Michael Hofer
University of Vienna

Jason K. Johnson
Future Cities Lab

Branko Kolarevic
University of Calgary

Gottfried Koppensteiner
ACIN, TU Vienna

Andrew Kudless
California College of the Arts

Christian Kühn
TU Vienna

Russell Loveridge
EPFL Lausanne

Alexis Meier
INSA Strasbourg

Andrew Payne
LIFT Architects

Fabian Scheurer
designtoproduction

Alexander Schiftner
Evolute

Kai Strehlke
Herzog & de Meuron

Sigrun Swoboda
IEMAR, TU Vienna

Markus Vincze
ACIN, TU Vienna

Gabriel Wurzer
IEMAR, TU Vienna

MIX
Papier aus verantwortungsvollen Quellen
Paper from responsible sources
FSC® C105338

If you have any concerns about our products,
you can contact us on
ProductSafety@springernature.com

In case Publisher is established outside the EU,
the EU authorized representative is:
Springer Nature Customer Service Center GmbH
Europaplatz 3, 69115 Heidelberg, Germany

Printed by Libri Plureos GmbH
in Hamburg, Germany